Statistical Theory and Methods for Evolutionary Genomics

Xun Gu

*Department of Genetics,
Development and Cell Biology,
Iowa State University, USA*

OXFORD

UNIVERSITY PRESS

OXFORD
UNIVERSITY PRESS

Great Clarendon Street, Oxford OX2 6DP

Oxford University Press is a department of the University of Oxford.
It furthers the University's objective of excellence in research, scholarship,
and education by publishing worldwide in

Oxford New York

Auckland Cape Town Dar es Salaam Hong Kong Karachi
Kuala Lumpur Madrid Melbourne Mexico City Nairobi
New Delhi Shanghai Taipei Toronto

With offices in

Argentina Austria Brazil Chile Czech Republic France Greece
Guatemala Hungary Italy Japan Poland Portugal Singapore
South Korea Switzerland Thailand Turkey Ukraine Vietnam

Oxford is a registered trade mark of Oxford University Press
in the UK and in certain other countries

Published in the United States
by Oxford University Press Inc., New York

© Xun Gu 2011

British Library Cataloguing in Publication Data

Data available

Library of Congress Cataloging in Publication Data

Data available

Typeset by SPI Publisher Services, Pondicherry, India
Printed in Great Britain
on acid-free paper by
CPI Antony Rowe, Chippenham, Wiltshire

ISBN 978–0–19–921326–9

3 5 7 9 10 8 6 4 2

To my family, Wei and Katie, and to my parents

Preface

Descending from a single ancestor, genomes became subject to Darwinian selection and adaptation. It is through mutations that functional innovation or loss of ancestral function could lead to an increase of genome complexity, and adaptation to the challenges of a continuously changing environment. Evolutionary genomics is a recently-emerged research field, with the ultimate goal of understanding the underlying evolutionary and genetic mechanisms for such processes. It stems from the links between high throughput data in functional genomics, statistical modeling and bioinformatics, and phylogeny-based analysis.

During the last decade, high throughput technologies in genomics have generated enormous genome sequence data in many organisms in all ranges of life forms. Indeed, these technological innovations have made evolutionary genomes a rapidly-growing field, witnessed by numerous research papers published in the literature. At this stage, inevitably these studies are not only highly heterogeneous in methodology, but also highly controversial in biological interpretation. The purpose of this book is to attempt to present a tentative framework of statistical theory and methods that are useful in the study of genomic evolution, and illustrate how to use them in real, large-scale data analyses. Since nowadays genomic data analyses are almost always software-based, our explanation of statistical methods is focused on biological foundations and model assumptions, rather than the details of computational algorithms and practical implementations.

The author wishes to unify the field of evolutionary genomics by building a theoretically inherent framework that can provide a basis for interpreting enormous genomic data by the means of evolution. This is a long-term, ambitious goal that cannot be accomplished within the volume of this book. This book discusses various models and methods, including those developed by the author and his collaborators. Some of them were investigated during the process of writing this book and have been published very recently or for the first time here. Preference was given to statistical models that are based on realistic biological assumptions and statistical methods that are practically useful. Since the author's laboratory has been engaged in this area of study since the emergence of genome sciences, this book includes a substantial portion of work conducted by his group. Yet, the author has acknowledged the impossibility of covering all important topics and references; that would be beyond his capability, as well as the mission of this book. Indeed, balance has to be made between personal preference and the different scientific views and approaches. For all these reasons, this book should be viewed as an effort from an individual scientist as a step toward the ultimate goal of synthesis.

This book is written for graduate students and researchers in the field of evolutionary genetics/genomics and related fields, as well as those scientists in mathematics, statistics, or computer sciences who are interested in computational biology and bioinformatics. If this book is used in the classroom, the author encourages the instructor to select the contents that best fit their teaching purposes. As this book is intended to introduce statistical models and methods, the author expects readers with the necessary training in statistics and mathematics. In addition, readers should have some basic knowledge of genome science and evolutionary biology. Readers who wish to know more on some important topics that are discussed only briefly may refer to many excellent books, such as Li (1997) and Nei and Kumar (2000) for molecular evolution and phylogeny, Yang (2006) for statistical methods of DNA sequence analysis, Lynch (2009c) for genomic evolution, and Evens and Grant (2005) for statistical bioinformatics.

In this book, the author introduces the basics of molecular evolutionary theory and bioinformatics tools so that readers can comprehend related issues discussed in this book. Then follows a range of topics, chapter by chapter, including functional divergence of gene families, expression divergence between duplicate genes, tissue-driven evolution, gene pleiotropy, and gene content evolution. These topics were organized by the model-driven approach, with a few well-selected examples of data analyses. The author acknowledges that there are a huge number of relevant publications in the literature that cannot all be cited here. In particular, this book does not discuss in detail the widely-used data-driven approach that takes advantage of high computational capacity and high throughput genomic data. Indeed, after developing integrated, comprehensive databases, one can utilize various efficient IT technologies such as clustering or machine-learning to conduct extensive data exploration, in an attempt to find biologically meaningful patterns that can be further tested by experimentation or independent datasets. In the author's view, the model-driven approach and the data-driven approach are complementary; both are indispensable for having a deep understanding of some fundamental biological problems. The last chapter of this book tackles several evolutionary issues that have not been completely resolved in systems biology. The intent was to explore the field coined 'evolutionary systems biology' for what has been accomplished, and what may potentially lie ahead. Hopefully these explorations will benefit research in this field. Finally, it should be noted that most statistical methods discussed in this book had computer programs available when they were published. For instance, previously, the author's laboratory published a computer program (DIVERGE) to predict amino acid sites that are responsible for functional divergence between two duplicate genes. Nevertheless, the author's group plans to develop an updated, user-friendly computer software package that can cover these methods.

The author has many people to thank: Jiazhen Tan (C. C. Tan), the founder of Chinese modern genetics who has influenced the author's career in genetics and evolution profoundly since he was an undergraduate student in Fudan University, China; Zhudong Liu, his master's degree supervisor in Fudan University, who provided him with a unique opportunity to learn mathematical population genetics; Wen-Hsiung Li, his Ph.D supervisor in the University of Texas, who guided him to the frontline of molecular evolution and phylogeny; and Masatoshi Nei, his postdoc supervisor in Penn State University, who encouraged him to develop his own research niche when genome science was emerging. The author would also like to thank all his friends, colleagues, and collaborators during his career development over the last two decades, especially Li Jin, Yunxin Fu, Ranajit Chakraborty, Sudhir Kumar, C-I Wu, Manyuan Long, Dan Graur, Mike Miyamoto, Ziheng Yang, Jianzhi Zhang, Zhenglong Gu, Gunter Wagner, Duane Enger, Dan Voytas, Tom Peterson, Patrick Schnable, Dan Nettleton, Karin Dorman, Hui-Hsien Chou, Xiaoqui Huang, Jonathan Wendel, Eric Gaucher, Yaping Zhang, Jun Yu, Ji Yang, and Yang Zhong. The author is grateful to former and current students, research associates, and collaborators in his laboratory, who have made great contributions to solving many challenging problems in the statistical study of genomic evolution, computer program development, and large-scale data analysis; especially to Yufeng Wang, Jianying Gu, Kent Vander Velden, Zhongqi Zhang, Shiquan Wu, Huaijun Zhou, Zhixi Su, Yong Huang, Yangyun Zou, Hongmei Zhang, and Xiujuan Wang. The author acknowledges support in all aspects from the Department of Genetics, Development and Cell Biology, Iowa State University, for giving him time to finish the book. The author also has special acknowledgement for his wife, Wei. Without her full support and encouragement over the past 17 years, this book would not have been possible.

Research activities conducted in the laboratory of the author have been partly supported by the National Institutes of Health, USA (NIH), the National Science Foundation, USA (NSF), the National Science Foundation in China (NSFC), DuPont Young Investigator Award, Iowa State University, and Fudan University. The author is grateful to these funding agencies for their generous support.

Xun Gu
Ames, Iowa USA
February, 2010

Contents

1
Basics in Molecular Evolution

Molecular evolution is the study of the process of evolution at the level of DNA, RNA, and proteins, in which the neutral or nearly neutral evolution model has provided the theoretical basis (Kimura 1968, 1983; Kimura and Ohta, 1971; Ohta 1973, 1993; Nei 1987; Li 1997). Yet, the role of positive selection at the molecular level remains a controversial issue (Gillespie 1991; McDonand and Kreitman 1991; Dean and Golding 1997; Messier and Stewart 1997; Zhang *et al.* 1998; Bustamante *et al.* 2000, 2005; Tanenbaum *et al.* 2005; Nielsen *et al.* 2007). Recent advances in genomics, including whole-genome sequencing, high-throughput protein characterization, and bioinformatics have led to a dramatic increase in studies in comparative and evolutionary genomics. In this chapter, we concisely introduce some widely-used methods in genomic analysis.

1.1 Evolutionary distance of DNA sequences

Evolutionary distance (d) is fundamental for the study of molecular evolution, which is usually measured by the number of nucleotide or amino acid substitutions per site between two homologous sequences (Li 1997; Nei and Kumar 2000). First, d has been widely used to reconstruct phylogenetic trees of genes and gene families. Second, d is a basic measure for studying the pattern and mechanism of DNA/protein evolution, for example, for testing the molecular clock hypothesis (Wu and Li 1985; Gu and Li 1992; Huang *et al.* 1998) and for detecting positive selection in sequence evolution (Hughes and Nei 1988). Third, with the assumption of constant rate and reliable fossil records, d can be used to date the divergent time of species (Kumar and Hedges 1998; Hedge and Kumar 2009), or gene/genome duplication events (Wang and Gu 2001; Gu *et al.* 2002b). However, d is generally not equal to the observed number of differences per site between two DNA sequences, because multiple substitutions at a given site may occur, especially when the sequence divergence is large. Therefore, to estimate d, a stochastic (Markov) model for DNA evolution is required. In the following we discuss these stochastic models and methods.

1.1.1 Jukes and Cantor's model: a tutorial

One of the simplest models of nucleotide substitution is that of Jukes and Cantor (1969). This model assumes that at each site a nucleotide changes to one of the three remaining nucleotides with the same substitution rate (r) per year. Let us

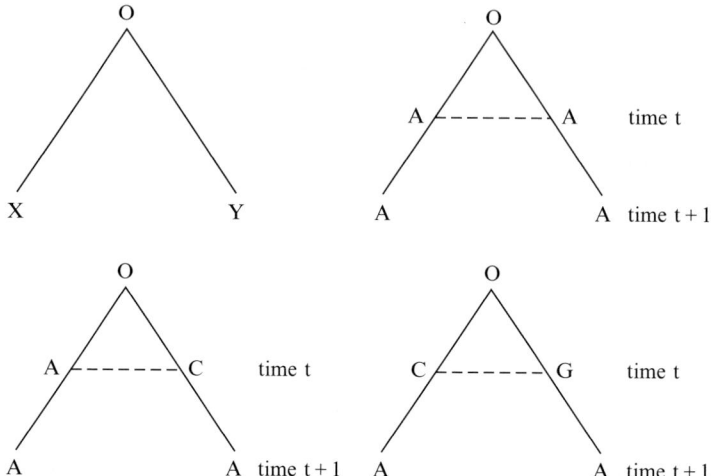

Fig. 1.1 Illustration for the derivation of Jukes–Cantor model in the case of two DNA sequences (X and Y) diverged from the common ancestor (O). For any site that has the same nucleotide (A) in both sequences at the time $t + 1$, at time t the nucleotide could be either the same, or different (one is A and the other is C). Nevertheless, double changes occurred in both lineages in the time interval between t and $t + 1$ are negligible

now consider two nucleotide sequences, X and Y, which diverged from the common ancestral sequence t years ago. We denote by q_t the probability of identical nucleotides between X and Y and by p_t ($= 1 - q_t$) the probability of different nucleotides. At time $t + 1$ (measured in years), the probability of identical nucleotides, q_{t+1}, can be derived as follows (Fig. 1.1).

(i) With a probability of q_t, two sequences have the same nucleotide at time t. At time $t + 1$, the chance that remains the same is $(1 - r)^2 \approx 1 - 2r$, that is, the probability of no change in both lineages. Note that double changes are negligible.

(ii) With a probability of $1 - q_t$, two sequences have different nucleotides at time t, the chance at time $t + 1$ that has the same nucleotide is $2 \times (1/3) \times r$; the factor of $1/3$ is for the change to a specific nucleotide, and the factor of 2 indicates that it could happen in either of two lineages. Double changes from t to $t + 1$ are negligible. Together with (i), we then have

$$q_{t+1} = (1 - 2r)q_t + \frac{2}{3}r(1 - q_t)$$

which can be written as

$$q_{t+1} - q_t = \frac{2r}{3} - \frac{8r}{3}q_t$$

(*iii*) Let us now use a continuous time model and represent $q_{t+1} - q_t$ by dq/dt. We then have the following differential equation

$$\frac{dq}{dt} = \frac{2r}{3} - \frac{8r}{3} q_t$$

Solution of this equation with the initial condition $q = 1$ at $t = 0$ gives

$$q = 1 - \frac{3}{4} \left(1 - e^{-8rt/3} \right)$$

(*iv*) Under the Jukes–Cantor (1969) model, the expected number of nucleotide substitutions per site (d) for the two sequences is $2rt$. Therefore, d can be estimated by

$$\hat{d} = -\frac{3}{4} \ln \left(1 - \frac{4}{3}\hat{p} \right) \tag{1.1}$$

where $\hat{p}$ is the proportion of sites with different nucleotides between X and Y.

However, the derivation of the evolutionary distance is tedious when the substitute model becomes sophisticated. In the next section we introduce a formal mathematical treatment.

1.1.2   Models of nucleotide substitution

Since at each site in a DNA sequence there are four possible nucleotides (A, T, C, and G), a model of nucleotide substitution is characterized by a 4×4 rate matrix **R**, which is also called the pattern of nucleotide substitution. This matrix can be represented as (Table 1.1): The ij-th element of **R**, denoted by r_{ij}, is the substitution rate from nucleotide i to nucleotide j, $i \neq j$; the diagonal elements are given by $r_{ii} = -\sum_{j \neq i} r_{ij}$, so that the sum of the elements in each row is zero. Thus, the most general model has 12 independent parameters (Table 1.1(a)), but this model is too complex to apply. Therefore, it is necessary to make some assumptions on **R** to develop a useful method for estimating d. As examples, the following are two simple but widely-used models.

The first is Jukes and Cantor's one-parameter model (Jukes and Canter 1969), in which the substitution rate is assumed to be the same for all nucleotides, i.e. $r_{ij} = \mu$, for all $i \neq j$, and $r_{ii} = -3\mu$; see Table 1.1(b) for the rate matrix **R**. The second is Kimura's two-parameter model (Kimura 1980), in which the rate of transitional substitution (i.e. changes between A and G or between C and T) may not be equal to that of transversional substitution (i.e. all the other types of changes); the two rates are denoted by s and v, respectively (Table 1.1(c)). This model is more realistic than the one-parameter model because transitions are generally more frequent than transversions in DNA evolution.

In general, let $P_{ij}(t)$ be the transition probability from nucleotide i to j after t time units. By Markovian theory, $P_{ij}(t)$ satisfies the linear differential equation

$$\frac{dP_{ij}(t)}{dt} = \sum_k r_{ik} P_{kj}(t) \tag{1.2}$$

Table 1.1 Models (R) of nucleotide substitution.

	A	T	C	G
(a)				
A	$-(a+b+c)$	a	b	c
T	d	$-(d+e+f)$	e	f
C	g	h	$-(g+h+i)$	i
G	j	k	l	$-(j+k+l)$
(b)				
A	-3μ	μ	μ	μ
T	μ	-3μ	μ	μ
C	μ	μ	-3μ	μ
G	μ	μ	μ	-3μ
(c)				
A	$-(2\nu+s)$	ν	ν	s
T	ν	$-(2\nu+s)$	s	ν
C	ν	s	$-2(\nu+s)$	ν
G	s	ν	ν	$-(2\nu+s)$
(d)				
A	r_{11}	$\pi_2\nu_1$	$\pi_3\nu_2$	$\pi_4 s_1$
T	$\pi_1\nu_1$	r_{22}	$\pi_3 s_2$	$\pi_4\nu_3$
C	$\pi_1\nu_2$	$\pi_2 s_2$	r_{33}	$\pi_4\nu_4$
G	$\pi_1 s_1$	$\pi_2\nu_3$	$\pi_3\nu_4$	r_{44}

where $i, j, k = A, G, T$ or C, with the initial condition $P_{ii}(0) = 1$ and $P_{ij}(0) = 0$ ($i \neq j$). Let matrix $\mathbf{P}(t)$ consist of $P_{ij}(t)$. In matrix form, Eq.(1.2) can be expressed as

$$\frac{d\mathbf{P}(t)}{dt} = \mathbf{R}\mathbf{P}(t) \tag{1.3}$$

with the initial condition $\mathbf{P}(0) = \mathbf{I}$, where $\mathbf{I}$ is the identity matrix. The solution of Eq.(1.3) is

$$\mathbf{P}(t) = e^{\mathbf{R}t} \tag{1.4}$$

Because we generally do not know the initial sequences, to study DNA sequence evolution we need to consider two homologous DNA sequences which diverged from O, the common ancestor, t time units ago (Fig. 1.1). The mean evolutionary rate (per year per site) is given by $\bar{r} = \sum_i f_i \sum_{j \neq i} r_{ij}$, where f_i is the frequency of nucleotide i in the ancestral sequence. Since $\sum_{j \neq i} r_{ij} = -r_{ii}$, it can be simplified as $\bar{r} = -\sum_i f_i r_{ii}$. Thus, the number of substitutions per site between the two sequences, the evolutionary distance (d), is expected to be

$$d = 2\bar{r}t = -2t \sum_{i=1}^{4} f_i r_{ii} \qquad (1.5)$$

where the factor 2 comes from the fact that d is the sum of the numbers of substitutions per site in the two lineages. One problem is that d depends on the initial frequencies f_i, which we do not know. However, if the nucleotide frequencies are stationary, that is, their expectations do not change with time, then the initial frequencies can be estimated by the average frequencies in the present sequences. The assumption of stationarity greatly simplifies the estimation problem.

We consider the case where the substitution process is stationary and time reversible (Gu and Li 1996b); this is called the SR model. Time reversibility means that $\mathbf{R}$ is restricted by the following condition:

$$\pi_i r_{ij} = \pi_j r_{ji} \qquad (1.6)$$

for any $i \neq j$, where π_i is the equilibrium frequency of nucleotide i; $i = 1, 2, 3, 4$ for A, T, C, and G, and $\pi_1 + \pi_2 + \pi_3 + \pi_4 = 1$ (Table 1.1(d)). Therefore, the SR model is a nine-parameter model, which includes the one-parameter model, the two-parameter model, and several other models (Tajima and Nei 1982, 1984; Tamura and Nei 1993) as special cases.

Consider the two-lineage scheme (Fig. 1.1). Since time reversibility implies that the substitution process from the common ancestor O to sequences X and Y is equivalent to that from X through O to Y (or from Y through O to X), the transition probability matrix from X to Y is given by

$$\mathbf{P}(2t) = e^{2t\mathbf{R}} \qquad (1.7)$$

By spectral decomposition, the diagonal elements of $\mathbf{R}$ can be expressed as

$$r_{ii} = \sum_{k=1}^{4} u_{ik} v_{ki} \lambda_k \qquad (1.8)$$

where λ_k ($k = 1, 2, 3, 4$) is the k-th eigenvalue of $\mathbf{R}$, one of which is zero, say $\lambda_4 = 0$; u_{ik} is the ik-th element of the eigenmatrix $\mathbf{U}$ and v_{ki} is the ki-th element of matrix $\mathbf{V} = \mathbf{U}^{-1}$. Putting Eq.(1.8) into Eq.(1.5) and setting $f_i = \pi_i$, we have

$$d = -2t \sum_{k=1}^{4} b_k \lambda_k \qquad (1.9)$$

where the constants $b_k = \sum_{i=1}^{4} \pi_i u_{ik} v_{ki}$; in particular $\lambda_4 = 0$ that means $b_4 = 0$.

In the following, we discuss how to derive the formula for estimating d under a specified substitution model. We will start with the one-parameter and two-parameter models, for which the eigenvalues of $\mathbf{R}$ can be obtained analytically. Then, we shall discuss the nine-parameter SR model, for which the eigenvalues of $\mathbf{R}$ cannot be obtained analytically. Besides, we show that all these methods can be extended to the case where the substitution rate varies among nucleotide sites.

1.1.3 One-parameter method

In the one-parameter model, the eigenvalues of $\mathbf{R}$ are given by $\lambda_1 = \lambda_2 = \lambda_3 = -4\mu$ and $\lambda_4 = 0$ and so $b_1 = b_2 = b_3 = 1/4$. From Eq.(1.9), the number of substitutions per site can be simplified as $d = 6\mu t$. From the eigenvalues and eigenmatrix of $\mathbf{R}$, the transition probability is given by

$$P_{ij}(t) = \begin{cases} \frac{1}{4} + \frac{3}{4}e^{-4\mu t} & \text{if } i = j \\ \frac{1}{4} - \frac{1}{4}e^{-4\mu t} & \text{if } i \neq j \end{cases} \tag{1.10}$$

Now consider the two-lineage scheme (Fig. 1.2). Let $I(t)$ be the probability that, at time t, the nucleotides at a given site in the two sequences are identical to each other. Since the probability that both sequences have nucleotide j is $\sum_{k=1}^{4} f_k P_{kj}^2(t)$, where f_k is the frequency of nucleotide k at the ancestor O, $I(t)$ is given by

$$I(t) = \sum_{k=1}^{4} f_k \left[P_{k1}^2(t) + P_{k2}^2(t) + P_{k3}^2(t) + P_{k4}^2(t) \right]$$

From Eq.(1.10) and after some simplifications, one can show that, regardless of the initial frequencies f_k, $I(t)$ is given by $I(t) = 1/4 + (3/4)e^{-8\mu t}$. Note that the probability that two sequences are different at a site at time t is $p = 1 - I(t)$, which can be estimated by the observed proportion of different nucleotides between the two sequences ($\hat{p}$). Therefore we have shown that the evolutionary distance $d = 6ut$ can be estimated by Eq.(1.1), that is,

$$d = -\frac{3}{4} \ln \left(1 - \frac{4}{3}\hat{p} \right)$$

1.1.4 Kimura's two-parameter method

Under Kimura's two parameter model (Kimura 1980), the eigenvalues of $\mathbf{R}$ are $\lambda_1 = \lambda_2 = -2(s + v)$, $\lambda_3 = -4v$ and $\lambda_4 = 0$; the constants are $b_1 = b_2 = 1/2$ and $b_3 = 1/4$. Thus, d is given by $d = 2(s + 2v)t$. On the other hand, the transition probability under this model is given by

$$P_{ij}(t) = \begin{cases} \frac{1}{4} + \frac{1}{4}e^{-4vt} + \frac{1}{2}e^{-2(s+v)t} & \text{for } i = j \\ \frac{1}{4} + \frac{1}{4}e^{-4vt} - \frac{1}{2}e^{-2(s+v)t} & \text{for transitions} \\ \frac{1}{4} - \frac{1}{4}e^{-4vt} & \text{for transversions} \end{cases} \tag{1.11}$$

Let P and Q be the probabilities that a nucleotide site between the two sequences differ by a transition and a transversion, respectively. One can show $P = 2\sum_k f_k(P_{kA}P_{kG} + P_{kT}P_{kC})$ and $P + Q = I(t)$. By Eq.(1.11) one can show

$$P = \frac{1}{4} + \frac{1}{4}e^{-8vt} - \frac{1}{2}e^{-4(s+v)t}$$

$$Q = \frac{1}{2} - \frac{1}{2}e^{-8vt}$$

Thus, $d = 2(s + 2v)t$ can be estimated by

$$d = -\frac{1}{2}\ln(1 - 2P - Q) - \frac{1}{4}\ln(1 - 2Q) \tag{1.12}$$

where P and Q can be estimated from the two sequences.

1.1.5   The general stationary and time-reversible model

So far, explicit solutions for estimating d have been obtained for up to six-parameter models (Tamura and Nei 1993; Li and Gu 1996). In general, however, it is difficult to derive an analytical formula for d because the eigenvalues of $\mathbf{R}$ cannot be expressed in an analytical form (Rodriguez *et al.* 1990). Under the SR (stationary and time-reversible) model, Gu and Li (1996b) provided a practically feasible solution. Let z_k be the k-th eigenvalue of $\mathbf{P}(2t)$. By matrix theory, Eq.(1.7) implies that $\mathbf{R}$ and $\mathbf{P}(2t)$ have the same eigenmatrix, and their eigenvalues have the following relationship

$$z_k = e^{2t\lambda_k} \qquad k = 1, \ldots, 4 \tag{1.13}$$

Note that $\lambda_4 = 0$ and $z_4 = 1$. Thus, the evolutionary distance under the SR model can be written as

$$d = -\sum_{k=1}^{3} b_k \ln z_k \tag{1.14}$$

where z_k and b_k can be estimated from sequence data (see below). Under the SR model, all eigenvalues z_k (or λ_k) are real.

To estimate z_k and b_k from sequence data, we must estimate the transition probability matrix $\mathbf{P}(2t)$ first. Let J_{ij} be the expected frequency of sites where the nucleotide is i in sequence X and j in sequence Y. Let matrix $\mathbf{J}$ consist of J_{ij}. It can be shown that matrix $\mathbf{J}$ is symmetric under the SR model. By the Markovian property, we have

$$J_{ij} = \sum_{k=1}^{4} \pi_k P_{ki}(t) P_{kj}(t) \tag{1.15}$$

$i, j = 1, \ldots, 4$. By time reversibility, i.e. $\pi_i P_{ij}(t) = \pi_j P_{ji}(t)$, we have

$$J_{ij} = \sum_{k=1}^{4} \pi_i P_{ik} P_{kj}(t) = \pi_i \sum_{k=1}^{4} P_{ik}(t) P_{kj}(t) = \pi_i P_{ij}(2t) \tag{1.16}$$

where $\sum_{k=1}^{4} P_{ik}(t) P_{kj}(t) = P_{ij}(2t)$ is a basic property of transition probabilities. Thus, Eq.(1.16) gives a simple method for estimating $P_{ij}(t)$ from sequence data J_{ij}:

(1) Count N_{ij}, the number of sites at which the nucleotide is i in sequence X and is j in sequence Y, and then compute $\hat{J}_{ij} = N_{ij}/N$, where N is the number of nucleotides in the sequence.

(2) Estimate the transition probability matrix $\mathbf{P}(2t)$ by

$$\hat{P}_{ij} = \frac{\hat{J}_{ij}}{\hat{\pi}_i}$$

($i, j = 1, \ldots, 4$). where $\hat{\pi}_i$ is the frequency of nucleotide i estimated by taking (simple) average between the two sequences. However, when $i \neq j$, the estimated frequency $\hat{J}_{ij}$ may not be equal to $\hat{J}_{ji}$, i.e. the estimated matrix $\hat{\mathbf{J}}$ may not be symmetric. Gu and Li (1996b) proposed a method to test whether $J_{ij} - J_{ji}$ ($i \neq j$) is significantly different from zero. If the null hypothesis is not rejected statistically, the deviation from symmetry can be regarded as sampling effects, and the ij-th and the ji-th elements of $\mathbf{J}$ are equally given by $[\hat{J}_{ij} + \hat{J}_{ji}]/2$.

(3) Then, the estimate of z_k ($k = 1, \ldots, 4$) can be obtained by solving the following characteristic equation

$$\det(\hat{\mathbf{P}} - \hat{z}\mathbf{I}) = 0$$

where $\hat{\mathbf{P}}$ consists of $\hat{P}_{ij}$ and $\mathbf{I}$ is the identity matrix; the corresponding eigenmatrix $\mathbf{U}$ and its inverse matrix $\mathbf{V}$ are also obtained simultaneously by a standard algorithm. Thus, d can be estimated according to Eq.(1.14) and the sampling variance can be approximately computed by the formula given by Gu and Li (1996b).

1.1.6 Estimation of d under variable rates

All the above methods for estimating d assume the same substitution rate for all sites. If this assumption is violated, d may be seriously underestimated, especially for divergent sequences. Because the substitution rate usually varies among sites in most genes, these methods need to be modified to take rate variation into account.

There is empirical evidence that the rate variation among sites follows a gamma distribution. This distribution is mathematically simple and is commonly used in the literature. Assume that the ij-th element of the rate matrix $\mathbf{R}$ is expressed by $r_{ij} = h_{ij}u$, where the constant h_{ij} represents the pattern of substitution rate and the random variable u varies among sites according to the following gamma distribution

$$\phi(u) = \frac{\beta^\alpha}{\Gamma(\alpha)} u^{\alpha-1} e^{-\beta u} \tag{1.17}$$

where the mean of u is given by $\bar{u} = \alpha/\beta$.

First, consider the one-parameter model; in this case, $h_{ij} = 1$ for all $i \neq j$. It follows that the mean of p, i.e. the mean proportion of differences between two sequences, is given by

$$\bar{p} = \int_0^\infty \frac{3}{4} \left(1 - e^{-8ut}\right) \phi(u) du = \frac{3}{4} \left[1 - \left(1 + \frac{8\bar{u}t}{\alpha}\right)^{-\alpha}\right] \tag{1.18}$$

Therefore, the average number of substitutions per site $d = 6\bar{u}t$ can be estimated by

$$d = \frac{3}{4}\alpha \left[\left(1 - \frac{4}{3}\bar{p}\right)^{-1/\alpha} - 1\right] \tag{1.19}$$

In the same manner, Jin and Nei (1990) extended Kimura's two-parameter method (1980) to the case where the rate varies according to a gamma distribution:

$$d = \frac{1}{4}\alpha \left[2 \left(1 - 2P - Q\right)^{-1/\alpha} + (1 - 2Q)^{-1/\alpha} - 3\right] \tag{1.20}$$

Finally, for the SR model, the average d is

$$d = \alpha \sum_{k=1}^3 b_k \left(z_k^{-1/\alpha} - 1\right) \tag{1.21}$$

where the eigenvalues z_k and the constants b_k can be estimated by the same approach as above for the SR model under the uniform rate model (Gu and Li 1998).

In summary, because the ancestral sequence is generally unknown, the most general model (i.e. 12 parameters) is difficult to apply so that some restrictions on the rate matrix **R** are necessary for developing useful methods for estimating d. As reviewed here, many methods have been developed for this purpose. As a general rule, methods based on more general models will have smaller estimation bias but larger sampling variances. Thus, when the sequence length (N) is large, more general methods are preferred. However, when N is small, simpler methods may be better. For example, Gu and Li's (1996b) simulation study suggested that, if N is less than 200 base pairs, the one-parameter method is on average better than the SR method; whereas if N is larger than 500, the SR method is on average better than the one-parameter method.

1.1.7 The LogDet distance

The stationarity of nucleotide frequencies is one of the most common assumptions made in estimating evolutionary distances (see Lanave *et al.* 1984; Zharkikh 1994; Gu and Li 1996b). It assumes that the expectations of nucleotide frequencies in a sequence do not change with time and are equal to those in the ancestral sequence. Therefore, to estimate the distance between two sequences, the nucleotide frequencies in the ancestral sequence are estimated by the averages of the nucleotide frequencies in the two extant sequences. If nucleotide frequencies vary with time so that stationarity does not hold, the estimated distance may not be accurate. As a consequence, a distance-matrix method for phylogeny reconstruction can be misleading, i.e. it tends to group sequences of similar nucleotide frequencies irrespective of the true evolutionary relationships (Hasegawa and Hashimoto 1993; Sogin *et al.* 1993; Steel 1994).

The LogDet (Lake 1994; Steel 1994; Lockhart *et al.* 1994; Gu and Li 1996a) distance has been proposed to deal with the nonstationarity problem. In spite of various versions, these methods are based on the most general model of nucleotide substitutions. Historically, these methods can be traced back to Barry and Hartingan (1987) and Cavender and Felseinstein (1987). In this section, we study some statistical properties of the LogDet distance.

Formula

Consider two sequences (denoted by X and Y, respectively) that evolved from O, their common ancestor, t time units ago (see Fig. 1.1). Let $\mathbf{J}$ be the data matrix whose ij-th element J_{ij} is the proportion of sites at which the nucleotide is i in sequence X and j in sequence Y. Then, the LogDet distance (between sequences X and Y) is defined as

$$d = -\frac{1}{4}\ln\det[\mathbf{J}] \tag{1.22}$$

where *det* means the determinant of a matrix. By using the delta method (Barry and Hartigan 1987), the sampling variance of the estimated LogDet distance $(\hat{d})$ is found to be

$$Var(\hat{d}) \approx \frac{1}{16L}\sum_{i=1}^{4}\sum_{j=1}^{4}(M_{ij}^2 J_{ij} - 1) \tag{1.23}$$

where L is the sequence length and M_{ij} is the ij-th element of $\mathbf{M} = \mathbf{J}^{-1}$. The estimation procedure for the LogDet distance is straightforward because the matrix $\mathbf{J}$ can be directly estimated from the sequence data.

Properties

The LogDet distance is potentially very useful in the study of DNA evolution because it has the following nice properties (Gu and Li 1996a):

(1) The LogDet distance is based on the most general model of nucleotide substitution, i.e. the 12-parameter model. Moreover, it is valid even if the rate matrix $\mathbf{R}$ differs among lineages.
(2) The LogDet distance is useful for phylogenetic reconstruction when nucleotide frequencies are nonstationary. It has been shown (e.g. Gu and Li 1996a) that for some distance matrix methods of phylogenetic reconstruction such as the neighbor-joining (NJ) method (Saitou and Nei 1987), the LogDet distance lead to the correct tree topology when the nucleotide frequencies vary considerably among sequences.
(3) Let μ_1 and μ_2 be the arithmetic mean of the evolutionary rate in lineage 1 (toward X) or lineage 2 (toward Y), respectively, that is, $\mu_1 = -\sum_{i=1}^{4} r_{ii}^{(1)}/4$ and $\mu_2 = -\sum_{i=1}^{4} r_{ii}^{(2)}/4$. The biological interpretation of the LogDet distance can be described as the following equation

$$d = 2\mu t - \frac{1}{4}\sum_{i=1}^{4} \ln f_{i,0} \qquad (1.24)$$

where $\mu = (\mu_1 + \mu_2)/2$ is the mean rate over the two lineages, and $f_{i,0}$ is the frequency of nucleotide i at the ancestral node O. Therefore, the LogDet distance is not only linear in time t, but also depends on the nucleotide frequencies at the ancestral node.

(4) The LogDet distance is useful for testing the molecular clock hypothesis under nonstationary frequencies because the relative-rate test (Wu and Li 1985) is not affected by the ancestral nucleotide frequencies.

On the other hand, the LogDet distance may have several disadvantages that need to be examined carefully in practice:

(1) The sampling variance of LogDet distance becomes large for short sequences. By simulation studies, it is recommended for use when the sequence length is >500 bp. Gu and Li (1996a) have demonstrated that the LogDet distance is, on average, overestimated especially when the sequences are short; the bias becomes trivial as the sequence length >2000 bp. Furthermore, Gu and Li (1996a) proposed an empirical bias-corrected LogDet distance as follows

$$\hat{d}_c = -\frac{1}{4}\ln\det[\hat{\mathbf{J}}] - 2Var(\hat{d}) \qquad (1.25)$$

Computer simulation showed that the statistical bias can be largely corrected by this formula.

(2) When $t = 0$, $d = -\sum_{i=1}^{4} \ln f_{i,0}/4 > 0$. In other words, the LogDet distance satisfies the non-negative condition but have a non-zero positive constant at the initial condition. One may modify the LogDet distance by

$$d = -\frac{1}{4}\ln\det[\mathbf{J}] - d_0$$

The problem is that it may cause the violation of non-negative condition in some cases. Note that adding $-d_0$ in the LogDet distance has no effect on the performance of tree-making but may affect the branch lengths estimation. In practice, one may calculate the effect of nucleotide frequencies $F = -\sum_{i=1}^{4} \ln f_i/4$ for all extant sequences and choose $d_0 = F_{min}$, the minimum value over all sequences to guarantee the non-negative nature of the evolutionary distance.

(3) Probably the theoretical challenge of the LogDet distance is the assumption of constant rate over nucleotide sites, which is obviously unrealistic. Indeed, when the rate varies among sites, the biological interpretation of the LogDet distance holds only approximately. In the estimation of evolutionary distance, how to solve the nonstationary problem and the rate variation among sites simultaneously remains an unsolved problem.

1.2 Evolutionary distance between protein-encoding sequences

1.2.1 Poisson distance of protein sequence

In the history of molecular evolution, study of the evolutionary change of proteins began with comparison of two or more amino acid sequences from different organisms. One simple measure of the extent of sequence divergence is the proportion of different amino acids between two sequences, also called the p-distance. However, the proportion of different amino acids (p) is not strictly proportional to the divergence time (t), as multiple amino acid substitutions start to occur at the same sites.

One simple way to estimate the protein distance more accurately is based on the Poisson process, which claims that the probability for no amino acid substitution occurred during t years at a site of a sequence is e^{-vt}, where v is the evolutionary rate. Hence, the probability (q) that neither of the homologous sites of the two protein sequences has undergone substitution is e^{-2vt}, which can be estimated by $q = 1 - p$. It follows that the expected number of amino acid substitutions per site for the two sequences, i.e. the evolutionary distance $d = 2vt$, is given by

$$d = -\ln(1 - p) \qquad (1.26)$$

It should be noted that the Poisson distance is approximate because backward mutations and parallel mutations (the same mutations occurring at the homologous amino acid sites in two different evolutionary lineages) are not taken into account. However, the effects of these mutations are generally very small unless p is large.

1.2.2 Amino acid substitution matrix

Empirical studies have shown that amino acid substitution occurs more often between amino acids that are similar in biochemical properties than between dissimilar amino acids (Dayhoff 1972). As a result, amino acid substitution is generally not random, and backward and parallel substitutions may occur quite often between similar amino acids. Some amino acids such as cysteine and tryptophan rarely change. To take into account these factors, Dayhoff *et al.* (1978) proposed the so-called PAM-based method of estimating evolutionary distance. The amino acid substitution matrix for a relatively short period of time is considered, and the relationship between the proportion of identical amino acids and the number of amino acid substitutions is derived empirically.

The amino acid substitution matrix Dayhoff *et al.* (1978) used was derived from empirical data for many proteins such as hemoglobins, cytochrome c, and fibrinopeptides. They first constructed an evolutionary tree for closely related amino acid sequences and then inferred the relative frequencies of substitutions between different amino acids. From these data, they constructed an empirical amino acid substitution matrix for the 20 amino acids. An element (m_{ij}) of this substitution matrix gives the empirical probability that the amino acid in row i changes to the amino acid in column j during one evolutionary time unit. The time unit used in the matrix is the time during which one amino acid substitution per 100 amino acid sites occurs on average. Dayhoff *et al.* (1978) measured the number of amino acid substitutions in terms of accepted point mutations (PAM) and represented one amino acid substitution per 100 amino acid sites.

While Dayhoff's substitution matrix is still widely used, Jones *et al.* (1992) constructed a new matrix based on a large amount of substitution data from many different proteins. Adachi and Hasegawa (1996) also produced a substitution matrix for 13 mitochondrial proteins in vertebrates. Theoretically, different protein families (such as globins or protein kinases) are expected to have different substitution matrices, so it is desirable to construct a substitution matrix for each group of proteins. Nevertheless, it has shown that the evolutionary distance under the empirical substitution matrix can be numerically approximated by some simple formulas. For instance, Kimura (1983) showed the following distance

$$d = -\ln(1 - p - 0.2p^2) \tag{1.27}$$

where p is the proportion of different amino acid sites, closely approached the Dayhoff's distance when $p < 0.8$.

1.2.3 Synonymous and nonsynonymous distances

Many (but not all) nucleotide substitutions at the third positions are silent and do not change amino acids, i.e. synonymous substitutions. Meanwhile, some silent substitutions may also occur at the first positions. Since synonymous substitutions are most likely free from natural selection (but see Chamary *et al.* 2006 for different opinions), the rate of synonymous substitution is often equated to the rate of neutral nucleotide substitution. By contrast, the rate of nonsynonymous substitution is generally much lower than that of synonymous substitution and varies extensively from gene to gene. (Kimura 1983). For some genes, on the other hand, nonsynonymous substitutions occur at a higher rate than synonymous substitutions (e.g. Hughes and Nei 1988; Lee *et al.* 1995). These nonsynonymous substitutions are apparently caused by positive selection, because under neutral evolution one would expect that the rates of synonymous and nonsynonymous substitution are equal to each other. For these reasons, estimation of the rates of synonymous and nonsynonymous substitution has become an important subject in the study of molecular evolution.

Nei–Gojobori Method

When the number of nucleotide substitutions between two DNA sequences is so small that there is no more than one nucleotide difference between any pair of homologous codons compared, the numbers of synonymous and nonsynonymous substitutions can be obtained by simply counting silent and amino acid altering nucleotide differences. However, when two or more nucleotide differences exist between a pair of codons, the distinction between synonymous and nonsynonymous substitutions is no longer simple, because of multiple evolutionary pathways between them. Nei and Gojobori (1986) developed an unweighted method to calculate the average numbers of synonymous and nonsynonymous substitutions over multiple evolutionary pathways (Perler *et al.* 1980; Miyata and Yasunaga 1980).

To estimate the synonymous (d_S) and nonsynonymous distances (d_N), one has to classify the synonymous and nonsynonymous sites: Let i be the number of possible synonymous changes at this site. This is counted as $i/3$ synonymous and $(1 - i/3)$ nonsynonymous. For instance, in the codon TTT (Phe), the first two positions are counted as nonsynonymous sites because no synonymous changes can occur at these positions. Meanwhile, the third position is counted as one-third synonymous and two-third nonsynonymous because one of three possible changes is synonymous. After obtaining the number of synonymous and nonsynonymous sites, it is straightforward to estimate d_S and d_N separately under the Jukes–Cantor (1969) model.

Li–Wu–Luo Method

Li *et al.* (1985) proposed an alternative way to averaging multiple evolutionary pathways, by classifying nucleotide sites into nondegenerate, twofold degenerate, and fourfold degenerate sites. A site is nondegenerate if all possible changes at this site are nonsynonymous, twofold degenerate if one of the three possible changes is synonymous, or fourfold degenerate if all possible changes are synonymous. This method then calculates the number of substitutions between two coding sequences for the three types of sites separately. Note that by definition all the substitutions at nondegenerate sites are nonsynonymous, and all the substitutions at fourfold degenerate sites are synonymous. At twofold degenerate sites, transitional changes (C/T or A/G) are synonymous and other changes (transversions) are nonsynonymous. For two exceptions in the universal genetic code (Arginine and Isoleucine), Li *et al.* (1985) suggested an ad hoc correction. Under the Kimura (1980) two-parameter model, transitional distance and transversional distance at each type of site can be calculated. Therefore, the synonymous distance (d_S) is the average of the evolutionary distance at the fourfold degenerate sites and the transitional distance at twofold degenerate sites. Similarly, the nonsynonymous distance (d_N) is the average of the evolutionary distance at the nondegenerate sites and the transversional distance at the twofold degenerate sites. To correct the bias in the original version of Li *et al.* (1985), Li (1993) proposed an unbiased method to estimate the rates of synonymous and nonsynonymous substitutions.

Codon substitution models

Goldman and Yang (1994) developed a likelihood method for estimating the rates of synonymous and nonsynonymous nucleotide substitution considering a nucleotide substitution model for 61 sense codons. (Three nonsense codons were eliminated.) Let us consider a pair of sequences of homologous codons and let π_j be the relative frequency of the j-th codon. They assumed that the instantaneous substitution rate (q_{ij}) from codon i to codon j is given by the following equations.

$$q_{ij} = \begin{cases} 0, & \text{if nucleotide change occurs at two or more positions} \\ \pi_j, & \text{for synonymous transversion} \\ k\pi_j, & \text{for synonymous transition} \\ \omega\pi_j, & \text{for nonsynonymous transversion} \\ \omega k\pi_j, & \text{for nonsynonymous transition} \end{cases} \tag{1.28}$$

where k is the transition/transversion rate ratio and ω is the nonsynonymous/synonymous rate ratio. Here k may be written as α/β if the rates of transitional and transversional changes are α and β, respectively.

There are 61 parameters for π_j, but if we assume that the codon frequencies are in equilibrium, they can be estimated by the observed codon frequencies when the number of codons used is large. Therefore, the only parameters to be estimated are k and ω, and these parameters can be estimated by using the maximum likelihood method (Goldman and Yang 1994).

1.3   Phylogenetics trees: an overview

Phylogenetic relationships of genes or organisms (generally referred as taxa) are usually presented in a tree-like form either with a root or without any root. They can be further classified into rooted tree or unrooted tree. The branching pattern of a tree is called a topology. If the number of taxa (m) is four, there are 15 possible rooted tree topologies and three possible unrooted tree topologies. However, the number of possible topologies rapidly increases with increasing m (millions of possible trees when $m > 15$). Therefore, it becomes a difficult task to find the true tree topology when m is large; the subject of study is called phylogenetic inference.

Using molecular data to reconstruct the phylogenetic tree can be traced back to Cavalli-Sforza and Edwards (1967), and Fitch and Margoliash (1967). Since a DNA sequence splits into two descendant sequences at the time of speciation or gene duplication, molecular phylogenetic trees are usually bifurcating. In an unrooted bifurcating tree of m taxa there are $2m - 3$ branches. Since there are m exterior branches connecting to m extent taxa, the number of interior branches is $m - 3$. The number of interior nodes is equal to $m - 2$. In a rooted tree, the numbers of interior branches and interior nodes are $m - 2$ and $m - 1$, respectively, and the total number of branches is $2m - 2$. However, when a relatively short sequence is considered, some interior branches may show no nucleotide substitution, so that a multifurcating node may appear. This type of tree is called a multifurcating tree.

In phylogenetic inference, a certain optimization principle, such as the maximum likelihood or the minimum evolution principle, is often used for choosing the most likely topology. There are many methods that have been developed for reconstructing phylogenetic trees from molecular data, which can be classified into four major groups: (1) distance methods, (2) parsimony methods, (3) likelihood methods, and (4) Bayesian methods. When the phylogeny is inferred, we need to evaluate the statistical reliability. For the distance, parsimony or likelihood methods, the bootstrapping approach (Felenstein 1985) has been widely used in practice. In the following sections we discuss each of them briefly. Due to the space limit, we will not address some advanced topics, such as phylogenetic inference under the nonhomogeneous model of DNA sequence evolution (Galtier and Gouy 1995; 1998), or maximum likelihood identification of coevolving protein residues (Pollock *et al.* 1999).

1.4 Distance method for phylogenetic inference

1.4.1 Principle: minimum-evolution (ME)

In distance methods, evolutionary distances are computed for all pairs of taxa, and a phylogenetic tree is constructed by considering the relationships among these distance values. There are many different methods of constructing trees from distance data. Here we discuss the method that has been widely used in molecular evolution: The principle of minimum-evolution (ME) and the neighbor-joining (NJ) algorithm.

In this method, the sum (S) of all branch length estimates in a given unrooted topology, i.e.

$$S = \sum_{i=1}^{2m-3} \hat{b}_i$$

where $\hat{b}_i$ is the estimated branch length of the i-th branch, is computed for all plausible topologies, and topology that has the smallest S value is chosen as the best tree. The theoretical foundation of the ME method is Rzhetsky and Nei's (1993) mathematical proof that when unbiased estimates of evolutionary distances are used, the expected value of S becomes the smallest for the true topology.

1.4.2 Algorithm: neighbor-joining (NJ) method

Although the ME method has nice statistical properties, it requires a substantial amount of computer time when the number of taxa compared is large. Saitou and Nei (1987) developed an efficient tree-building method that actually is based on the minimum evolution principle. This method does not examine all possible topologies, but at each stage of taxon clustering a minimum evolution principle is used. This method is called the neighbor-joining (NJ) method. Currently, there are a number of versions that may improve the performance of NJ algorithm, e.g. the BIONJ (Gascuel 1997).

Construction of a tree by the NJ method begins with a star tree, which is produced under the assumption that there is no clustering of taxa (Fig. 1.2). Then we estimate the branch lengths of the star-tree and compute the sum of all branches (S_0). Since the star-tree is generally incorrect, this sum (S_0) should be greater than the sum (S_F) for the final NJ tree. In practice, since we do not know which pair of taxa are true neighbors, we consider all pairs of taxa as a potential pair of neighbors and compute the sum of branch lengths (S_{ij}) for the i-th and j-th taxa using a topology similar to that given in Fig. 1.2. We then choose taxa i and j that show the smallest S_{ij} value. Once a pair of neighbors is identified, they are combined into one composite taxon, and this procedure is repeated until the final tree is produced.

Mathematically, S_0 for the star-tree is given by

$$S_0 = \sum_{i=1}^{m} L_{iX} = \sum_{i<j}^{m} d_{ij}/(m-1) = T/(m-1)$$

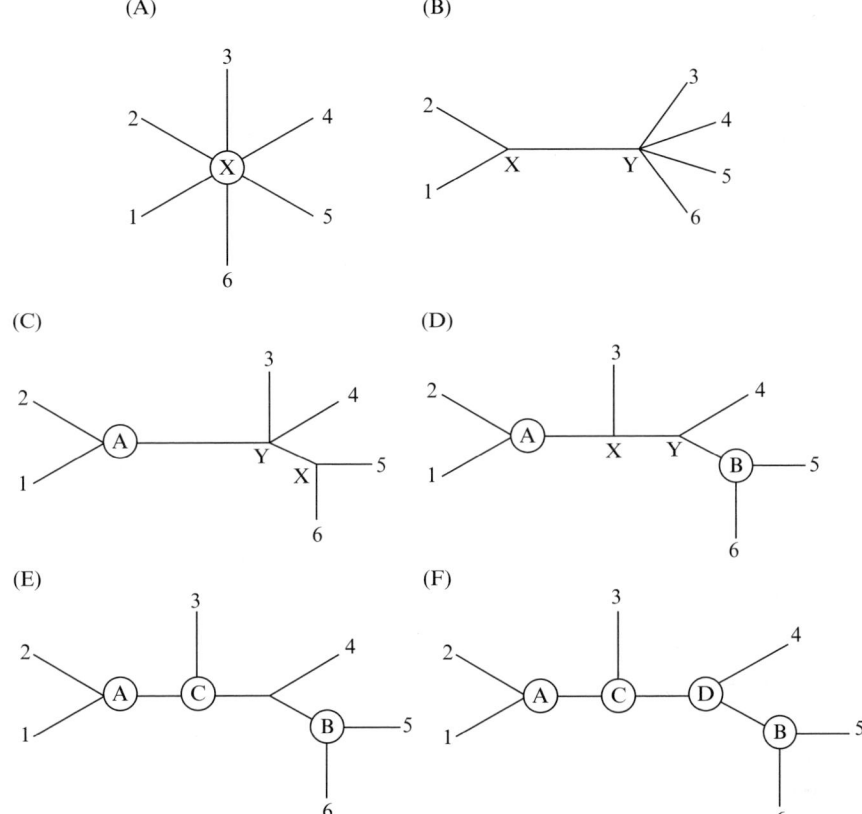

Fig. 1.2 Illustration of the computational procedure in the neighbor-joining (NJ) method. The NJ method begins with a start tree (*A*). Once a pair of neighbors (taxa 1 and 2) (*B*), they are combined into one composite taxon called taxon A (*C*). For the rest of the four original taxa plus composite taxon A, suppose taxa 5 and 6 are the neighbors so that they are again combined to be composite taxon B (*D*). The final two steps (*E* and *F*) solve the tree-making problem in a similar iteration. Modified from Nei and Kumar (2000).

where L_{iX} is the branch length estimate between nodes i and X, and $T = \sum_{i<j} d_{ij}$. In the star-tree, i stands for the i-th exterior node and X the interior node. Suppose, as shown in Fig. 1.2, that taxa 1 and 2 are neighbors such that S_{12} is given by the sum of the following terms

$$S_{12} = L_{1X} + L_{2X} + L_{XY} + \sum_{i=3}^{m} L_{iY}$$

where $L_{1X} + L_{2X} = d_{12}$. Similar to the star-tree for the rest of $m - 2$ taxa, we have $(m - 3)\sum_{i=3}^{m} L_{iY} = \sum_{3 \leq i < j} d_{ij}$. Moreover, Saitou and Nei (1987) have shown

$$L_{XY} = \frac{1}{2(m-2)} \left[\sum_{i=3}^{m} (d_{1i} + d_{2i}) - (m-2)d_{12} - 2 \sum_{i=3}^{m} d_{iY} \right]$$

Therefore, we have

$$S_{12} = \frac{1}{2(m-2)} \sum_{i=3}^{m} (d_{1i} + d_{2i}) + \frac{1}{2}d_{12} + \frac{1}{m-2} \sum_{3 \leq i < j} d_{ij} \qquad (1.29)$$

Obviously, S_{ij} can be computed in the same way if we replace 1 and 2 by i and j, respectively, in the above equation. Once the smallest S_{ij} is determined, we can create a new node (A) that connects taxa i and j. The branch lengths (b_{A_i} and b_{A_j}) from this node to taxon i and j are given by

$$b_{Ai} = \frac{1}{2(m-2)} \left[(m-2)d_{ij} + R_i - R_j \right]$$

$$b_{Aj} = \frac{1}{2(m-2)} \left[(m-2)d_{ij} - R_i + R_j \right]$$

where $R_i = \sum_{k=1}^{m} d_{ik}$ and $R_j = \sum_{k=1}^{m} d_{jk}$. These values are known to be the least-squared (LS) estimated for the topology under consideration (Saitou and Nei 1987).

The next step is to compute the distance between the new node (A) and the remaining taxa ($k \neq i, j$) (Fig. 1.2). This distance is given by

$$d_{Ak} = (d_{ik} + d_{jk} - d_{ij})/2$$

If we compute all the distances using this equation, we have a new $(m-1) \times (m-1)$ matrix. From this matrix, we can compute a new S_{ij} matrix. To find the new pair of 'neighbors', we choose a pair with the smallest S'_{ij} value. A new node B is then created for this pair of taxa, and a new $(m-2) \times (m-2)$ distance matrix is computed. This procedure is repeated until all taxa are clustered in a single unrooted tree. The final tree obtained in this way is called the NJ tree.

1.4.3 Four-point condition and NJ algorithm

Consider a tree for four taxa and assume that taxa 1 and 2 are a pair of true neighbors and also taxa 3 and 4 (Fig. 1.3). For an additive tree, we obviously have the following inequalities

$$d_{12} + d_{34} < d_{13} + d_{24}$$

$$d_{12} + d_{34} < d_{14} + d_{23} \qquad (1.30)$$

This four-point condition has been used by Sattath and Tversky (1977) and Fitch (1981) to reconstruct the topology of a tree. It should be noted that the NJ method and these two methods require the same condition for obtaining the correct topology for the case of four taxa. To show this we compare the difference between S_{13} (the sum of branch lengths after assuming taxa 1 and 3 are the pair of neighbors) and S_{12} (the sum after assuming taxa 1 and 2 are the pair of neighbors). It has been shown that

$$S_{13} - S_{12} = \frac{1}{2(m-2)} \sum_{k=4}^{m} [(d_{13} + d_{2k}) - (d_{12} + d_{3k})] = \frac{1}{2(m-2)} \sum_{k=4}^{m} U_{12,3k}$$

where $U_{12,3k} = d_{13} + d_{2k} - (d_{12} + d_{3k})$ is the score of four-point condition. On the right hand of this equation are the sum of four-point conditions between taxa 1, 2, 3, and any of $4 \leq k \leq m$. Therefore, the NJ criterion for choosing taxa 1 and 2 as a pair neighbors, i.e. $S_{13} > S_{12}$, is the same condition that the sum of four-point condition scores is positive.

1.4.4 The Q-score of Studier and Keppler

Studier and Keppler (1988) rewrote the formula of S_{12} in Eq. (1.29) as follows

$$S_{12} = \frac{2T - R_1 - R_2}{2(m-2)} + \frac{d_{12}}{2}$$

where $R_i = \sum_{k=1}^{m} d_{ik}$ and $R_j = \sum_{k=1}^{m} d_{jk}$. Since T is the same for all pairs, S_{12} can be replaced by

$$Q_{12} = (m-2)d_{12} - R_i - R_j$$

for the purpose of computing the relative value of S_{ij} (Studier and Keppler 1988). In fact, most computer programs use Q_{12} rather than S_{12} to facilitate the computation.

1.5 Parsimony methods for phylogenetic inference

Maximum parsimony (MP) methods were originally developed for morphological characters (Hennig 1966). Here we discuss how MP can be useful for analyzing molecular data. These MP methods consider four or more aligned nucleotide (or amino acid) sequences ($m \geq 4$). The smallest number of nucleotide (or amino acid) substitutions that explain the entire evolutionary process for the topology is computed. This computation is done for all potential topologies, and the topology that requires the smallest number of substitutions, also called the shortest tree length, is chosen to be the best tree. The theoretical basis of this method is William Ockham's philosophical idea that the best hypothesis to explain a process is the one that requires the smallest number of assumptions.

Homoplasy and long-branch attraction If there are no backward and no parallel substitutions (no homoplasy) at each nucleotide/amino acid site, MP methods are expected to produce the correct tree. However, nucleotide/amino acid sequences are often subject to backward and parallel substitutions (high homoplasy). In this case, MP methods tend to give incorrect topologies. Felsenstein (1978) has shown that when the rate of nucleotide substitution varies extensively among branches, MP methods may generate incorrect topologies. In this case, long branches of the true tree tend to join (or attract) together in the MP inferred tree. Therefore, this phenomenon is called the long-branch attraction.

Unweighted MP and weighted MP methods In unweighted MP methods, all types of nucleotide or amino acid substitutions are assumed to occur with nearly-equal probability. However, certain substitutions, such as transitions, occur more often than transversion substitutions. It is therefore reasonable to assign different weights to different types of substitutions, as the weighted MP methods.

Informative sites In MP methods, nucleotide or amino acid sites that have the same type of nucleotide or amino acid sites (invariable sites) are eliminated from the analysis. Note that not all variable sites are useful. For instance, any nucleotide site at which only unique nucleotides exist (called a singleton site) can be explained by the same substitution pattern in all topologies. For a nucleotide site to be informative for constructing an MP tree, there must be at least two different kind of nucleotides, each represented in at least two sequences (taxa). These sites are called informative sites, or precisely, parsimony-informative sites. In the MP tree reconstruction, it is sufficient to consider only parsimony-informative sites.

Consistency index and homoplasy index It is important to have many informative sites to obtain reliable MP trees. However, when the extent of homoplasy (backward and parallel substitutions) is high, MP trees would not be reliable because these informative sites tend to be inconsistent. For this reason, people have proposed several measures to measure the extent of homoplasy, a widely used measure is the consistency index (CI) computed for all informative sites, and the homoplasy index $HI = 1 - CI$. When there is no backward and parallel substitutions, we have $CI = 1$ and $HI = 0$. In this case, the topology is uniquely determined by the principle of parsimony.

Searching for MP trees When the number of sequences or taxa (m) is small, say, $m < 10$, it is possible to compute the tree lengths of all possible trees and determine the MP tree. This type of search for MP trees is called the exhaustive search. While the number of topologies (m) rapidly increases, it is virtually impossible to examine all topologies if m is large. However, if we know clearly incorrect taxa, we can simply compute only for potentially correct trees. This type of search is called the specific-tree search.

There are two ways of obtaining MP trees when $m > 10$. One is to use the branch-and-bound method (Hendy and Penny 1982). In this method, the trees that obviously have a tree length longer than that of a previously examined tree are all ignored, and the MP tree is determined by evaluating the tree lengths for a group of trees that potentially have shorter tree lengths. This method guarantees finding all MP trees, although it is not an exhaustive search. However, even this method becomes very time-consuming if m is large. In this case, one has to use another approach called the heuristic search. In this method, only a small portion of all possible trees is examined, and there is no guarantee that the MP tree will be found. Yet, it is possible to enhance the probability of obtaining the MP tree by using several heurestic algorithms.

1.6 Maximum-likelihood (ML) methods for phylogenetic inference

Felsenstein (1981) has established the basic framework to infer the phylogenetic tree based on the principle of maximum likelihood. Later, Yang (1993, 1994a, 1994b, 1997) has extended this approach to tackle a number of issues in molecular evolution. In the following we use a simple example to illustrate the maximum likelihood (ML) methods for phylogenetic inference.

1.6.1 Likelihood function

Consider a simple tree of four taxa (DNA sequences with n nucleotides and no deletions/insertions) given in Fig. 1.3. Denote the observed nucleotides (A, T, C, or G) for sequences 1, 2, 3, and 4 at a given site by x_1, x_2, x_3, and x_4, respectively. The unobserved nucleotides at root or internal nodes 0, 5, and 6 are denoted by x_0, x_5, and x_6, respectively.

Let $P_{ij}(t)$ be the transition probability that nucleotide i at the initial time becomes nucleotide j at time t at a given site. Here i and j refer to any of A, T, C, and G. In the ML method, the rate of substitution (r) is allowed to vary from branch to branch, so that it is convenient to measure evolutionary time in terms of the expected number of substitutions ($v = rt$). In the following, we denote the expected number of substitutions for the i-th branch by $v_i = r_i t_i$. According to the Markov chain property, the likelihood function for a nucleotide site (the k-th) is then given by

$$l_k = \pi_{x_0} P_{x_0 x_5}(v_5) P_{x_0 x_6}(v_6) P_{x_5 x_1}(v_1) P_{x_5 x_2}(v_2) P_{x_6 x_3}(v_3) P_{x_6 x_4}(v_4)$$

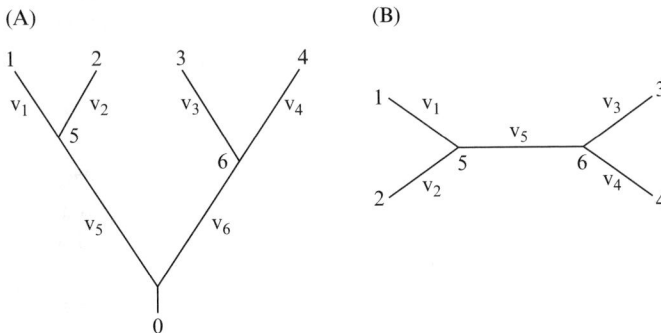

Fig. 1.3 A hypothetical phylogenetic tree (rooted and unrooted, respectively) that has been used to illustrate the building of likelihood function in the text. In both cases, two internal nodes are numbered 5 and 6, while four external branch lengths are denoted by v_1, v_2, v_3, and v_4, respectively. In the rooted tree, there are two internal branch lengths toward the ancestor O, with branch lengths v_5 and v_6, respectively, while these two branches are merged into one in the unrooted tree.

where π_{x_0} is the probability that node 0 (root) has nucleotide x_0, which is often set to be the same as the relative frequency of nucleotide x_0 in the entire set of sequences. In practice we do not know x_0, x_5, and x_6, so the likelihood will be the sum of the above quantity overall possible nucleotides at root 0 and internal nodes 5 and 6. That is,

$$L_k = \sum_{x_0}\sum_{x_5}\sum_{x_6} \pi_{x_0} P_{x_0 x_5}(v_5) P_{x_0 x_6}(v_6) P_{x_5 x_1}(v_1) P_{x_5 x_2}(v_2) P_{x_6 x_3}(v_3) P_{x_6 x_4}(v_4) \quad (1.31)$$

In ML methods, one must consider all nucleotide sites including invariable sites. Since the likelihood (L) for the entire sequence is the product of L_k's for all sites, the log likelihood of the entire tree becomes

$$\ln L = \sum_{k=1}^{n} \ln L_k$$

1.6.2 Time-reversibility and the root problem

To know $P_{ij}(v)$ explicitly, we have to use a specific substitution model. For instance, Felsenstein (1981) used the equal-input model, in which $P_{ii}(v)$ and $P_{ij}(v)$ ($i \neq j$) are given by

$$P_{ii}(v) = \pi_i + (1 - \pi_i)e^{-v}$$
$$P_{ij}(v) = \pi_j(1 - e^{-v})$$

where π_i is the relative frequency of the i-th nucleotide. When $\pi_i = 1/4$ and $v = 4rt$, the above equations become identical with those for the Jukes–Cantor (1969) model.

In particular, we consider a class of substitution models called time-reversible. This is because if we use a reversible model of nucleotide substitution for defining $P_{ij}(v)$, there is no need to consider the root (Fig. 1.3). A reversible model means that the process of nucleotide substitutions between time 0 and time t remains the same whether we consider the evolutionary process forward or backward in time. Mathematically, the reversibility condition is given by $\pi_i P_{ij}(v) = \pi_j P_{ji}(v)$, for all i and j. One can easily verify that the equal-input model satisfies this condition. When a reversible model is used, the number of nucleotide substitutions ($v_5 + v_6$) between nodes 5 and 6 of tree A remains the same irrespective of the location of the root 0. Therefore, we designate $v_5 + v_6$ in tree A by v_5 in tree B. Assuming that evolutionary change starts from some point of the tree, say, from node 5 for convenience, the likelihood function L_k can be rewritten in the following way

$$L_k = \sum_{x_5}\sum_{x_6} \pi_{x_5} P_{x_5 x_6}(v_5) P_{x_5 x_1}(v_1) P_{x_5 x_2}(v_2) P_{x_6 x_3}(v_3) P_{x_6 x_4}(v_4) \quad (1.32)$$

1.6.3 Search strategies for ML trees

In practice we have to consider all nucleotide sites. Since the likelihood (L) for the entire sequence is the product of all sites, the log-likelihood for a topology in general can be written as

$$\ln L = \sum_{k=1}^{n} \ln L_k = f(\mathbf{x}, \theta)$$

where $\mathbf{x}$ is a set of observed nucleotide sequences and θ is a set of parameters such as branch lengths, nucleotide frequencies, and substitution rates. In ML methods, the likelihood of observing a given set of sequence data for a specific substitution model is maximized for each topology, and the topology that gives the highest maximum-likelihood is chosen as the final tree.

Since the search for an ML tree is time-consuming, various heuristic methods for finding the ML tree have been proposed (e.g. Felsenstein 1981; Adachi and Hasegawa 1996). Though many of these algorithms, in principle, are similar to those used for obtaining minimum evolution or parsimony trees, their efficiencies in obtaining the correct topology are not necessarily the same.

For distantly-related protein-coding genes, the DNA likelihood may have some problems, because the synonymous substitution may have been saturated, suggesting that the stationary model of nucleotide substitution is no longer valid. In this case, the evolutionary change of protein sequences may be more appropriate. Kishino *et al.* (1990) proposed a protein-likelihood method, in which Dayhoff *et al.*'s (1978) empirical transition matrix for 20 different amino acids was used. Later, Adachi and Hasegawa (1996) used various transition matrices including the Poisson model, Jones *et al.*'s (1992) empirical transition matrix for nuclear proteins, and their own matrix for mitochondrial proteins.

1.7 Bayesian methods for phylogenetic inference

Bayesian methods provide a computationally efficient approach to infer the phylogenetic tree, by calculating the posterior distribution of phylogenetic trees (Huelsenbeck *et al.* 2001). Given the multiple alignment of sequence data, $\mathbf{D}$, the Bayes rule states that the posterior probability of the i-th possible tree topology denoted by T_i is given by

$$P(T_i|\mathbf{D}) = \frac{P(T_i)P(\mathbf{D}|T_i)}{\sum_T P(T)P(\mathbf{D}|T)} \tag{1.33}$$

where $P(T_i|\mathbf{D})$ is the probability of tree T_i given the sequence data $\mathbf{D}$, $P(\mathbf{D}|T_i)$ is the probability or likelihood of the data given tree T_i, and $P(T_i)$ is the prior probability of T_i. The denominator sums the probabilities over all possible trees. Moreover, for each of these possible trees, $P(\mathbf{D}|T_i)$ should be integrated over all possible values of the branch lengths of the tree and over the parameters of the model of sequence evolution. Let $\mathbf{t}$ be a vector of the branch lengths of the tree and $\mathbf{m}$ a vector of the parameters of the model of sequence evolution, then we have

$$P(\mathbf{D}|T_i) = \int_{\mathbf{t}} \int_{\mathbf{m}} P(\mathbf{D}|T_i, \mathbf{t}, \mathbf{m})P(\mathbf{t})P(\mathbf{m})d\mathbf{t}d\mathbf{m} \tag{1.34}$$

where $P(\mathbf{t})$ and $P(\mathbf{m})$ are the prior probabilities of the branch lengths and the parameters of the model.

Since the number of possible unrooted topologies for n species is $(2n-5)!/2^{n-3}(n-3)!$, the summation in the denominator is over a high number of topologies even for ten sequences. This problem can be solved via the MCMC sampling algorithm (Hastings 1970; Metropolis *et al.* 1953). Under the MCMC methods, a Markov chain is constructed, the states of which are different phylogenetic trees (Huelsenbeck *et al.* 2001; Larget *et al.* 1999; Mau *et al.* 1999; Rannala and Yang 1996; Yang and Rannala 1997). At each step in the chain a new tree is proposed by altering the topology, or by changing branch lengths, or by changing the parameters of the model of sequence evolution. The Metropolis–Hastings algorithm is then used to accept or reject the new tree. A newly proposed tree that improves upon the previous tree in the chain is always accepted, otherwise it is accepted with probability proportional to the ratio of its likeness to that of the previous tree in the chain. If such a Markov-chain has been run long enough, it reaches a stationary distribution.

The argument for Bayesian inference of phylogeny can be concisely stated as follows. At stationarity, the Metropolis–Hastings sampling algorithm ensures that the Markov chain wanders through the universe of trees, sampling better and worse trees, rather than inexorably moving toward 'better' trees as an optimizing approach would do. A properly constructed chain samples trees from the posterior density of trees in proportion to their frequency of occurrence in the actual density. Hence, the Markov chain draws a sample of trees that can be used to approximate the posterior distribution. In the current implementation, the stationary distribution simultaneously samples the posterior density of trees, the posterior distribution of the branch lengths, and parameters of the model of sequence evolution. For a desirable degree of precision, in practice the chain should be allowed to run perhaps hundreds of thousands or million of trees.

1.8 Ancestral sequence inference

Ancestral sequence inference has been shown to be useful to predict the ancestral function of a gene family, to detect positive selection, and to reconstruct the ancestral genome (Golding and Dean 1998; Soyer and Bonhoeffer 2006). There are two major types: the parsimony approach and the probabilistic approach, which are briefly discussed below.

1.8.1 The maximum parsimony approach

The idea of maximum parsimony is to identify the ancestral states at each node of a tree that minimize the number of character (nucleotide or amino acid) changes needed to explain the observed differences among the extant sequences. We use the original algorithm developed by Fitch (1971) as an example (for nucleotides) to show the principle. To exemplify the Fitch algorithm, we consider a simple five-taxon tree in Fig. 1.4. For the character illustrated, the data observed are $X_1 = A$, $X_2 = C$, $X_3 = G$, $X_4 = C$, and $X_5 = T$. At node X_6, the intersection of the sets of its two descendants X_1 and X_2 is $(A) \cap (C) = \phi$ (here ϕ means empty). If it is empty, the union set must be assigned to X_6: $(A) \cup (C) = (A, C)$. Likewise, the union set of X_3 and X_4 is assigned

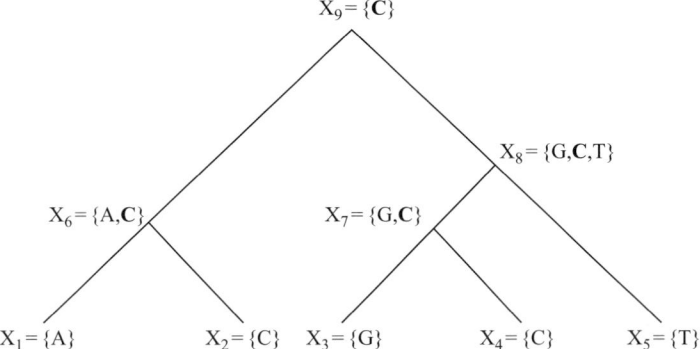

Fig. 1.4 Given the current observed nucleotide pattern in five sequences at a site (denoted by X_1 to X_5), the ancestral state of each internal node (X_6 to X_9) can be inferred by the parsimony algorithm or the probabilistic method.

at node X_7 with $(G) \cup (C) = (G, C)$. Now the set at node X_8 can be determined, since the interaction of sets X_5 and X_7 is again empty, the union of these sets (G, C, T) is assigned. Finally the set in the root X_9 is the intersection of the sets X_8 and X_6: $(A, C) \cap (G, C, T) = (C)$. Three union operations were needed, suggesting a minimum of three changes is needed for this reconstruction.

In the next step, the ancestral states are determined (marked in bold type in Fig. 1.4) by traversing the tree in pre-order (from the root to the extant taxon, or leaves). First the state C is determined at the root, the state at X_8 is also set to C because this state is in the ancestral state in the parent node (X_9) and is also a member of set at that node (X_8). Similarly, the state at X_6 is C and the state at X_7 is also C. Because the Fitch algorithm penalizes equally any change among the four nucleotide states, the procedure can result in multiple ancestral state inferences with equal parsimony. The Sankoff (1975) algorithm is a generalization of Fitch's original version, allowing different costs for different character changes.

1.8.2 The probabilistic (Bayesian) approach

Though the parsimony method is simple and intuitive, the major shortcoming is the lack of statistical evaluation for the reconstructed ancestral states. For instance, the parsimony method is not statistically meaningful for discriminating among equally parsimonious reconstructions. To solve this problem, a number authors (e.g. Schluter 1995; Yang *et al.* 1995; Koshi and Goldstein 1996; Pupko *et al.* 2000, 2002) developed efficient algorithms using the probabilistic models.

The probabilistic (Bayesian) approach for ancestral state inference has two steps in principle. We use Fig. 1.4 for illustration. First, it calculates the likelihood (the joint probability) of the observed nucleotide pattern $X_1, \ldots, X_5 = (A, C, G, C, T)$ under the given phylogeny. Denote four nucleotides by 1, 2, 3, and 4, respectively, it formally can be written as

$$P(X_1, \ldots, X_5) = \sum_{x_9=1}^{4} f_{x_9} \sum_{x_8=1}^{4} \sum_{x_7=1}^{4} \sum_{x_6=1}^{4} P_{x_9 x_6} P_{x_9 x_8} P_{x_6 x_1} P_{x_6 x_2} P_{x_8 x_5} P_{x_8 x_7} P_{x_7 x_4} P_{x_7 x_5}$$

where f_{x_9} is the nucleotide frequency at the root (X_9); $P_{x_i x_j}$ is the transition probability from nodes X_i to X_j; here branch lengths are omitted for simplicity. Note that in most cases, the probabilistic model for ancestral sequence inference is time-reversible. As a result, the position of a root does not affect the likelihood so that the ancestral root state cannot be reconstructed. Next we rewrite $P(X_1, \ldots, X_5)$ as

$$P(X_1, \ldots, X_5) = \sum_{x_8=1}^{4} \sum_{x_7=1}^{4} \sum_{x_6=1}^{4} P(X_1, \ldots, X_5; x_6, x_7, x_8)$$

where $P(X_1, \ldots, X_5; x_6, x_7, x_8) = \sum_{x_9=1}^{4} f_{x_9} P_{x_9 x_6} P_{x_9 x_8} P_{x_6 x_1} P_{x_6 x_2} P_{x_8 x_5} P_{x_8 x_7} P_{x_7 x_4} P_{x_7 x_5}$. That is, the likelihood $P(X_1, \ldots, X_5)$ is a sum of $4 \times 4 \times 4 = 64$ different terms, each corresponding to a specific ancestral sequence assignment at nodes X_6, X_7 and X_8. Therefore, the joint nucleotide assignment (x_6, x_7, and x_8) which contributes the most to the above likelihood is called the ancestral state inference, which is explictly given by the expression

$$\text{argmax}_{x_6, x_7, x_8} \left[P(X_1, \ldots, X_5; x_6, x_7, x_8) \right]$$

The maximum of the above expression is the likelihood of the joint reconstruction. In essence, the posterior probability of x_6, x_7 and x_8 given the dataset ($X_1, \ldots, X_5$) is

$$P(x_6, x_7, x_8 | X_1, \ldots, X_5) = \frac{P(X_1, \ldots, X_5; x_6, x_7, x_8)}{P(X_1, \ldots, X_5)}$$

In other words, one may call the joint nucleotide assignment (x_6, x_7 and x_8) which has the highest posterior probability (given the data) the ancestral state inference.

In practice, particularly when the number of sequences is large, the computational time required for joint likelihood (or posterior probability) becomes huge. One feasible approach is to use the single-node inference. This algorithm considers the nucleotide assignment for each internal node separately. For instance, one may write the likelihood with respect to node X_8

$$P(X_1, \ldots, X_5) = \sum_{x_8=1}^{4} \left[\sum_{x_7=1}^{4} \sum_{x_6=1}^{4} P(X_1, \ldots, X_5; x_6, x_7, x_8) \right]$$

Hence, the nucleotide assignment at node X_8 which contributes the most to the likelihood is given by

$$\text{argmax}_{x_8} \left[\sum_{x_7=1}^{4} \sum_{x_6=1}^{4} P(X_1, \ldots, X_5; x_6, x_7, x_8) \right]$$

It has been shown that the assignment of characters (nucleotides or amino acids) at ancestral nodes could be affected by the substitution model of nucleotides or amino acids. For instance, two almost equal likelihood assignments under the simple Juke–Cantor (1969) model may have different outcomes under more sophisticated models. Nevertheless, ancestral sequence inference is fairly robust against other parameter estimations such as branch lengths or the rate variation among sites, though the statistical evaluation for the inference reliability could vary; see Yang *et al.* (1995) and Pupko *et al.* (2002) for detailed discussions.

1.8.3 Deletions and insertions

None of the models described thus far consider deletions and insertions (called indels or gaps). A common technique in phylogenetic inference is to exclude all sites in which at least one sequence contains a gap. However, for ancestral sequence inference, the goal is to infer the most likely, nearly complete, ancestral sequence. To this end, it is essential to determine whether a character (nucleotide or amino acid) or gap is the ancestral state. This problem has not been well solved. In the following we introduce some tentative approaches.

Gap as additional character: One approximation to escape this problem is to represent a gap by adding an additional character to the model (thus creating an alphabet of size 21 for amino acids, or five for DNA/RNA). In spite of easy implementation, there are two main difficulties with this approach: First, the probabilities of such transitions from each character state to a gap and *vice versa* are unclear. Second, this approach assumes independence among sites. Thus, an insertion or deletion of two sites will be considered as two independent so-called character-to-gap transitions, rather than the biologically more reasonable explanation of a single two-site indel event.

Edward-Shields method: The algorithm developed by Edward-Shields (2005) first approximates the probabilities of gaps at each site and internal node, using a two-state character model (0 is a gap site and 1 for a non-gap site). Once the ancestral state (0/1) for each node is determined, the nongapped sites are estimated in an informal likelihood approach using probabilities derived from empirical substitution matrices.

An integrated framework: Recently, we proposed (unpublished results) an integrated approach to infer a nearly complete ancestral sequence. The novelty of this approach is in dividing the ancestral sequence inference algorithm into several separate tasks:

(1) For nucleotide or amino acid sites without any gap, follow the previous probabilistic method to assign the most likely character states at internal nodes.
(2) Given the phylogeny, count the number distribution of indels with the gap length. Assume that the change of indels follows a Poisson process with the parameter $v_k T$ (v_k is the rate of indels with gap length k, and T is the total evolutionary time of the tree). According to the study of Gu and Li (1995), we use a log-link function $v_k = v_0/(1 + b \ln k)$ ($k \geq 1$) to measure the length-dependent evolutionary rate of indels. Apparently, v_k is inversely related to the

gap length. The unknown parameters, v_0 and b, can be estimated from the size distribution of indels.

(3) For a given site with indels, we adopt the additional character approach with an important extension. That is, the rate from other characters to the gap or *vice versa* is proportional to the gap-length factor $v_0/(1 + b \ln k)$.

We have conducted some preliminary analyses and found that the performance is, in general, satisfactory, as long as the multiple alignment is reliable.

1.9 Rate variation among sites

It is well known that different amino acid residues of a protein may have different functional constraints such that the substitution rate varies among sites. Although this phenomenon was first described over 20 years ago (Uzzel and Corbin 1971), its importance for molecular evolutionary study has not been recognized until recently (Nei and Kumar 2000; Yang 1993; Gu 1999; 2001b; 2007a, 2007b). In particular, the gamma distribution has been widely used for modeling the rate variation among sites (Yang 1993; Gu *et al.* 1995). Under this model, the variation of substitution rate (λ) among sites can be described as follows:

$$\phi(\lambda) = \frac{\beta^\alpha}{\Gamma(\alpha)} \lambda^{\alpha-1} e^{-\beta\lambda}$$

where the shape parameter α is important because it describes the degree of rate variation, and β is a scalar. Since $1/\sqrt{\alpha}$ is the coefficient of variation of λ, the larger α is, the weaker the rate variation is, and $\alpha = \infty$ means a uniform rate among sites.

Several methods have been developed for estimating α from sequence data, these methods can be classified into two groups. The first group is the maximum-likelihood (ML) approach, which was constructed under the framework of Felsenstein (1981) (e.g. Yang 1993; Gu *et al.* 1995). However, the algorithms developed by these authors for maximizing the likelihood function are time consuming. This problem has been solved by the approximate method of discrete-gamma distribution (Yang 1994b). The second group of methods for estimating α is usually called the parsimony method, and has been widely used because it is computationally fast (e.g. Uzzel and Corbin 1971, Holmquist *et al.* 1983; Tamura and Nei 1993; Sullivan *et al.* 1995; Tourasse and Gouy 1997). In these methods, the principle of parsimony (Fitch 1971) was used to infer the (minimum required) number of substitutions. Since the parsimony method tends to underestimate the number of substitutions, it is known that the shape parameter (α) can be seriously overestimated: in other words, the degree of rate variation among sites can be underestimated (Wakeley 1993).

Since rate variation among sites has been shown to be important for evolutionary functional analysis of protein families (Gu 1999, 2001b, 2006), as well as the estimation of gene pleiotropy (Gu 2007a), implementation of a statistically unbiased and computationally fast method is desirable. Gu and Zhang (1997) proposed a simple ML method that has two steps: (1) At each site, the expected number of substitutions corrected for multiple hits is estimated by a likelihood approach, based on the phylogeny and inferred ancestral sequences; and (2) the ML estimate of α is obtained under a negative binomial distribution (Uzzel and Corbin 1971) using the estimated number of substitutions.

1.9.1 Number of substitutions at a site

The number of substitutions at a site cannot be observed from present-day sequences, so it has to be inferred. The traditional method of inference invokes the parsimony principle (Fitch 1971), which tends to underestimate the true number of substitutions (Gu and Zhang 1997; Zhang and Gu 1998). To understand the bias caused by parsimony, it is important to distinguish between the number of substitutions (k) and the number of branches on which the amino acids (or nucleotides) at the two ends of a branch are different (m); in the following, m is also concisely called the number of changes. For given sequence data with a known tree, the difference between m and k is due to multiple substitutions that may occur when the branch is long, resulting in $m \leq k$. It should be noted that the minimum-required number of substitutions is actually an inference of m rather than k, because the possibility of multiple hits is completely neglected.

Gu and Zhang (1997) developed an asymptotically unbiased method to estimate k. For simplicity, we only discuss amino acid sequences here; it is virtually the same for nucleotide sequences. Suppose we have a protein dataset with n homologous sequences, whose phylogenetic tree (topology) is known or can be inferred. It is known that the total number of branches for an unrooted tree is $M = 2n - 3$, or $M = 2n - 2$ for a rooted tree. At a given site, we assume that k along the tree follows a Poisson distribution, whose expectation is written by $\bar{k} = uB$, where B is the total branch length of the tree and u is the evolutionary rate at this site. It is noteworthy that k under the Poisson model is a random variable. Our purpose is to estimate the expected number of substitutions ($\bar{k}$), which will be used for estimating α.

The number of substitutions at a given site that occur on branch i also follow a Poisson distribution with the expectation ub_i, where u is the site-specific rate and b_i is the length of branch i. Because the expected number of substitutions at this site along the phylogeny is $\bar{k} = uB$, we have $ub_i = \bar{k}b_i/B$. Thus, the probability of no change on branch i (i.e. the amino acids at the two ends of this branch are the same) is given by

$$p_i = \exp\{-\bar{k}b_i/B\}$$

and the probability of a change (i.e. the amino acids are different at the two ends of the branch) is

$$q_i = 1 - p_i = 1 - \exp\{-\bar{k}b_i/B\}$$

For a given site, the branches along the tree can be divided into two groups. The first group, denoted by G_1, includes the branches on which (amino acid) changes occur, and the second group, denoted by G_0, includes the branches on which no changes occur. Obviously, the total number of branches in G_1 is equal to m at the site, and the total number of branches on G_0 is therefore given by $M - m$. Then, when the information about groups G_1 and G_0 at a site is known, the (conditional) likelihood function can be written as

$$L = \prod_{i \in G_1} q_i \prod_{j \in G_0} p_j = \prod_{i \in G_1} \left[1 - \exp\{-\bar{k} b_i / B\} \right] \prod_{j \in G_0} \exp\{-\bar{k} b_j / B\} \qquad (1.35)$$

The subscripts under the product signs mean that branch i belongs to group G_1 and branch j belongs to group G_0, respectively.

After re-numbering branches in group G_1 from 1 to m and branches in group G_0 from $m + 1$ to M, Gu and Zhang (1997) have shown that the ML estimate equation with respect to $\bar{k}$ can be concisely expressed by

$$\sum_{i=1}^{m} \frac{b_i / B}{1 - e^{-\hat{k} b_i / B}} = 1 \qquad (1.36)$$

The ML estimate of the expected number of substitutions ($\hat{k}$), which is the positive solution of the above equation, depends on m and the estimated branch lengths. If for every branch, b_i / B is so small that $1 - e^{-\hat{k} b_i / B} \approx \hat{k} b_i / B$, then $\hat{k} \approx m$. This result is consistent with the intuition that the expected number of substitutions estimated approaches the number of changes when all branch lengths in the tree are short. On the other hand, if all branch lengths are the same, i.e. $b_i = b$ and so $B = Mb$ (M is the number of branches), it can be simplified as follows

$$\hat{k} = -M \ln \left(1 - \frac{m}{M} \right) \qquad (1.37)$$

In the above formulation, we assume that ancestral sequences are known. Thus, under a given phylogenetic tree, it is easy to classify a branch into G_0 or G_1 by simply comparing the amino acids at the two ends of the branch at each site, and count the number of changes along the tree (m). In practice, ancestral sequences have to be inferred by the Bayesian-based methods (e.g. Schluter 1995; Yang *et al.* 1995; Koshi and Goldstein 1996). In this approach, the amino acid assignment that has the highest (posterior) probability is chosen to represent the inferred ancestral amino acids at this site.

1.9.2   Estimation of α

If amino acid (or nucleotide) substitutions at each site follow a Poisson process, and the substitution rate (λ) varies among sites according to the gamma distribution $\phi(\lambda)$, the number of sites with the occurrence of k substitutions follows a negative binomial distribution, i.e.

$$f(k) = \frac{\Gamma(\alpha + k)}{k! \Gamma(\alpha)} \left(\frac{D}{D + \alpha} \right)^k \left(\frac{\alpha}{D + \alpha} \right)^\alpha \qquad (1.38)$$

where D is the average number of substitutions per site along the tree.

The ML approach for estimating the parameter α from a negative binomial distribution was clearly discussed by Johnson and Kotz (1969); it was also used by Sullivan *et al.* (1995) and Tourasse and Gouy (1997) for the rate variation among sites. In our case, the difference from the standard algorithm is that the number of substitutions at a site is replaced by its expectation $\hat{k}$. Therefore, the log-likelihood function can be written as

$$\ln L = \sum_{i=1}^{N} \ln f(\hat{k}_i)$$

where N is the total number of sites and $\hat{k}$ is the estimate of the expected number of substitutions at site i, which is not necessarily an integer. One can easily show that the ML estimate of D is the same as that for the normal case, which is given by $\hat{D} = \sum_{i=1}^{N} \hat{k}_i/N$. There is no simple solution for the estimate of α, but it can be numerically obtained, and the sampling variance of α can also be approximately obtained.

2
Basics in Bioinformatics and Statistics

2.1 Bioinformatic resources for evolutionary genomics

The available web-based genome resources provide great opportunities for biomedical scientists to identify functional elements in a particular genome region or explore the evolutionary pattern of genome dynamics. Because of the broad scope of comparative genomics, it is difficult to address all its aspects here. We have therefore selected several important topics and give a brief review for the availability of web-based database and software, as shown in Table 2.1 (Gu and Su 2005).

Genome databases for comparative genomics: Though researchers always find these valuable for obtaining a wealth of useful information, it should be noted that genome-wide databases change rapidly both in their internal implementation and in the dataset updating. For instance, GALA is a database of genome alignments and annotation (Giardine *et al.* 2003), which provides access to information on genes (known and predicted), gene ontology, expression patterns, genome alignments, and conserved transcription factor binding sites predicted by the TRANSFAC weight matrix (Wingender *et al.* 2001). Hence, given a set of genes expressed in a particular tissue, GALA is able to identify all the predicted binding sites for one or more transcription factors of interest that are both conserved in mammals. Another example is EnsMart, a branch of the Ensembl project (Kasprzyk *et al.* 2004), which integrates data from Ensembl and several other resources, using a 'warehouse star-schema' with central biological objects (e.g. genes or SNPs) connected to a set of satellite tables, such as disease, transcript, and Protein FAMily (PFAM) attributes. Thus, EnsMart provides users with fast and effective access to deep data in and around genes.

Multi-genome alignment and gene prediction: Genome-wide alignment servers for two closely related species are available on the web. The BLAST suite of tools (Altschul *et al.* 1990; 1997), implemented at NCBI, is the most frequently used one. Several servers were specially designed to align two or more long genomic sequences at high sensitivity while detecting common rearrangements or duplications, e.g. Pip-Maker (Schwartz *et al.* 2000), MultiPipMaker (Schwartz *et al.* 2003), and zPicture (Ovcharenko *et al.* 2004), VISTA, and MAVID (Mayor *et al.* 2000; Bray *et al.* 2003). These servers are suitable for species such as those from different mammalian orders. Several pipelines have been designed for mammalian genome alignment (Brudno *et al.*

Table 2.1 Some examples of websites for tools and databases useful for comparative genomics.

Tool or database	Website
NCBI	http://www.ncbi.nlm.nih.gov
EMSEMBL	http://www.ensembl.org
UCSC Genome Browser	http://genome.uscs.edu/
EnsMart	http://www.ensembl.org/Multi/martview
NCBI BLAST	http://www.ncbi.nlm.nih.gov/BLAST/
WU-BLAST	http://blast.wustl.edu/
GALA	http://gala.cse.psu.edu/
PipMaker and MultiPipMaker	http://bio.cse.psu.edu/pipmaker/
zPicture	http://zpicture.dcode.org/
VISTA	http://www-gsd.lbl.gov/vista/
MAVID	http://baboon.math.berkeley.edu/mavid/
MEME	http://meme.sdsc.edu
GLASS and Rosetta	http://crossspecies.lcs.mit.edu/
SGP2	http://genome.imim.es/software/sgp2/
TWINSCAN	http://genes.cs.wustl.edu/query.html
GeneID	http://www1.imim.es/geneid.html
DOUBLESCAN	http://www.sanger.ac.uk/Software/analysis/doublescan/
TRED	http://rulai.cshl.edu/TRED
RNAdb	http://research.imb.uq.edu.au/rnadb/
NONCODE	http://noncode.bioinfo.org.cn
PAML	http://abacus.gene.ucl.ac.uk/software/paml.html
DIVERGE	http://xgu.zool.iastate.edu
Mgenome	http://xgu.zool.iastate.edu
GRIMM	http://www-cse.ucsd.edu/groups/bioinformatics/GRIMM/
GRAPPA	http://www.cs.unm.edu/~moret/GRAPPA/
TRANSFAC	http://www.gene-regulation.de/
FootPrinter and PhyME	http://bio.cs.washington.edu/software.html
MSARi	http://theory.csail.mit.edu/MSARi/
RNAz	http://www.tbi.univie.ac.at/~wash/RNAz/

2003; Couronne *et al.* 2003; Schwartz *et al.* 2003). For more distant species, or ancient paralogous genes, different alignment methods should be recommended. One major application is to look for common motifs in the upstream regions of coexpressed genes. Two examples of these approaches are MEME and Gibbs sampling (Thompson *et al.* 2003).

One application of multi-genome alignment is to improve the efficiency of gene finding. ROSETTA reconstructs colinear gene structures from global alignments and defines exons as subsequences bounded by splice sites (Batzoglou *et al.* 2000). SGP1 reconstructs genes from a collection of local alignments between two sequences (Wheeler *et al.* 2002), while SGP2 assesses the reliability of gene models predicted by GENEID (Parra *et al.* 2003), a conventional gene predictor (Guigo 1998). Similarly, TWINSCAN represents a direct extension of the Genscan algorithm that integrates conservation information between two sequences (Korf *et al.* 2001; Burge and Karlin 1997). DOUBLESCAN uses a Pair Hidden Markov Model (Pair HMM) to reconstruct gene structures from a series of local alignments created with BLAST (Meyer and Durbin 2002).

Evolutionary approaches for protein function detection: There are many software packages available for molecular evolutionary analysis. Here we only illustrate a few examples. MEGA (Kumar *et al.* 2008) is a widely-used software package that is user-friendly. PAML is a software package that includes a wealth of methods for statistically testing the evolutionary pattern of coding sequences, which can be used for positive selection detection (Yang 1997). For instance, PAML is able to estimate the ratio of nonsynonymous rate to synonymous rate at each amino acid residue along the lineages of a given phylogenetic tree. DIVERGE is a program to study functional divergence of a protein family by detecting site-specific change in evolutionary rate using a multiple alignment of amino acid sequences for a given phylogenetic tree (Gu 1999; Gu and Vander Velden 2002). It first conducts a statistical test for site-specific rate shifts along the tree, and then predicts candidate amino acid residues responsible for functional divergence based on posterior analysis. These results can be mapped on the 3D protein structure if available.

Multiple genome rearrangement by signed reversal: For comparative gene mapping, it is important to reconstruct the ancestral gene orders for a given set of current genomes. Mathematically, it becomes the problem of signed reversals, that is, how the genomes evolve from a common ancestral genome based on signed reversal of genes or gene sets. Since this problem is NP-hard (Caprara 1999), most work has focused on heuristic algorithms for reconstructing the gene order of ancestral genomes. Sankoff *et al.* (1996) searched for the optimal ancestral genome for a median problem upon a grid. Bourque and Pevzner (2002) designed MGR algorithm to reconstruct ancestral genomes by a greedy-split strategy. Wu and Gu (2002, 2003) improved the searching accuracy by a nearest path search algorithm, and developed a neighbor-perturbing algorithm to reconstruct optimal gene order of ancestral genomes.

Identification of functional noncoding elements by comparative genomics: Although the majority of eukaryote genomes are noncoding regions and used to be regarded as 'junk DNA', recent studies have indicated that the noncoding region harbors important functional elements such as cis-regulatory modules (CRMs) (Dermitzakis *et al.* 2002; Gibbs 2003). Computational detection of these functional noncoding elements has been extremely challenging. It has been recognized that comparative genomics may be a promising approach to solve this problem. For example, 'Phylogenetic footprinting' focuses on the discovery of novel regulatory elements based on the sequence conservation among a set of orthologous noncoding regions. Using this method, many successful motif discovery programs, such as, Gibbs sampler (Lawrence *et al.* 1993), MEME (Bailey and Elkan 1995), Consensus (Hertz and Stormo 1999), AlignAce (Roth *et al.* 1998), ANN-Spec (Workman and Stormo 2000), FootPrinter (Blanchette and Tompa 2003), and PhyMe (Sinha *et al.* 2004) have been developed. For noncoding RNA (ncRNA) elements, many tools have been developed to identify the evolutionary conservation of secondary structures serving as compelling evidence for biologically relevant ncRNAs function. Some examples are QRNA (Rivas and Eddy 2001), DDBRNA (di Bernardo *et al.* 2003), MSARI (Coventry *et al.* 2004), and RNAZ (Washietl *et al.* 2005). In addition, a number of databases about functional noncoding elements were also available, for example, TRED (Zhao *et al.* 2005), RNAdb (Pang *et al.* 2005) and NONCODE (Liu *et al.* 2005).

Given that substantial resources are available, the challenge in fact turns on how to transfer the explosion in genomic data to biological knowledge. The internet has facilitated the transition process, but the progress depends on the development of new ideas and analysis pipelines that combine many approaches.

2.2   Basic statistics for homologous search

Once genomic scientists obtain a genomic sequence by contig assembly, the next step is to find out what the long stretch of nucleotides means. This is the subject of sequence annotation. A pivotal method of sequence annotation is based on known genes in sequence databases with homologous search. Two widely-used homologous search tools are BLAST (Altschul *et al.* 1990; Altschul *et al.* 1997), and FASTA (Pearson and Lipman, 1988). Conventional methods for homologous search are based on local sequence alignment using dynamic programming (Smith and Waterman 1981). For a given scoring scheme, such methods will guarantee the finding of the optimal alignment, but are computationally slow. FASTA and BLAST have used heuristic methods for the similarity search. Though they may miss some homologous sequences, both are fast enough for a large-scale genome analysis. In the following, we discuss the statistical meaning of the so-called E-score, which plays a critical role to select a cutoff that is statistically appropriate.

Let P_A, P_C, P_G, and P_T be the nucleotide frequencies of A, C, G, and T in a target (database) sequence, respectively. For a query sequence (Q) with the sequence length (L_Q), at random, the probability of the query sequence having a perfect match of the target sequence D with the sequence length (L_D) is given by

$$p = P_A^{n_A} P_C^{n_C} P_G^{n_G} P_T^{n_T} \qquad (2.1)$$

where the number of each type of nucleotide is denoted by n_A, n_C, n_G, or n_T, respectively. Since the number of possible perfect matching operations between the query and target sequences is $n = L_D - L_Q + 1$, the probability distribution of the number of matches (x) follows (approximately) a binomial distribution

$$P(x) = \frac{n!}{x!(n-x)!}p^x(1-x)^{n-x} \tag{2.2}$$

where p is given by Eq.(2.1). When $np < 1$ and n is large, the binomial distribution can be approximated by the Poisson distribution with the mean and variance equal to $u = np$, that is,

$$P(x) \approx \frac{u^x}{x!}e^{-u}$$

If we assume $P_A = P_C = P_G = P_T = 0.25$, from Eq.(2.1) the probability of finding an exact match of at least L consecutive letters is $p = 0.25^L$. The match of L consecutive letters between Q and D can happen at $m = L_Q - L + 1$ positions on Q and at $n = L_D - L + 1$ positions on D. In BLAST, m refers as the *effective length of query* and n as the *effective length of database*. In total there are $m \times n$ possible matching operations, each with a probability of 0.25^L of getting a match of L consecutive letters. Putting together, the expected number of matches with length at least L is therefore given by $E = mn \times 0.25^L$. In BLAST and FASTA literature, E can be written as a general form

$$E = mne^{-\lambda L} \tag{2.3}$$

to take the unequal nucleotide frequencies into account. The BLAST output for nucleotide sequences typically reports $\lambda = 1.37$ which is based on computer simulations of more realistic sequences (Altschul 1996; Altschul *et al.* 1997; Pearson 1998; Waterman and Vingron, 1994).

The E-value can be used as the u parameter in the Poisson distribution for calculating the probability of having $0, 1, \ldots, x$ matches that are as good or better than the reported match. In particular, the probability of having at least one match (i.e. $x \geq 1$) that is as good or better than the reported match, is given by

$$P(x > 1) = 1 - P(x = 0) = 1 - e^{-E} = 1 - exp\left\{-mn \times e^{-\lambda L}\right\} \tag{2.4}$$

which is a special form of the extreme value distribution (EVD). In fact, the EVD is used in BLAST and FASTA to calculate statistical significance to a match score between two sequences.

The E-value is the expected number of random matches with match scores as equally good as the reported one, or even better. It is not a probability. Nevertheless, when E is very small, it can be approximately interpreted as the probability of finding one match that is no worse than the reported one, as $1 - e^{-E} \approx E$ when $E \to 0$.

2.3   Sequence alignment

In bioinformatics, a sequence alignment is to identify sequence similarity that may be the result of functional, structural, or evolutionary relationships between the sequences (Needleman and Wunsch 1970). If two sequences in an alignment share a common ancestor, mismatches can be interpreted as point mutations and gaps as indels (insertion or deletion mutations) introduced in one or both lineages. In the protein sequence alignment, the degree of similarity between amino acids occupying a particular position roughly indicates how conserved a particular sequence region. Indeed, absence of substitutions in a particular region of the sequence may suggest that this region has structural or functional importance. Hence, a reliable sequence alignment is critical for almost all comparative genomic analyses.

2.3.1 Pairwise alignment

Given two strings $S = s_1 s_2 \dots s_n$ and $T = t_1 t_2 \dots t_m$, a pairwise alignment of S and T is defined as an ordered set of pairings of (s_i, t_j) and of gaps $(s_i, -)$ and $(-, t_j)$. An optimal alignment is operationally defined as the pairwise alignment with the highest alignment score for a given scoring scheme. The dynamic programming (Smith and Waterman 1981) guarantees that the resulting alignment is the optimal alignment or one of the equally optimal alignments, when the penalty score is specificed. Apparently, the alignment is reduced to the two original strings when all gaps in the alignment are deleted. Readers may find a detailed introduction about the dynamic programming in bioinformatics textbooks (Ewens and Grant 2005). Here we focus on the score scheme that may practically affect the performance of sequence alignment. A simple scheme assumes the same mismatch score for any type of mismatch. More complicated score schemes are discussed below.

Similarity matrix: nucleotide sequences

There are two types of nucleotide changes: the transitions (i.e. substitutions between nucleotides A and G and between C and T) and the transversions (when A or G is replaced by C or T). Since transitions generally occur more frequently than transversions, transitions should be penalized less than transversions. Moreover, there are often ambiguous bases in the input of sequences, e.g. R for A or G and Y for C or T. An A-R pair is neither a strict match nor a strict mismatch, but has a probability, say, 0.5, being a match or a transition. Xia (2001) proposed the 'transition bias matrix' that has different scores of transition/transversion and ambiguous nucleotides.

Similarity matrix: protein sequences

As amino acid can differ from each other in volume, charge, polarity, and many other ways, amino acid replacements between very different amino acids are generally against the purifying selection due to strong functional constraints. In practice this effect can be measured by an empirical amino acid substitution matrix. Frequently used substitution matrices for protein sequence alignment are the classical PAM matrix (Dayhoff *et al.* 1978) and the BLOSUM matrix (Henikoff and Henikoff 1992) that is now widely adopted. Both PAM and BLOSUM matrices are derived from sequence alignment related proteins. The notation BLOSUM-xx matrix is based on sequence blocks with no less than xx per cent divergence. The default matrix in BLAST is BLOSUM 62, which is calculated from comparisons of sequences with no less than 62 per cent divergence.

Gap penalty function

An important extension of the simple scoring scheme is to introdue the gap penalty function instead of a constant gap penalty. A simple linear gap function is

$$G(x) = a + bx$$

where x is the length of the gap, and a and b are the gap open and gap extension penalties, respectively. The gap penalty increases linearly with the length of the gap. The linear gap penalty function, which is used in BLAST (Altschul *et al.* 1990; Altschul *et al.* 1997), has one particular advantage, that is, it allows the alignment to be completed in time proportional to MN, where M and N are the length of the two sequences to be aligned.

From a biological point of view, the alignment with two independent gaps $(x = 1)$ is less likely than the one with only one gap of length $x = 2$. The gap open penalty catches this intuition. On the other hand, it has been observed that the linear increase of the gap penalty with the gap length may result in fragmented alignment when there is a large gap. It is therefore suggested that a log-linear gap length penalty may be more appropriate, that is

$$G(x) = a + b \ln x$$

(Gu and Li 1995).

Profile alignment

Profile alignment aligns one sequence (designated T) against a set of already aligned sequences in the form of a profile (designated S), or align two profiles S_1 and S_2. As shown below, this is the key step for multiple sequence alignment algorithms such as CLUSTAL. Among various approaches to profile alignment, the simplest one is to get a consensus sequence from S and align with T by using the pairwise alignment method. In case it inserts a gap, this gap must correspond to all sequences in S.

A reasonable extension of this consensus approach is to represent S with a site-specific frequency profile. That is, S can always be represented by a site-specific profile in the form of a $N \times L$ matrix where N is the number of symbles (e.g. $N = 21$ for 20 amino acids plus gap), and L is the sequence length. The next step is to perform the dynamic programming based on a special scoring scheme for comparing two profiles rather than two sequences.

2.3.2 Multiple alignment with a guide tree: Clustal

Though it is an extension of pairwise alignment, the multiple sequence alignment is computationally difficult and the most formulations of this problem lead to NP-complete combinatorial optimization problems. Nevertheless, a number of bioinformatic tools with heuristic algorithms are suitable for aligning a high number of sequences in practice.

The most well-known representative of this approach is the Clustal family of programs (Higgins and Sharp 1988; Thompson *et al.* 1994; Higgins *et al.* 1996). Multiple alignment of N sequences in Clustal consists of three major steps. First, perform all $N(N-1)/2$ pairwise alignments by dynamic programming. For each pairwise alignment, calculate the alignment score. Second, construct a guide tree by converting these pairwise alignment scores into distance measures of sequence similarity, and then use the neighbor-joining method (Saitou and Nei 1987) to build the guide tree. And third, traverse the node to align sequences by pairwise alignment and profile alignment.

The first two steps are straightforward. The final step is based on the guide tree. The multiple alignment starts from the most similar sequences, say, seq1 and seq2, with the internal node-(1, 2). So we first move to the internal node-(1,2) and align seq1 and seq2. A profile is then created to represent the aligned Seq1 and Seq2. Treating the aligned Seq1 and Seq2 as a new single (combined) sequence, the number of sequence for multiple alignment is reduced to $N-1$. The next-round alignment is to find the most similar sequences in these $N-1$ (profiled) sequences.

Note that starting with the most similar sequences can minimize the effect of an alignment error to be propagated in the subsequent alignment. A wrong guide tree could bias the subsequent alignment. Nevertheless, though a guide tree built from alignment scores in the Clustal is typically poor, the final output of the multiple alignment seems to be fairly robust.

2.4   Microarrays and statistics

2.4.1   Types of microarray data

Microarray is a technology for measuring the mRNA levels for thousands of genes simultaneously. Due to the highly parallel nature of micorarray data, the statistical methods used to analyze these data are usually sophisticated. Nevertheless, the basic question which microarrays data are used to address is what genes are expressed in a given sample, and which genes are differentially expressed between different samples. There are two major types:

Spotted arrays: Spotted microarrays are either glass slides or nylon filters which are printed with thousands of spots, where each spot contains a set of identical probes for a particular gene. The length of the probes on spotted arrays is, at minimum, around 70 nucleotides. Since many genes may share some common features at the sequence level, the requirement of probe selection is to avoid ambiguously representing more than one gene so that the effect of 'cross-hybridization' can be reduced. There are two major approaches to selecting probes. The first one is to choose them from libraries of cDNA clones, such as EST databases. The second method is to start with the genomic sequence, which is now becoming available for many species. These probes are chosen to be as unique to that gene as possible, to minimize cross-hybridization.

Oligonucleotide probles are then synthesized representing the chosen sequences and spotted onto the array. To control high noise from the variation of spots, a reference sample is often labelled with a different dye, for example, Cy3 (green) can be used for the sample and Cy5 (red) for the reference. These so-called 'two-channel' array experiments are among the most common types of array in use.

Affymetrix arrays: The Affymetrix approach is to use short probes, called 'match' probes, on the order of 20 bases long, and to use a set of 10 to 20 different probes for each gene, each matching a different small segment of the mRNA. A probe and its mismatch are called a 'probe pair' (denoted by PM and MM for perfect match and mismatch), and generally there are 10 to 20 probe pairs per gene.

2.4.2 Sources of noises

Biological variations: Biological variation is the natural variation of biological samples from different populations of cells. Therefore, one should have sufficient biological replicates to reduce this type of variation. Recent technology improvemments have reduced the cost and increased the efficiency significantly. Yet, the biological replicate, which usually ranges from 3 to 20, remains a small sample size problem from the view of statistics.

Experimental variations: Experimental variation is the variation deriving from the technical aspects of the procedure: (i) Array-specific effects: It could be random array-to-array or pin-specific variations. (ii) Gene-specific effects: Hybridization conditions may vary considerably among thousands of genes in the array. (iii) Dye-specific effects: Incorporation of fluorescent dye varies. Two dyes might be different in incorporation for the same transcript. (iv) Background noise, artifacts, and preparation effects: There is always a low level background glow on an array, such as well dust, scratches, smears, etc.

Normalization for correcting systematic bias: Some types of noise/variation cause systematic biases that can be corrected by data normalization. However, that many non-systematic sources cannot be eliminated completely. For instance, those from the level of background noise may preclude the possibility of using microarrays to detect low levels of expression. Furthermore, since the two-channel procedure deals with ratios, in those spots whose intensity levels are in the background level, the calculation procedure would introduce a large amount of artifact of variability into these ratios.

2.4.3 Multiple gene problem

Microarrays are not used to determine whether a single gene is differentially expressed between two conditions, because they generate much noisier and less unreliable data than other low throughput methods such as RT-PCR. Instead, microarray technology is suitable to predict a set of genes that are differentially expressed between the two conditions. To this end, the statistical challenge is in defining a procedure which achieves acceptable control of false positive and false negative rates.

Some basic concepts

Ranked list   A simple approach to help find differentially expressed genes is to produce a ranking of genes. For instance, genes could be ranked in the decreasing order of absolute values of expression differences. A statistically sound approach, however, is to rank the gene by the *P*-values of the statistics. It should be noted that, however, these estimated *P*-values are rough, because some assumptions underlying the statistical test may bot be valid.

***t*-Statistics**   If for each gene the (log-transformed) expression level is normally distributed and has equal variance in the two conditions, the standard *t*-test is appropriate. That is, the gene with the higher *t*-statistic (absolute value), is the gene with the more significant result. In practice, however, the normality and equal variance may not hold. Yet the *t*-statistic is still widely used, as it is generally robust to the violation of these assumptions. Several studies have been made to adjust the *t*-statistic. For instance, Storey and Tibshirani (2003) introduce a correction factor into the denominator of the *t*-statistic to minimize the gene dependency.

Family-wise error rate (FWER) and false discovery rate (FDR)   In the single gene case the null hypothesis usually means no difference of the expression levels between two experimental conditions. We will refer to the null hypotheses for the individual genes as the gene-wise null hypotheses. In the multiple gene case we are interested in procedures that predict which genes are differentially expressed by controlling the number of false positive cases. There are two prevailing approaches to defining false positive rates: The first is the family-wise error rate (FWER), which is the probability of having at least one false positive in the predicted set. The simplest method of FWER is the well-known Bonferroni correction. The second, which is widely used in the microarray analysis, is the false discovery rate (FDR). Introduced by Benjamini and Hochberg (1995), it is based on the expected proportion of the predicted set that consists of false predictions. Most procedures that control the FWER or FDR start with gene-by-gene *P*-values and then adjust them accordingly.

2.4.4   False discovery rate (FDR)

Definition

In practice, microarrays have been used to predict some of differentially expressed genes that might be as low as, say 1 per cent of the whole gene set. In such case, calculation of FDR for a predicted gene set is needed. It is important to keep in mind that an FDR is fundamentally different from a *P*-value, and is for a very different purpose. A *P*-value is generally used to assess the significance of the result, which must have been rigorously defined, such as less than 0.05. On the other hand, an FDR is generally used as a selecting tool, depending on the purpose of the analysis.

Consider the predicted gene set for which the null hypothesis is rejected. Let *V* be the number of these genes for which the null hypothesis is true (or falsely rejected), and S be the number of which the null hypothesis is false (or correctly rejected). Let $R = V + S$. The quantity Q is defined as

$$Q = V/R \qquad (2.5)$$

if $R > 0$, or $Q = 0$ if $V = R = 0$.

The false discovery rate FDR defined by $E(Q)$ could also be written as $E(V/R|R > 0)Prob(R > 0)$, Storey (2002) introduced the 'positive false discovery rate', denoted by pFDR, as

$$pFDR = E(V/R|R > 0)$$

This differs from the FDR by not including the term $prob(R > 0)$, but by the condition on the event $R > 0$ instead. Though there are some theoretical advantages of using pFDR, they are virtually the same in the real data analysis.

Benjamini and Hochberg's method

The original method proposed to control the FDR was given by Benjamini and Hochberg (1995), starting with gene-by-gene P-values derived from g tests, under g individual null hypotheses. It is assumed that these P-values are independent. Let g_0 be the number of tests for which the null hypothesis is true. For those tests where the null hypothesis is true, the individual P-value follows a uniform distribution. Without loss of generality, let $H_1, \ldots, H_{g_0}$ be the true null hypotheses and $H_{g_0+1}, \ldots, H_g$ the false null hypotheses.

On the other hand, these g independent hypothesis tests result in individual test P-values denoted by $P_1 \ldots, P_g$, respectively. Each P_i is a random variable, whose observed value is given by p_i. Let $P_{(i)}$ be the i-th smallest of these P-values such that $P_{(1)} \leq, \ldots, \leq P_{(g)}$. Let $H_{(i)}$ be the hypothesis corresponding to $P_{(i)}$. Then, the Benjamini–Hochberg testing procedure is as follows: Let q_i be

$$q_i = \frac{i}{g}\beta, i = 1, \ldots, g$$

where β is the desired false discovery rate. Let k be maximum i such that $p_{(i)} \leq q_i$, where $p_{(i)}$ is the observed value of $P_{(i)}$. If there is no value i such that $p_{(i)} \leq q_i$, we accept all g null hypothesis. If $k \geq 1$, we reject the null hypothesis $H_{(1)}, \ldots, H_{(k)}$ and accept all others. Moreover, Benjamini and Hochberg (1995) have shown that

$$E(Q) = \frac{g_0}{g}\beta \qquad (2.6)$$

Hence, $E(Q) \leq \beta$ regardless of the value of g_0.

Statistical analysis of microarrays (SAM) method

Benjamini and Hochberg's method has several drawbacks. First, these estimated P-values are usually not accurate. Second, the procedure assumes that these tests are independent, an assumption which is unlikely to hold in the microarray analysis. Third, the true FDR (β) cannot be estimated since this procedure controls the mean value of Q to be less or equal to $(g_0/g)\beta$. Some recent studies (Tusher *et al.* 2001; Storey and Tibshirani 2003) have discussed how these problems might be addressed by using permutation methods. As indicated above, see Eq.(2.5), the FDR is defined as the mean ratio of V/R. Although bootstrap resampling can give an estimate of the distribution of R, estimation of the distribution of the ratio V/R has been shown to be a difficult task. Instead, statisticians considered $E(V)/R$. The mean of V can be roughly estimated by the permutation, while R is treated as known from the testing procedure.

A popular permutation method for estimating the FDR is called the SAM approach, proposed by Tusher *et al.* (2001). It computes, for gene $i = 1, 2 \ldots, g$, a t-like statistic $d(i)$ by adding a positive quantity s_0 to the denominator of the standard t-statistic. These g genes are then ranked according to their respective $d(.)$ values, and the notation is changed so that the gene with the largest $d(.)$ is now called gene-1, the gene with the second largest $d(.)$ value is now called gene-2, and so on. The follow-up procedure is similar to the original Benjamini and Hochberg method.

2.4.5   ANOVA analysis of many genes

A number of ANOVA models have beeen proposed for the microarray analysis. It is generally accepted that microarray expression levels do not have a normal distribution. In practice the logarithm (X) of expression level has been used routinely, under the assumption that these logarithms have a normal distribution, or at least to a sufficiently close approximation. For illustration, Kerr and Churchill (2001)'s model can be concisely represented as

$$X_{ijkg} = \mu + A_i + \delta_j + \tau_k + \gamma_g + B_{ig} + \psi_{kg} + e_{ijkg} \qquad (2.7)$$

where X_{ijkg} is the logarithm of a gene expression level for array i, dye j, treatment k, and gene g; A_i is the random effect due to array i; δ_j is a fixed effect due to dye j; τ_k is a fixed effect due to treatment k; γ_g is a fixed effect due to gene g; B_{ig} and ψ_{kg} are the random array×gene and (fixed) treatment×gene interactions, respectively; and finally, random variable e_{ijkg} is the error term.

Any ANOVA model in microarray analysis focuses on the treatment×gene interaction. If the F-ratio testing for this interaction is significant, we have strong evidence that some genes are expressed differently in different tissues. Numerious microaarray ANOVA designs are to be found in the literature.

2.5   Markov chain Monte Carlo (MCMC)

Until recently, the application of many computational tools had been prevented by computational difficulties. In the case of the Bayesian method as a general framework in bioinformatic analysis, the prior and the likelihood (the probabilistic model) are easy to claculate, but the normalizing constant, which usually involves high dimensionsal integrals, is hard to calculate. The development of Markov chain Monte Carlo (MCMC) algorithms provides a powerful method for achieving the Bayesian computation.

2.5.1 Metropolis-Hastings algorithm

The goal of the algorithm proposed by Metropolis *et al.* (1953) is to generate a Markov chain. The states of this Markov-chain are represented by the parameters θ, and the steady-state (stationary) distribution is

$$\pi(\theta) = f(\theta|D).$$

The right hand of the equation is the posterior distribution of θ given the data D. For the current state of the Markov chain denoted by θ, the Mestropolis algorithm proposes a new state θ^* through a specificed density or jumping kernal $q(\theta^*|\theta)$, which is symmetrical, i.e. $q(\theta^*|\theta) = q(\theta|\theta^*)$. A simple jumping kernal is a uniform distribution around θ, which can be written as $q(\theta^*|\theta) = U(\theta - w/2, \theta + w/2)$. The window size w controls the size of steps taken. The next step is to make a decision whether the candidate θ^* is acceptable. If the new state θ^* is accepted, the chain moves to θ^*. If it is rejected, the chain stays at the current state θ. The probability of the acceptance is

$$\alpha = \min\left[1, \frac{\pi(\theta^*)}{\pi(\theta)}\right]$$

Apparently, a random step in a random direction of states is chosen. If the step is 'uphill', i.e. $\pi(\theta^*) > \pi(\theta)$, it is always taken with a probability of 1. However, if the step is downhill, i.e. $\pi(\theta^*) < \pi(\theta)$, it is accepted with a probability of $\alpha = \pi(\theta^*)/\pi(\theta)$, or rejected with a probability of $1 - \pi(\theta^*)/\pi(\theta)$.

The procedure is repeated for many interactions, including both acceptations and rejections, generating a so-called Markov chain because these values satisfy the Markovian property that 'given the present, the future is independent of the past'. Metropolis *et al.* (1953) have shown that this Markov chain has $\pi(\theta)$ as the stationary distribution as long as the proposed jumping kernal $q(.|.)$ specifies an irreducible and aperiodic chain. In other words, it should allow the chain to reach all possible states starting from any state, and the chain should not be a period.

Hastings (1970) extended the Metropolis algorithm to allow the use of asymmetric proposal densities as jumping kernals, that is, $q(\theta^*|\theta) \neq q(\theta|\theta^*)$. This involves a simple correction in calculation of the acceptance probability.

$$\alpha = \min\left[1, \frac{\pi(\theta^*)}{\pi(\theta)} \times \frac{q(\theta|\theta^*)}{q(\theta^*|\theta)}\right]$$

2.5.2 Calculation of posterior distribution

When the MCMC algorithm is used to approximate the posterior distribution of parameters θ, we have

$$\pi(\theta) = f(\theta|D) = \frac{f(\theta)f(D|\theta)}{f(D)}$$

so that

$$\frac{\pi(\theta^*)}{\pi(\theta)} = \frac{f(\theta^*)f(D|\theta^*)}{f(\theta)f(D|\theta)}$$

Importantly the normalizing constant $f(D)$ cancels. Consequently, the acceptance probability is thus

$$\alpha = \min\left[1, \frac{f(\theta^*)}{f(\theta)} \times \frac{f(D|\theta^*)}{f(D|\theta)} \times \frac{q(\theta|\theta^*)}{q(\theta^*|\theta)}\right]$$

$$= \min\left[1, \text{ prior ratio } \times \text{likelihood ratio } \times \text{proposal ratio }\right]$$

In typical applications of MCMC algorithms to the Bayesian analysis, the prior ratio $f(\theta^*)/f(\theta)$ is easy to calculate. The likelihood ratio $f(D|\theta^*)/f(D|\theta)$ is often easy to calculate as well, even though computationally expensive. The proposal ratio $q(\theta|\theta^*)/q(\theta^*|\theta)$ greatly affects the efficiency of the MCMC algorithm. So much effort is spent on developing good proposal algorithms.

3

Functional Divergence after Gene Duplication: Statistical Modeling

Many organisms have undergone genome-wide or local chromosome duplication events during their evolution (Ohno 1970; Holland *et al.* 1994; Wolfe and Shields 1997; Gu and Nei 1999; Dermitzakis and Clark 2001; Gu *et al.* 2002b; Su *et al.* 2006; Xu *et al.* 2009). As a result, many genes are represented as several paralogs in the genome with related but distinct functions (gene families). Since gene duplication is thought to have provided the raw materials for functional innovations, it is desirable to identify amino acid sites that are responsible for functional divergence from the sequence analysis of a gene family. Several computational methods have been proposed (e.g. Casari *et al.* 1995; Lichtarge *et al.* 1996; Livingstone and Barton 1996; Landgraf *et al.* 2001). As most amino acid substitutions are not related to the functional divergence but only represent the neutral evolution, it becomes crucial how to statistically distinguish between these two possibilities. To this end, Gu (1999; 2001a; 2001b; 2006) has developed a series of statistical models, based on the principle that functional divergence between duplicate genes is highly correlated with the change of evolutionary rate after the gene duplication. In this chapter, we discuss these statistical and computational methods. Case studies for applying these methods in biological problems will be presented in the next chapter.

3.1 Modeling functional divergence

Consider a multiple alignment of a gene family with two duplicate gene clusters 1 and 2, respectively (Fig. 3.1). Although various terminologies have been used in previous literatures, amino acid sites of a gene family can be tentatively classified into four types:

(1) Type-0: represents amino acid patterns that are universally conserved through the whole gene family, implying that these sites are important for the common function shared by duplicate genes.
(2) Type-I: represents amino acid patterns that are very conserved in duplicate gene cluster 1 but highly variable in cluster 2, or *vice versa*, implying that these sites have experienced shifted functional constraints.
(3) Type-II: represents amino acid patterns that are very conserved in both duplicate gene clusters but their biochemical properties are very different, e.g. charge

(A) 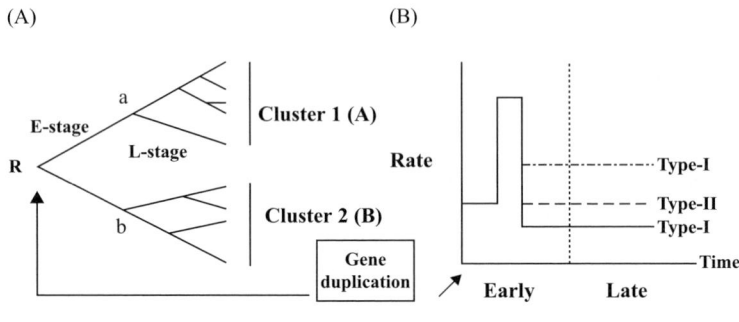 (B)

(C)

	Sequence	Type-0	Type-1	Type-2	Type-U
	1	CR	WQLV	RV	KTLI
	2	CR	WQIV	RV	RVLI
	3	CR	WQVG	RV	KIIV
Gene 1	4	CR	WQVG	RV	NVLL
	5	CR	WQAT	RV	DMLL
	6	CR	WQAT	RV	IKLL
	7	CL	WQVI	RV	EKLI
	8	CR	WQIT	RV	DLVL
	9	CR	LTFD	DR	LKLM
	10	CR	ITFD	DR	QLLV
	11	CR	ITFD	ER	RLVV
	12	CR	YSFD	DK	LHVV
Gene 2	13	CR	LEFD	DR	KMAL
	14	CL	LEFE	DR	KLLI
	15	CR	LEFD	DR	KLLL
	16	CR	VGFD	DK	ELII
	17	CR	VTFD	DR	RLII

Fig. 3.1 (A) Two gene clusters after gene duplication. E and L are early- and late-stages of gene clusters 1 and 2, respectively. (B) Type-I and type-II functional divergences after gene duplication. In the early-stage, the evolutionary rate (say, in cluster 1) may increase for functional divergence-related change, but in the late-stage it may be higher (or lower) than its original rate, resulting in shifted functional constraints between clusters 1 and 2, or type-I functional divergence. If the rate in the late-stage is back to the same as the original one, no shifted functional constraints between clusters 1 and B2 can be observed, or type-II functional divergence. (C) A hypothetical multiple-alignment to show universally conserved sites (type-0), type-I and type-II amino acid patterns, and U-type sites (unclassified). Figure modified from Gu (1999) and Gu (2001b).

positive vs negative, implying that these sites may be responsible for functional specifications.

(4) Type-U: represents amino acid patterns that cannot be classified into the three above types, for instance, amino acid sites that are highly variable in both clusters.

According to the evolutionary view that functional divergence between duplicates occurred in the early stage after the gene duplication, one can formulate two basic mechanisms for the functional divergence.

(1) *Type-I functional divergence*: postulates that functional divergence between duplicate genes results in shifted functional constraints (i.e. different evolutionary rate) at some sites, or type-I amino acid patterns. If a site is related to the type-I functional divergence, called F_1-site, we assume that the evolutionary rate at this site becomes independent between duplicate clusters. In other words, functional constraint of an F_1-site in one duplicate gene contains no information (high or low) for predicting the constraint in the other one.

(2) *Type-II functional divergence*: postulates that functional divergence between duplicate genes at some sites results in typical type-II amino acid patterns. If a site is related to type-II functional divergence, called F_2-site, we assume that the evolutionary rate at this site has been decoupled between the early and late stages after the gene duplication. While radical changes has been allowed in type-II functional divergence, both duplicate genes may retain a similar level of sequence conservation.

In the following we discuss the statistical methods for detecting functional divergence between duplicate genes. In particular, we address two important issues: Whether functional divergence between duplicate genes is statistically significant and if this is the case, how can we predict amino acid sites that are largely responsible for these functional divergences?

3.2 Poisson model for type-I functional divergence

3.2.1 Two-state model

Consider the case of a gene family containing two duplicate gene clusters (Fig. 3.1). For each duplicate gene cluster, it is usually assumed that orthologous genes within the cluster are functionally equivalent, implying that the evolutionary rate of a site remains virtually constant though it may vary among different sites. Since a molecular clock is not assumed, lineage-specific factors, such as generation-time effect (Wu and Li 1985), will not affect our results. Without loss of generality, the evolutionary rates in gene cluster 1 and gene cluster 2 are simply denoted by λ_1 and λ_2, respectively.

An ideal case is that we already know exactly which sites are related to type-I functional divergence. Then, all sites can be classified into either of two states:

(1) F_0 (functional divergence unrelated): The evolutionary rate (λ) of an F_0-site is the same between duplicate gene clusters, indicating no shift in functional constraint, that is, $\lambda_1 = \lambda_2 = \lambda$.

(2) F_1 (type-I functional divergence related): The evolutionary rate of an F_1-site has no correlation between duplicate gene clusters, as such the site has experienced shifted functional constraint. Consequently, λ_1 and λ_2 are independent.

However, in practice we do not know to which state each site belongs. This problem can be solved by implementing a two-state probabilistic model: an amino acid site is in the state of F_1 with a probability of $P(F_1)$, or the state of F_0 with a probability of $P(F_0)$. We denote $\theta_I = P(F_1)$, called *the coefficient of type-I functional divergence*. As θ_I increases from 0 to 1, the functional divergence between two clusters increases from very weak to extremely strong. Intuitively, based on θ_I one may formulate a statistical test to evaluate the significance of type-I functional divergence after the gene duplication, e.g. the null hypothesis $\theta_I = 0$ versus the alternative $\theta_I > 0$. To this end, we shall develop a statistical model for gene family evolution.

3.2.2 The Poisson-gamma model for protein sequence evolution

A simple model for protein sequence evolution is the Poisson process. At a given site, the number of amino acid changes in each cluster, denoted by X_1 and X_2 for gene clusters 1 and 2, respectively, follows a Poisson distribution. That is, the probability for $X_i = k$ changes is given by

$$p(X_i|\lambda_i) = \frac{(\lambda_i T_i)^{X_i}}{X_i!} e^{-\lambda_i T_i}, \quad i = 1, 2 \tag{3.1}$$

where T_1 and T_2 are the total evolutionary times of clusters 1 and 2, respectively.

To apply Eq. (3.1) in our study, we need to know the number of changes at each site for each gene cluster. Since these numbers (X_1 and X_2) cannot be directly observed from the multiple alignment, a conventional solution is to use the number of minimum-required changes (m) as an approximation, which can be inferred by the parsimony under a known phylogenetic tree (Fitch 1971). However, m is a biased 'estimate' for the true number of changes because it does not consider the possibility of multiple hits (Wakeley 1993). This parsimonious bias can be corrected by the method developed by Gu and Zhang (1997); also see section 1.9. Under a known phylogeny, they showed that the expected number of changes ($X = X_1$ or X_2) at a given site is the non-negative solution of the likelihood equation

$$\sum_{i=1}^{M} \frac{\delta_i b_i}{1 - e^{-\hat{X} b_i / B}} = 1 \tag{3.2}$$

where B is the total branch length of the gene cluster, b_i is the i-th branch length, $i = 1, \ldots, M$ (M is the total number of branches); $\delta_i = 1$ if there is an amino acid change in the i-th branch, otherwise $\delta_i = 0$. Computer simulation has shown that the estimate of corrected number of changes is asymptotically unbiased. Here we mention two interesting special cases: (1) $\hat{X} \approx m$ for short branch lengths, and (2) $\hat{X} = -M \ln(1 - m/M)$ for equal branch lengths.

Because amino acid sites differ in functional constraints, the evolutionary rate (λ) is expected to vary among amino acid sites (Kimura 1983). Since the rate λ at each site is usually unknown, a common practice is to assume that the rate variation among sites follows a specified distribution (Uzzel and Corbin 1971; Yang 1993; Gu *et al.* 1995), e.g. the widely-used gamma distribution

$$\phi(\lambda) = \frac{\beta^\alpha}{\Gamma(\alpha)} \lambda^{\alpha-1} e^{-\beta\lambda} \tag{3.3}$$

where $\lambda = \lambda_1$ or λ_2, respectively. The shape parameter α describes the degree of rate variation among sites, whereas β is only a scalar. Since $1/\sqrt{\alpha}$ is the coefficient of variation of λ, the larger the α value is, the weaker the rate variation is, and $\alpha = \infty$ means a uniform rate among sites.

3.2.3   The likelihood function

Our goal is to obtain $P(X_1, X_2)$, the joint distribution of the numbers of amino acid changes in clusters 1 and 2. We use the hierarchical modeling approach, which starts from a single site where the evolutionary rate in cluster 1 or 2 is denoted by λ_1 or λ_2, respectively. Since the two gene clusters are monophyletic, or phylogenetically independent (Fig. 3.1), it is reasonable to assume that the two Poisson processes of amino acid substitutions at this site, $p(X_1|\lambda_1)$ and $p(X_2|\lambda_2)$, are independent, and so the joint distribution of X_1 and X_2 is given by

$$P(X_1, X_2|\lambda_1, \lambda_2) = p(X_1|\lambda_1)p(X_2|\lambda_2)$$

The next step is to integrate out the evolutionary rates λ_1 and λ_2, both of which vary among sites according to the gamma distribution $\phi(\lambda)$ in Eq. (3.3). However, the correlation between λ_1 and λ_2 differs in different states under the two-state model of type-I functional divergence. First consider the state of F_0, under which the site is functional divergence unrelated, implying the same rate $\lambda_1 = \lambda_2 = \lambda$ holds. It follows that the joint distribution of X_1 and X_2 under F_0 is given by

$$P(X_1, X_2|F_0) = \int_0^\infty p(X_1|\lambda)p(X_2|\lambda)\phi(\lambda)d\lambda$$

Let D_1 and D_2 be the mean number of amino acid changes per site in duplicate clusters 1 and 2, respectively. From Eq. 3.1 to Eq. 3.3, Gu (1999) has shown that for $X_1 = i$ and $X_2 = j$, $P(X_1, X_2|F_0) = K_{12}(i, j)$ where

$$K_{12}(i, j) = \frac{\Gamma(i + j + \alpha)}{i!j!\Gamma(\alpha)} \left(\frac{D_1}{D_1 + D_2 + \alpha}\right)^i \left(\frac{D_2}{D_1 + D_2 + \alpha}\right)^j \left(\frac{\alpha}{D_1 + D_2 + \alpha}\right)^\alpha \tag{3.4}$$

Secondly, we consider the state of F_1, under which the amino acid site is functional divergence related, implying λ_1 and λ_2 are independent. This immediately results in the joint distribution of X_1 and X_2 under F_1 as follows

$$P(X_1, X_2 | F_1) = P(X_1 | F_1) \times P(X_2 | F_1)$$

For each cluster, the distribution of the number of changes ($X = X_1$ or X_2) under F_1 is given by

$$P(X | F_1) = \int_0^\infty p(X | \lambda) \phi(\lambda) d\lambda$$

Similar to the derivation of Eq. (3.4), for $X_1 = i$ one can show $P(X_1 | F_1) = Q_1(i)$ where

$$Q_1(i) = \frac{\Gamma(i + \alpha)}{i! \Gamma(\alpha)} \left(\frac{D_1}{D_1 + \alpha} \right)^i \left(\frac{\alpha}{D_1 + \alpha} \right)^\alpha \tag{3.5}$$

Similarly, for cluster 2 ($X_2 = j$) we have $P(X_2 | F_1) = Q_2(j)$, where $Q_2(j)$ is the same form of $Q_1(i)$ except that D_1 is replaced by D_2.

Putting all above together, under the two-state model of functional divergence, one can write the joint distribution of X_1 and X_2 as

$$P(X_1, X_2) = P(F_0)P(X_1, X_2 | F_0) + P(F_1)P(X_1, X_2 | F_1)$$

Since $P(F_1) = \theta_I$ and $P(F_0) = 1 - \theta_I$, we obtain the analytical result

$$P(X_1, X_2) = (1 - \theta_I)K_{12} + \theta_I Q_1 Q_2 \tag{3.6}$$

which provides the statistical foundation for estimating the coefficient of type-I functional divergence.

3.2.4   Maximum-likelihood estimation (MLE)

Let $P(X_1 = i_k, X_2 = j_k)$ be the probability of $X_1 = i_k$ and $X_2 = j_k$ at site k. Then, given the observed number of changes in two clusters along the sites, the likelihood function can be written as

$$L = \prod_k P(X_1 = i_k, X_2 = j_k)$$

There are four unknown parameters, D_1, D_2, α, and θ_I, which can be numerically estimated by a standard maximum likelihood approach. Using appropriate initial values, the ML estimates of D_1, D_2, α, and θ_I, as well as their approximate sampling variances, can be obtained numerically.

Apparently, we are mostly interested in whether θ_I, the coefficient of type-I functional divergence, is significantly greater than 0, which can be accessed by the likelihood ratio test (LRT): The null hypothesis is $H_0 : \theta_I = 0$ vs the alternative $H_A : \theta_I > 0$. Based on the likelihood ratio LR, it is known that the statistic $\delta = -2 \ln(LR)$ asymptotically follows a $\chi^2_{[1]}$ distribution with one degree of freedom.

An Example: Transferrins are iron-binding transport proteins which can bind two atoms of ferric iron Fe^{3+}. They are responsible for the transport of iron from sites

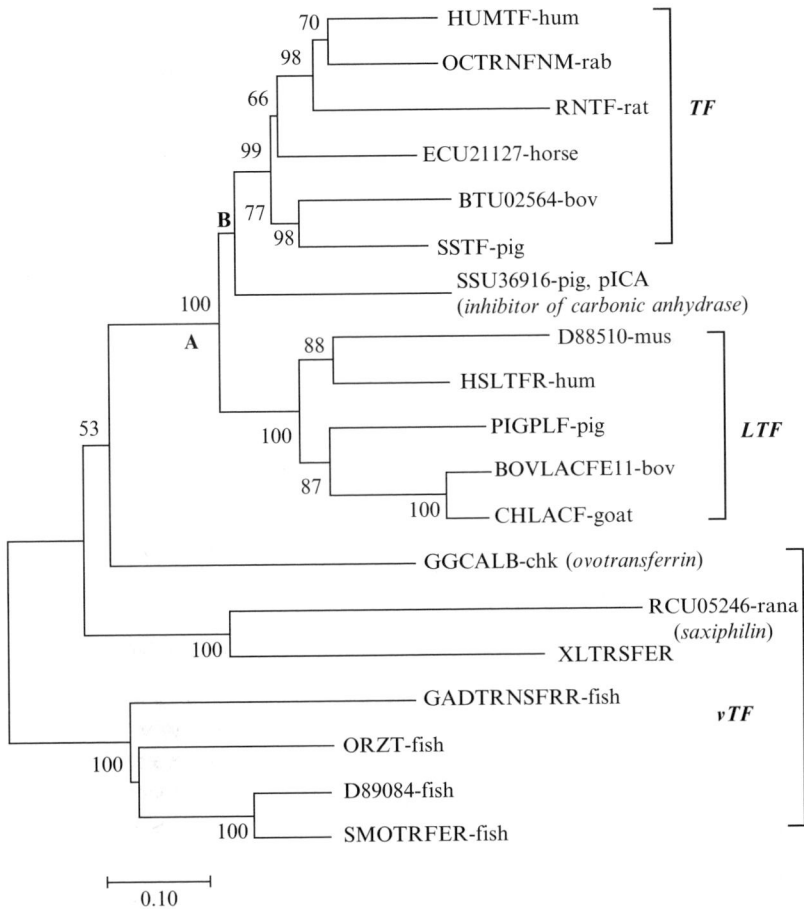

Fig. 3.2 The phylogenetic tree of the transferin gene family, which was inferred by the neighbor-joining method, using amino acid sequences with Poisson distance. Bootstrapping values of more than 50 per cent are presented. Figure modified from Gu (1999).

of absorption and heme degradation to those of storage and utilization. There is only one gene in non-mammalian vertebrates. In mammals, two close-linked tissue-specific genes are found, which encode serum transferrin (TF) and lactotransferrin (LTF), respectively. Figure 3.2 shows the phylogenetic tree of the transferrin gene family by the neighbor-joining method (Saitou and Nei 1987). Apparently, the TF/LTF duplication event occurred before the radiation of mammals but after the divergence between birds and mammals, resulting in two duplicate gene clusters with a high bootstrap (100 per cent) value. The expected number of changes (X_1 or X_2) at each site in TF or LTF cluster was shown in Fig. 3.3. Thus, we obtain the ML estimate $\theta_I = 0.19 \pm 0.07$ ($P < 0.01$). Hence, we conclude that type-I functional divergence is statistically significant between mammalian TF and LTF duplicate genes.

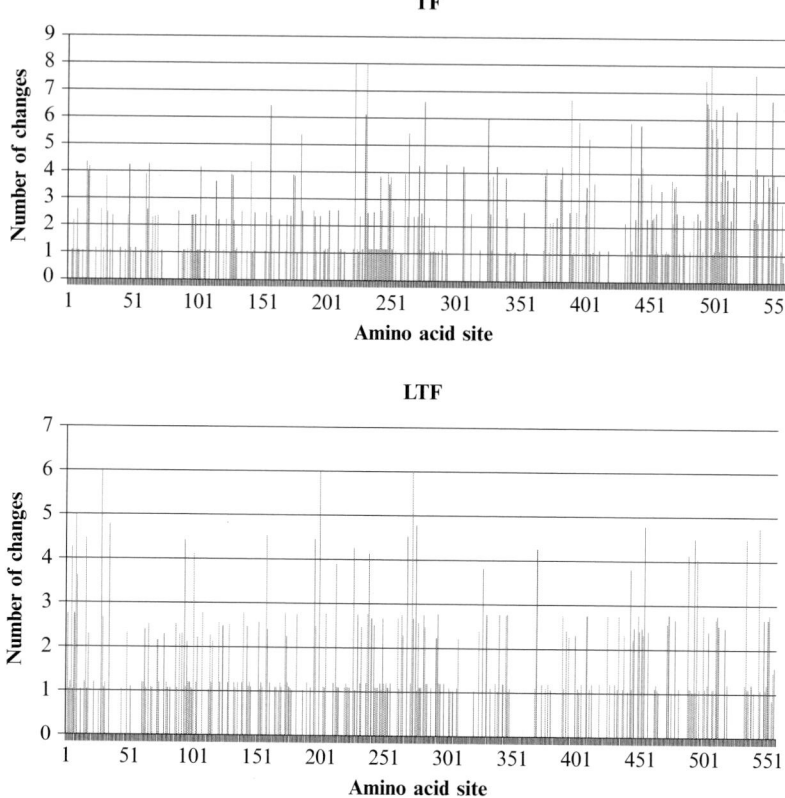

Fig. 3.3 The expected number of changes at each amino acid site in TF or LTF, respectively, estimated by Gu and Zhang's (1997) method, based on the phylogeny given in Fig. 3.2. Figure modified from Gu (1999).

3.2.5 Predicting critical amino acid residues

If the likelihood ratio test provides strong statistical evidence for the type-I functional divergence after gene duplication (i.e. $\theta_I > 0$), it is of great interest to (statistically) predict which sites are likely to be responsible for these functional differences. Gu (1999) has addressed this issue, developing a site-specific profile to calculate the posterior probability that an amino acid site is functional divergence related, given the observed numbers of amino acid changes in two duplicate clusters.

In the two-state model of functional divergence, each site has two possible states, F_0 and F_1. From the Bayesian view, the probability $P(F_1)$ and $P(F_0)$ are treated as priors. In other words, the coefficient of type-I functional divergence $\theta_I = P(F_1)$ is interpreted as the prior probability of the state of functional divergence related.

Therefore, to provide a statistical basis for predicting which state (F_0 or F_1) is more likely at a given site, we need to compute the posterior probability of state F_1 at this site with X_1 and X_2 changes in clusters 1 (and 2), respectively, denoted by $P(F_1|X_1, X_2)$. According to the Bayesian law, we have

$$P(F_1|X_1, X_2) = \frac{P(F_1)P(X_1, X_2|F_1)}{P(X_1, X_2)} = \frac{\theta_I Q_1 Q_2}{(1 - \theta_I)K_{12} + \theta_I Q_1 Q_2} \quad (3.7)$$

and $P(F_0|X_1, X_2) = 1 - P(F_1|X_1, X_2)$. For conciseness, let q_k be the posterior probability $P(F_1|X_1, X_2)$ at site k, $k = 1, \dots, L$, where L is the number of sites. From Eq. (3.7), one can further show the following relationship

$$\theta_I = \sum_{X_1, X_2} P(X_1, X_2) \times P(F_1|X_1, X_2) \approx \sum_{k=1}^{L} q_k/L \quad (3.8)$$

Hence, those probabilities for single sites being functional divergence related can be viewed as a (weighted) spectrum of the coefficient of type-I functional divergence. Approximately, θ_I is the average of posterior probabilities over all sites.

Examples: In practice, we use the ML estimates to replace the corresponding unknown parameters, an approach usually called 'empirical Bayesian'. When the site-specific profile (q_k) is calculated, one may identify these amino acid sites that may be responsible for the type-I functional divergence, given a cut-off value. Gu (1999) analyzed the transferrin (TF/LTF) and Myc (N-myc/C-myc) gene families. For example, site-specific profile of posterior ratio [$R_k = q_k/(1 - q_k)$] for predicting critical amino acid sites responsible for type-I functional divergence between N-myc and C-myc is plotted against the amino acid position (Fig. 3.4); in this case the ML estimate $\hat{\theta}_I = 0.39 \pm 0.08$. As expected, most amino acid sites have low values, indicating that their functional roles have not been changed after the gene duplication. In contrast, it shows that functional divergence between N-myc/C-myc has been only affected by a small number of amino acid residues. Table 3.1 lists these candidate sites under the cutoff $R_k > 2.5$ or $q_k > 0.7$. Indeed, these sites have dramatically different X_1 and X_2 changes between N-myc and C-myc duplicate gene clusters, e.g. one of them has no change at all, whereas another one has many changes. These predicted residues can be used as targets for further experimentation to determine their functional roles.

Statistical evaluation for predicted amino acid sites A simple cutoff, say, the posterior probability $q_k > 0.7$, can be used to identify amino acid sites that are responsible for type-I functional divergence. Since such cutoff is somewhat arbitrary, a statistical evaluation is necessary. Let L_c be the number of sites predicted under a cutoff $q_K \geq c$. Thus, the false discovery rate $FDR(c)$ can be approximtaely calculated by

$$FDR(c) = 1 - \sum_{k \in C} q_k/L_c$$

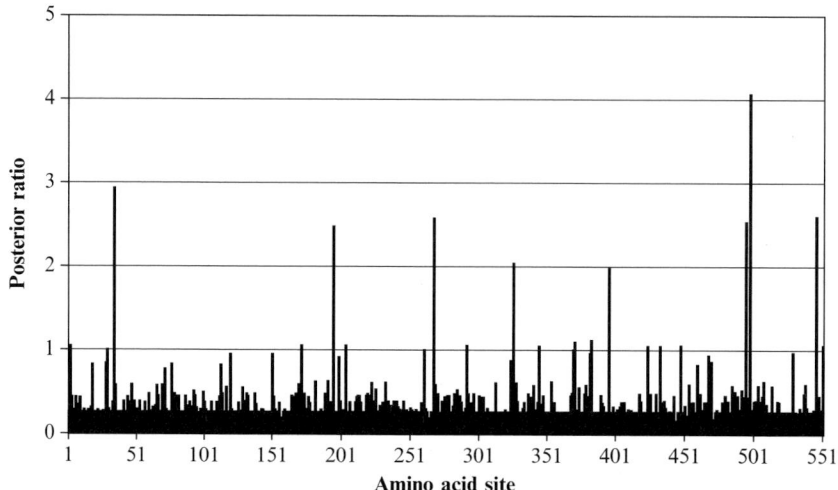

Fig. 3.4 The site-specific profile (R_k) for predicting critical amino acid sites responsible for the (type-I) functional divergence between TF and LTF, measured by the posterior probability ratio. Figure modified from Gu (1999).

where C is the set for all sites such that $q_k \geq c$. Roughly, FDR is the proportion of predicted sites under the cutoff c that are not related to the type-I functional divergence. Similarly, the false negative rate $FNR(c)$ is defined by

$$FNR(c) = \sum_{k \in C^*} q_k / (L - L_c)$$

where C^* is the set for all sites such that $q_k < c$. It roughly represents the proportion of unselected sites that are actually related to the type-I functional divergence. Table 3.1 shows both statistical measures for the top 17 sites in the N-myc/c-myc case. While FDR is highly dependent of the number of sites selected (after the ranking based on q_k or R_k), FNR is much less sensitive. This observation can explained as follows. Given the sample size of the sequence data, only a small number of sites can be effectively detected as F_1-sites. Indeed, after removing these sites as listed in Table 3.1, θ_I between N-myc and c-myc becomes nonsignificant. Consequently, FNR cannot be considerably reduced except for a large number of sites selected, which may be biologically unreasonable. Hence, we suggest that one may choose a cutoff c under a given false positive rate. For instance, one may select these 17 sites (Table 3.1) as candidate F_1-sites under the criterion of $FDR = 0.2$.

The most liberal prediction is to take all sites as F_1-sites, with the maximum false discovery rate. In this case, we have $FDR_{max} = 1 - \sum_k q_k / L \approx 1 - \theta_I$. Obviously any nontrivial prediction can reduce FDR, or increase the prediction accuracy. Tentatively, we propose the following index

$$PA_c = \frac{1 - FDR(c)}{1 - FDR_{max}}$$

Table 3.1 Statistical evaluation of predicted amino acid sites for type-I functional divergence between C-myc and N-myc genes, add ($\theta_1 = 0.39$).

	position	X_1	X_2	R_k	FDR	FNR	PA
1	253	7.5	0	23.6	0.041	0.388	2.46
2	245	7.0	0	18.5	0.046	0.386	2.45
3	50	0	7.0	14.0	0.053	0.384	2.43
4	176	5.1	0	7.8	0.068	0.382	2.39
5	179	0	4.6	5.1	0.090	0.380	2.33
6	95	0	4.6	5.1	0.102	0.379	2.30
7	244	0	4.5	5.0	0.112	0.377	2.28
8	149	0	4.4	4.8	0.119	0.375	2.26
9	243	3.9	0	4.6	0.126	0.373	2.24
10	118	3.7	0	4.1	0.133	0.372	2.22
11	48	0	3.6	3.5	0.141	0.370	2.20
12	247	7.6	1.3	3.2	0.149	0.369	2.18
13	97	0	3.4	3.2	0.156	0.367	2.16
14	37	0	3.4	3.1	0.162	0.366	2.15
15	56	1.1	7.5	3.1	0.168	0.364	2.13
16	135	7.2	1.3	2.8	0.174	0.363	2.12
17	89	1.1	7.0	2.7	0.179	0.361	2.10

to measure by how many folds the prediction accuracy has been improved under the cutoff c. In the case of N-myc/c-myc, Table 3.1 shows PA is about 2–2.5 fold for any cut off $q_k > 0.7$. If one selects only one site with the highest posterior probability denoted by q_{max}, we have the upbound of the prediction power $PA_{max} = q_{max}/\theta_I \leq 1/\theta_I$, as $q_{max} \leq 1$ always holds.

3.2.6 Reduced rate correlation between duplicate genes: an alternative view of θ_I

Consider the case of two duplicate gene clusters (Fig. 3.1). If all amino acid sites have experienced no functional divergence after the gene duplication, the two duplicate genes have no shifted functional constraints so that the evolutionary rates λ_1 and

λ_2 are virtually equal over all sites. In other words, the coefficient of correlation between λ_1 and λ_2, denoted by r_λ, could be very close to 1. Obviously, shifted functional constraints at some sites can result in different evolutionary rates, reducing the rate correlation between them, i.e. $r_\lambda < 1$. Hence, the shifted functional constraints between two gene clusters, or type-I functional divergence, can be measured by r_λ, the coefficient of rate correlation between λ_1 and λ_2. Under the Poisson model for amino acid substitutions, Gu (1999) has shown that r_λ can be calculated by the following formula

$$r_\lambda = \frac{\sigma_{12}}{\sqrt{(V_1 - D_1)(V_2 - D_2)}} \tag{3.9}$$

where D_1 and V_1 (or D_2 and V_2) are the mean and variance of the number (X_1 or X_2) of changes (over sites) in cluster 1 (or cluster 2), respectively, and σ_{12} is the covariance (over sites) between them.

Under the two-state model of type-I functional divergence, the evolutionary rates λ_1 and λ_2 follow the same distribution. Hence, it is easy to show that the variances of λ_1 and λ_2 are the same, i.e. $Var(\lambda_1) = Var(\lambda_2) = Var(\lambda)$, and the covariance between λ_1 and λ_2 is given by $(1 - \theta_I)Var(\lambda)$. This simple argument directly leads to the following simple relationship between r_λ and θ_I

$$r_\lambda = 1 - \theta_I \tag{3.10}$$

Therefore, the coefficient of type-I functional divergence can also interpreted as the reduced coefficient of rate correlation between duplicate clusters. Moreover, it provides a 'model-free' approach to estimate θ_I by $1 - r_\lambda$, because it does not rely on a specified model for the functional divergence. In this case, Gu (1999) derived an approximate formula for the sampling variance

$$Var(\hat{\theta}_I) \approx \frac{1}{L - 3}\left(\frac{1 - r_X^2}{r_M}\right)^2 \tag{3.11}$$

where L is the number of sites, $r_X = \sigma_{12}/\sqrt{V_1 V_2}$ and $r_M = \sqrt{(1 - D_1/V_1)(1 - D_2/V_2)}$. For example, see Table 3.2 for the result of this model-free analysis between TF and LTF duplicate clusters.

3.3 Markov chain model for type-I functional divergence

3.3.1 The Markov chain model

Gu (2001b) has formulated the Markov chain model for type-I functional divergence, a standard framework in molecular phylogenetic analysis (Felsenstein 1981; Kishino *et al.* 1990). Under this model, the transition probability matrix for a given time period t can be computed as $\mathbf{P} = e^{\lambda \mathbf{R} t}$, where the rate matrix $\mathbf{R}$ represents the pattern of amino acid substitutions, which can be empirically determined by, for example, the Dayhoff model (Dayhoff *et al.* 1978). The evolutionary rate (λ) may vary among sites

Table 3.2 Type-I functional divergence between TF and LTF gene families based on the model-free analysis.

D_1	1.17
D_2	0.86
V_1	2.87
V_2	1.49
σ_{12}	0.76
r_X	0.37
r_M	0.50
θ_I	0.26 ± 0.08
p-value	$< 10^{-3}$

Note: Cluster 1 is for TF and cluster for LTF. The significance level (p-value) is computed by the method of Fisher's transformation (Gu 1999).

because of different functional constraints. Usually λ is treated as a random variable, which follows a gamma distribution $\phi(\lambda)$; see Eq. (3.3).

Consider the phylogenetic tree in Fig. 3.5. Let $X = (x_1, x_2, x_3, x_4)$ and $Y = (y_1, y_2, y_3, y_4)$ be the observed amino acid pattern of a site for clusters 1 and 2, respectively. For the (unrooted) subtree for cluster 1 or 2, the conditional probability of observing X or Y at a site can be written as follows, respectively

$$f(X|\lambda) = \sum_{x_5=1}^{20} \sum_{x_6=1}^{20} b_{x_5} P_{x_5 x_1} P_{x_5 x_2} P_{x_5 x_6} P_{x_6 x_3} P_{x_6 x_4}$$

$$f(Y|\lambda) = \sum_{y_5=1}^{20} \sum_{y_6=1}^{20} b_{y_5} P_{y_5 y_1} P_{y_5 y_2} P_{y_5 y_6} P_{y_6 y_3} P_{y_6 y_4}$$

where $P_{ij} = P_{ij}(v_{ij})$ is the transition probability from node i to node j, v_{ij} is the branch length between them; b_i is the frequency of amino acid i. By integrating out the random variable λ, the probability of observing X at a site is given by

$$p(X) = \int_0^\infty f(X|\lambda)\phi(\lambda)d\lambda = E[f(X|\lambda_1)]$$

where E means taking expectation. Similarly, we have $p(Y) = E[f(Y|\lambda_2)]$.

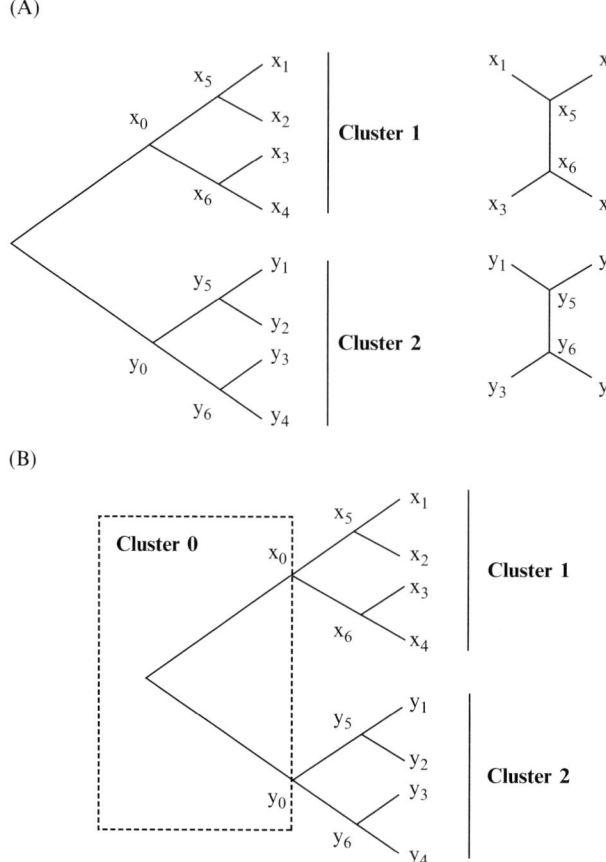

Fig. 3.5 A gene family tree with two duplicate gene clusters used for the illustration of type-1 and type-2 functional divergence analyses. Modified from Gu (2001b). (A) Two clusters 1 and 2 under the gene family and the corresponding unrooted trees. (B) Illustration of (ancestral) cluster 0.

Under the two state model of type-I functional divergence, the evolutionary rates (λ_1 and λ_2) at an F_1-site are statistically independent, whereas they are completely correlated ($\lambda_1 = \lambda_2 = \lambda$) at an F_0-site. Therefore, the joint probability of subtrees conditional on F_0 or F_1 is given by

$$f(X, Y | F_0) = \int_0^\infty f(X | \lambda) f(Y | \lambda) \phi(\lambda) d\lambda = E[f(X | \lambda) f(Y | \lambda)]$$

$$f(X, Y | F_1) = p(X) p(Y) = E[f(X | \lambda_1)] \times E[f(Y | \lambda_2)]$$

Similar to the derivation of Eq. 3.6, the joint probability of two subtrees can be written as follows

$$p(X, Y) = (1 - \theta_I) f(X, Y | F_0) + \theta_I f(X, Y | F_1) \tag{3.12}$$

Thus, under the assumption of site-independence, the likelihood function over all sites (gaps excluded) is given by

$$L(\mathbf{x} | data) = \prod_k p(X^{(k)}, Y^{(k)}) \tag{3.13}$$

where k runs for sites, and $\mathbf{x}$ is the set of unknown parameters. Gu (2001b) developed a practically feasible procedure to obtain the ML estimate of θ_I under the given phylogeny.

(1) The phylogenetic tree of the gene family is inferred by the standard tree-making methods.
(2) Given the inferred topology, the branch lengths ($\mathbf{v}$) are estimated by a least-square method, and the gamma shape parameter (α) is estimated by Gu and Zhang's (1997) method. The discrete gamma distribution approximation (Yang 1994b) is implemented for computing the likelihood functions.
(3) Regarding all these parameters as constants, the maximum likelihood estimate (MLE) of θ_I can be obtained by $\partial \ln L / \partial \ln \theta_I = 0$.
(4) A numerical iteration such as simplex method can be implemented to find the final ML estimates of $\mathbf{v}$, α and θ_I under the given phylogeny.

After obtaining these ML estimates, the likelihood ratio test (LRT) can be constructed under the null hypothesis $H_0 : \theta_I = 0$ vs. $H_A : \theta_I > 0$. Apparently, rejection of H_0 significantly provides statistical evidence for type-I functional divergence after the gene duplication. Moreover, according to the Bayesian law, the posterior probability of F_1 for a given site when the amino acid pattern (X, Y) is observed can be written as

$$P(F_1 | X, Y) = \frac{\theta_I f(X, Y | F_1)}{p(X, Y)} = \frac{\theta_I \, p(X) \, p(Y)}{p(X, Y)} \tag{3.14}$$

which can be computed in the algorithm described above.

3.3.2 Case study: COX (cyclooxygenase) gene family

The cyclooxygenase (COX) enzymes catalyze a key step in the conversion of arachidonate to PGH2, the immediate substrate for a series of cell prostaglandin and thromboxane synthases. Prostaglandins play critical roles in numerous biological processes, including the regulation of immune function, kidney development, reproductive biology, and gastrointestinal integrity (Williams *et al.* 1999). There are two tissue-specific isoforms in mammals: COX-1 and COX-2. Molecular cloning of COX-2 lead to a major investment by pharmaceutical companies in the development of selective inhibitors (Wallace 1999). The central tenets are that prostaglandins that contribute to inflammation are derived from COX-2, whereas prostaglandins that are involved in normal physiological processes are derived from the constitutively expressed isoform

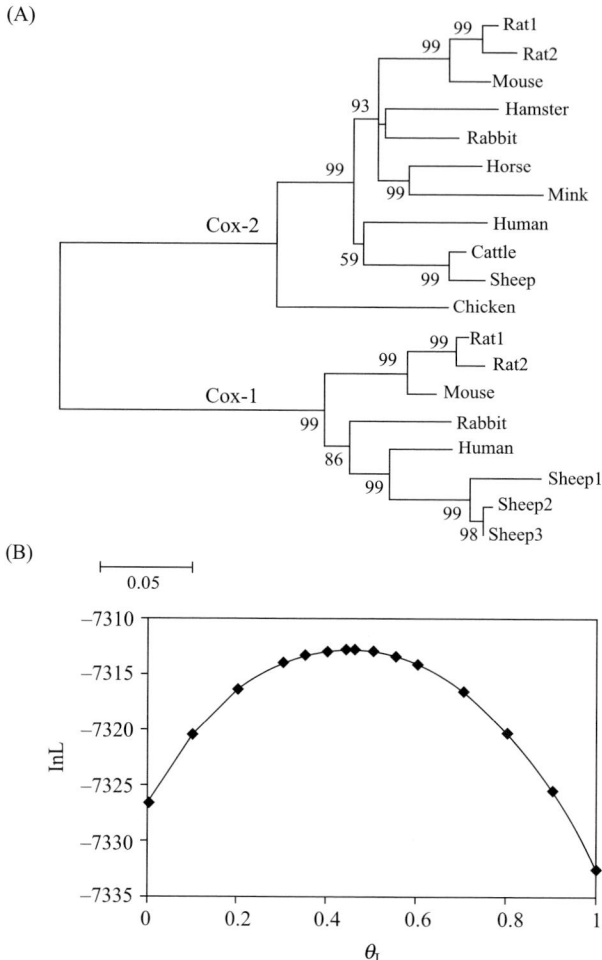

Fig. 3.6 (A) The phylogenetic tree of the COX gene family, inferred by the neighbor-joining method, using amino acid sequences with Poisson distance. Bootstrap values of more than 50 per cent are presented. (B) The log of likelihood value of the COX gene family amino acid sequences plotting against the coefficient of type-I functional divergence (θ_I). Figure modified from Gu (2001b).

COX-1. Therefore, investigating the pattern of functional divergence between COX-1 and COX-2 amino acid sequences is helpful for the drug design.

Figure 3.6(A) shows the phylogenetic tree of the COX gene family, inferred by the neighbor-joining method (Saitou and Nei 1987). It is clear that these two isoforms were generated in the early stage of vertebrates. Given this topology, the ML estimate of the coefficient of type-I functional divergence between COX-1 and COX-2 is $\hat{\theta}_I = 0.44 \pm 0.09$. Figure 3.6(B) shows the shape of the log of likelihood with respect to θ_I. We have conducted the likelihood ratio test (LRT) for the null hypothesis $H_0 : \theta_I = 0$. The logarithm of likelihood under the functional divergence is $\ln L = -7312.70$, and that under H_0 is $\ln L = -7326.51$, resulting in $\delta \ln L = -7312.70 - (-7326.51) = 13.81$. Thus, assuming that $2\delta \ln L$ approximately follows a χ^2 distribution, we conclude that the null hypothesis (no functional divergence) can be rejected statistically ($p < 0.001$).

3.3.3 Comparisons between the Poisson model and the Markov chain model

For comparison, Gu's (1999) method has been applied for the same dataset, which is based on a simple model but is computationally fast. Interestingly, the results from Gu's (1999) method is very close to what we have obtained by the current maximum likelihood approach, i.e. $\hat{\theta}_I = 0.46 \pm 0.08$. The correlation of posterior predictions between these two methods has been studied. Gu (2001b) showed that these two site-specific profiles are quite similar ($R^2 = 0.96$). Indeed, amino acid residues responsible for functional divergence can be predicted by both methods.

3.4 Statistical method for type-II functional divergence

3.4.1 Modeling type-II functional divergence

Early and late stages after gene duplication In principle, the evolution of protein sequences of duplicate genes can be divided into two stages, the early (E) stage after gene duplication, and the late (L) stage (Fig. 3.1). We assume that type-II functional divergence between duplicate genes has occurred in the E-stage, while in the late (L) stage, the purifying selection plays a major role to maintain related but distinct functions of two duplicate genes. Accordingly, Gu (2006) specified the two-state model to the type-II functional divergence:

(1) In the early (E) stage, an amino acid site can be in either of two states: F_0 (functional divergence-unrelated) and F_2 (type-II functional divergence-related). The probability of a site being under F_2 is $P(F_2) = \theta_{II}$, called *the coefficient of type-II functional divergence*.
(2) In the late (L) stage, an amino acid site is always under the state of F_0, indicating no further functional divergence. Amino acid substitutions in this stage are mainly under the purifying selection.

Substitution models under F_0 and F_2 The pattern of amino acid substitutions during the evolution, or the substitution model, relies on the states of functional divergence (F_0/F_2). The F_0-substitution model largely reflects the conserved evolution of protein sequences, which can be empirically determined by the Dayhoff model (Dayhoff *et al.* 1978), or the JTT model (Jones *et al.* 1992). In contrast, under F_2, radical amino acid substitutions may occur more frequently, apparently due to the functional divergence between duplicate genes (Lichtarge *et al.* 1996). To avoid over-parameterization, we propose a simple substitution model that can distinguish between the *radical* and *conserved* amino acid substitutions.

(1) Tentatively, classify 20 amino acids into 4 groups: charge positive (K, R, H), charge negative (D, E), hydrophilic (S, T, N, Q, C, G, P), and hydrophobic (A, I, L, M, F, W, V, Y). An amino acid substitution is called radical (denoted by **R**) if it changes from one group to another; otherwise it is called conserved, i.e. within the group, denoted by **C**. The status of no substitution is denoted by **N**.

(2) Under the state of F_0, the transition probability for a radical, conserved, or no substitution, follows an extended Poisson model, that is,

$$P(R|F_0) = \pi_R(1 - e^{-\lambda t})$$
$$P(C|F_0) = \pi_C(1 - e^{-\lambda t})$$
$$P(N|F_0) = e^{-\lambda t} \tag{3.15}$$

respectively, where t is the evolutionary time, λ is the substitution rate, and π_R (π_C) is the proportion of radical (or conserved) substitutions in the total substitutions; $\pi_R + \pi_C = 1$. Based on the Dayhoff PAM matrix, one can empirically determine $\pi_R = 0.312$ and $\pi_C = 0.688$. Indeed, without any functional divergence, conserved amino acid substitutions are more likely to occur, as anticipated by the theory of neutral evolution (Kimura 1983).

(3) Consider the transition probabilities under F_2 in the early stage, denoted by $P(Y|F_2)$ for $Y = N, R, C$. By definition, an amino acid site that has no change in the early stage is essentially unrelated to the type-II functional divergence, implying that $P(N|F_1) = 0$. Further, after assuming that the process of functional divergence is a fast process, we obtain

$$P(R|F_1) = a_R$$
$$P(C|F_1) = a_C$$
$$P(N|F_1) = 0 \tag{3.16}$$

That is, a_R (or a_C) is the (F_2)-proportion of radical (or conserved) substitutions. Moreover, the F_2-radical amino acid substitution (a_R) can be much higher than that under F_0 (π_R).

Evolutionary link between early and late stages   The evolutionary link between early and late stages depends on the status of type-II functional divergence. Let λ_E and λ_L be the evolutionary rates in the early (E) and late (L) stages, respectively. The statistical framework we developed is under the following assumptions:

(1) A random variable u, called the rate component, varies among sites according to a standard gamma distribution, where the shape parameter α describes the strength of rate variation among sites (Gu *et al.* 1995).

(2) Under F_0, the evolutionary rates in the early (λ_E) and late (λ_L) stages share the same rate component u. That is, $\lambda_E = u$ and $\lambda_L = u$.

(3) F_2-amino acid substitutions in the early stage is independent of the rate component u. In other words, F_2-amino acid substitutions have escaped from the ancestral functional constraint on the protein sequence.

3.4.2 Two clusters by gene duplication

Consider the typical case of two clusters generated by a gene duplication event, each of which consists of several orthologous genes (Fig. 3.1). Let X be the amino acid pattern of the late stage, a column (site) in the multiple alignment of the sequences that includes two clusters A an B. Let $Y = (a, b)$ be the amino acid pattern of the early stage, the ancestral sequences of two internal nodes a and b. From assumption (2), the joint probability of X and Y under F_0 is given by

$$P(X, Y|F_0) = \int_0^\infty P(X|Y)P(Y|F_0)\phi(u)du$$

where $P(Y|F_0)$ is determined by Eq.(3-16) for $Y = N, C$ or R, respectively, and $P(X|Y)$ is the likelihood of the subtrees of two clusters A and B, conditional on the ancestral states a and b, which can be constructed according to the Markov chain property under a known phylogeny (Felsenstein 1981; Gu (2001b)). Similarly, from assumption (3), under F_2 we have

$$P(X, Y|F_2) = P(Y|F_2) \times \int_0^\infty P(X|Y)\phi(u)du$$

where $P(Y|F_1)$ is given by Eq.(3-16). Remembering that the probability of a site being under F_2 is given by $P(F_2) = \theta_{II}$, the coefficient of type-II functional divergence, we have the joint probability for X and Y as follows

$$P(X, Y) = (1 - \theta_{II})P(X, Y|F_0) + \theta_{II}P(X, Y|F_2) \tag{3.17}$$

Direct application of Eq. (3.17) for estimating θ_{II} may face some difficulties because the amino acid pattern of early-stage (Y) is unobservable. A straightforward solution is to invoke the ancestral sequence inference, e.g. Yang *et al.* (1995). Treating the ancestral sequences as inferred observations, the standard procedure for the likelihood analysis of protein sequences can be applied. In spite of nice statistical properties, it requires a detailed description of the model and is sensitive to the statistical uncertainty in ancestral sequence inference. To solve this problem, we thus propose a simple but robust method that is computationally efficient, allowing genome-wide proteomic analysis.

3.4.3 Poisson model in the late-stage

Testing type-II functional divergence between two clusters (the early-late stages) utilizes the within-cluster amino acid patterns to examine the conservation in the late-stage. Therefore, a Poisson-based model that counts the number (k) of substitutions may be sufficient for this purpose, where smaller values of k of substitutions in a gene cluster indicate high conservation. Formally, at a given amino acid residue, the number of substitutions in each cluster (A or B) follows a Poisson process, e.g. for cluster A, we have

$$p_A(k) = \frac{(\lambda_A T_A)^k}{k!} e^{-\lambda_A T_A}$$

with the same applying to $p_B(k)$, where T_A (or T_B) is the total evolutionary time of cluster A (or B), and λ_A (or λ_B) is the evolutionary rate of cluster A (or B), respectively. Because of the Poisson property, we claim that

$$P(X = (i,j)|Y) = p_A(i)p_B(j)$$

which is independent of the early stage Y, where i or j is the number of substitutions in cluster A or B.

Hence, under the Poisson model, Gu (2006) has derived the analytical form of early-late joint distribution $f_{ij,Y} = P(X = (i,j), Y)$. Let $Z = \alpha/(D_A + D_B + \alpha)$, $Z_A = D_A/(D_A + D_B + \alpha)$ and $Z_B = D_B/(D_A + D_B + \alpha)$; D_A and D_B are the total branch lengths of clusters A and B, respectively, and α is the gamma shape parameter. Define $W = \alpha/(D_A + D_B + d + \alpha)$, $W_A = D_A/(D_A + D_B + d + \alpha)$ and $W_B = D_B/(D_A + D_B + d + \alpha)$. For $Y = N, R$ or C, $f_{ij,Y}$ is given by

$$f_{ij,N} = (1 - \theta_{II})M_{ij}$$

$$f_{ij,R} = (1 - \theta_{II})(Q_{ij} - M_{ij})\pi_R + \theta_E \, a_R Q_{ij}$$

$$f_{ij,C} = (1 - \theta_{II})(Q_{ij} - M_{ij})\pi_C + \theta_E \, a_C Q_{ij} \qquad (3.18)$$

where Q_{ij} and M_{ij} are given by

$$Q_{ij} = \frac{\Gamma(i + j + \alpha)}{i! \, j! \, \Gamma(\alpha)} Z^\alpha Z_A^i Z_B^j$$

$$M_{ij} = \frac{\Gamma(i + j + \alpha)}{i! \, j! \, \Gamma(\alpha)} W^\alpha W_A^i W_B^j \qquad (3.19)$$

3.4.4 Maximum-likelihood estimation

Let $n_{ij,Y}$ be the number of sites with the late-stage pattern $X = (i,j)$ and the early-stage pattern $Y = N, Y$ or C. Thus, the likelihood function can be written as

$$L = \prod_{i,j,Y} f_{ij,Y}^{n_{ij,Y}}$$

Gu (2006) implemented practical feasible algorithms to estimate unknown parameters, which is concisely discussed below. It is always assumed that the phylogenetic tree of the gene family is known or can be reliably inferred.

First, one can verify that the distribution of late-stage, i.e. the probability of a site being i and j substitutions in the two clusters, is given by

$$P(X = (i,j)) = f_{ij,N} + f_{ij,C} + f_{ij,R} = Q_{ij}$$

which depends on three (late-stage) parameters D_A, D_B, and α. Thus, the likelihood method of Gu and Zhang (1997) can be used to obtain the ML estimates of these parameters, with some technical modifications. Note that the algorithm of Gu and Zhang (1997) corrected the parsimony bias in counting the number of substitutions.

Second, after replacing three unknown late-stage parameters by their ML estimates, Gu (2006) developed a likelihood approach to estimate the early-stage parameters θ_{II}, a_R/a_C, and d, based on the inferred ancestral sequences of the early stage. We use the Bayesian algorithm of Yang *et al.* (1995). In particular, we find a simplified method called the U-likelihood is useful, which utilizes amino acid sites that are universally conserved in both clusters, i.e. $i = j = 0$. Let n_{00Y} be the number of sites with $Y = N$ (the U-type), R, or C, respectively. Let $n_{00} = n_{00N} + n_{00R} + n_{00C}$, and $f_{00} = f_{00N} + f_{00R} + f_{00C}$. Then, the U-likelihood can be written as

$$L = (1 - f_{00})^{N-n_{00}} \times \prod_{Y=N,R,C} f_{00,Y}^{n_{00,Y}}$$

Let $\hat{f}_{00N} = n_{00N}/N$. It has been shown that the U-ML estimates of θ_{II} and d are given by

$$\theta_{II} = 1 - \hat{f}_{00,N} \left[1 + \frac{\hat{D}_A + \hat{D}_B + d}{\hat{\alpha}} \right]^{\hat{\alpha}}$$

$$d = -\ln(1-p) + \ln(1 - \theta_{II}) \tag{3.20}$$

where p is the proportion of different amino acid sites between the ancestral nodes of two clusters. Since the U-method largely relies on the universally conserved sites, it seems robust against the inaccuracy of ancestral sequence inference and sequence alignment.

3.4.5 Predicting critical amino acid residues: empirical Bayesian approach

The identification of which sites are responsible for these type-II functional differences is of great interest, if the coefficient of functional divergence (θ_{II}) between early- and late-stages is significantly larger than 0. Here we develop a method of predicting such sites, which indeed can be further tested by experimentation, using molecular, biochemical, or transgenic approaches.

We wish to know the probability of state F_2 in the early-stage at a site, i.e. $P(F_2|X,Y)$. According to the Bayesian law, we have

$$P(F_2|X,Y) = \frac{P(F_2)P(X,Y|F_2)}{P(X,Y)}$$

where the prior probability of F_2 in the early-stage is given by $P(F_2) = \theta_{II}$. Under the Poisson-based model, $P(X = (i,j), Y|F_2)$ and $P(X = (i,j), Y|F_0)$, and $P(X = (i,j), Y)$ are given by Eq. (3.19). Noting that $a_Y = 0$ if $Y = N$, one can show

$$P(F_2|X,Y) = 0 \qquad\qquad \text{if} \qquad Y = N$$

$$P(F_2|X,Y) = a_C \theta_{II} Q_{ij}/f_{ij,Y} \qquad \text{if} \qquad Y = C$$

$$P(F_2|X,Y) = a_R \theta_{II} Q_{ij}/f_{ij,Y} \qquad \text{if} \qquad Y = R \qquad (3.21)$$

One may find it is simple to use the posterior probability ratio of F_2 to F_0, i.e. $R(F_2|F_0) = P(F_2|X,Y)/P(F_0|X,Y)$. After some algebras, we obtain

$$R(F_2|F_0) = 0 \qquad\qquad\qquad\qquad \text{if} \qquad Y = N$$

$$R(F_2|F_0) = \frac{\theta_{II}}{1-\theta_{II}} \frac{a_C}{\pi_C} \frac{1}{1-(1-h)^{i+j+\alpha}} \qquad \text{if} \qquad Y = C$$

$$R(F_2|F_0) = \frac{\theta_{II}}{1-\theta_{II}} \frac{a_R}{\pi_R} \frac{1}{1-(1-h)^{i+j+\alpha}} \qquad \text{if} \qquad Y = R \qquad (3.22)$$

where $h = d/(D_A + D_B + d + \alpha)$.

An important observation is that the posterior ratio $R(F_2|F_0)$ reaches its maximum if there is no amino acid substitution in each gene cluster but the amino acid is different between them, i.e. $i = j = 0$ and $Y \neq N$. As usually observed, and assuming that the proportion of radical changes under F_2 is higher than that under F_0 such that $a_R/a_C > \pi_R/\pi_C$, we have

$$R(F_2|F_0)_{max} = \frac{\theta_{II}}{1-\theta_{II}} \frac{a_R}{\pi_R} \frac{1}{1-(1-h)^{\alpha}}$$

Hence, a typical cluster-specific site indeed will receive a highest score for the type-II functional divergence, consistent with the intuitive biological interpretation. However, it should also be indicated that a high score could be statistically meaningless if θ_{II} is not significantly larger than 0. Finally, we note that $R(F_2|F_0)_{max} \to \infty$ if $h \to 0$. This means that greater accuracy is achieved as more sequences are analyzed (i.e. increasing D_A or D_B). In practice, one may use this property to determine how many sequences are sufficient to achieve the statistical resolution of site prediction.

An Example   We analyzed the cyclooxygenase (COX) enzymes catalyze which is a key step in the conversion of arachidonate to PGH2, the immediate substrate for a series of cell prostaglandin and thromboxane synthases. There are two tissue-specific isoforms in mammals: COX1 and COX2. Molecular cloning of COX2 led to a major investment by pharmaceutical companies in the development of selective inhibitors.

We estimated that the coefficient of type-II functional divergence between COX1 and COX2 duplicate genes was $\theta_{II} = 0.159 \pm 0.036$, which is statistically significant ($p < 0.001$). More detailed analysis shall be presented in Chapter 4.

3.5 A unifying model for type-I and -II functional divergences

The statistical methods described above have been shown to be useful for estimating type-I and type-II functional divergence of protein sequences after the gene duplication. To be feasible in practice, additional assumptions are needed in these methods, for instance, the ancestral sequence inference. Moreover, the model of type-I functional divergence assumes no type-II functional divergence, and *vice versa*. Therefore, it is desirable to build a likelihood function that takes these two types of functional divergence into consideration simultaneously.

The likelihood function Gu (2001b) solved this problem by considering the internal branch between two duplicate clusters (i.e. the early-stage) as ancestral cluster 0 (Fig. 3.5). Let λ_1 and λ_2 be the evolutionary rates in clusters 1 and 2, respectively, and λ_0 be the evolutionary rate in the internal branch (cluster 0), each of which follows a gamma distribution $\phi(\lambda)$; see Eq. 3.3. In each cluster, a given site has two possible states, F_0 (functional divergence unrelated) and $F = F_1$ or F_2 (functional divergence related). Therefore, we have $2^3 = 8$ possible combined states that can be reduced to 5 functional divergence patterns, denoted by $S_0, \ldots, S_4$. Under each S_j, $j = 0, \ldots, 4$, the relationship between λ_0, λ_1, and λ_2 is shown in Table 3.3. Let π_j ($j = 0, \ldots, 4$) be the probability of a site being under S_j, i.e. $\pi_j = P(S_j)$. By the Markov chain property, the conditional probability for observing X and Y at a site is given by

$$f(X, Y | \lambda) = \sum_{x_0=1}^{20} \sum_{y_0=1}^{20} b_{x_0} P_{x_0 y_0}(v | \lambda_0) f(X | \lambda_1; x_0) f(Y | \lambda_2; y_0) \qquad (3.23)$$

where $f(X | \lambda_1; x_0)$ and $f(Y | \lambda_2; y_0)$ are the likelihood functions for clusters 1 and 2, conditional of the roots x_0 and y_0, respectively, and v is the internal branch length. When the phylogeny is given in Fig. 3.5, one can write down

$$f(X | \lambda; x_0) = \sum_{x_5} \sum_{x_6} P_{x_0 x_5} P_{x_5 x_1} P_{x_5 x_2} P_{x_0 x_6} P_{x_6 x_3} P_{x_6 x_4}$$

$$f(Y | \lambda; y_0) = \sum_{y_5} \sum_{y_6} P_{y_0 y_5} P_{y_5 y_1} P_{y_5 y_2} P_{y_0 y_6} P_{y_6 y_3} P_{y_6 y_4}$$

To be concise, we use the expectation notation for any function $u(\lambda)$ for

$$E[u(\lambda)] = \int_0^\infty u(\lambda) \phi(\lambda) d\lambda$$

Table 3.3 Combined states (functional divergence patterns) for the unifying model with two gene clusters.

State (S_i)	Pattern	$P(S_i)$	Rate-independence[a]	Functional divergence[b]
S_0	(F_0, F_0, F_0)	π_0	$\lambda_0 = \lambda_1 = \lambda_2$	no
S_1	(F_1, F_0, F_0)	π_1	$\lambda_0, \lambda_1 = \lambda_2$	type II
S_2	(F_0, F_1, F_0)	π_2	$\lambda_0 = \lambda_2, \lambda_1$	type I
S_3	(F_0, F_0, F_1)	π_3	$\lambda_0 = \lambda_1, \lambda_2$	type I
S_4	including:	π_4		
	(F_0, F_1, F_1)		$\lambda_0, \lambda_1, \lambda_2$	type I
	(F_1, F_0, F_1)		$\lambda_0, \lambda_1, \lambda_2$	type I
	(F_1, F_1, F_0)		$\lambda_0, \lambda_1, \lambda_2$	type I
	(F_1, F_1, F_1)		$\lambda_0, \lambda_1, \lambda_2$	type I

[a] Rate-independence under each state can be illustrated by the following example: $\lambda_0, \lambda_1 = \lambda_2$ means that λ_0 is independent of λ_1 or λ_2.
[b] It indicates the type of functional divergence under each state.

Thus, according to the property of five functional divergence patterns (Table 3.3), we obtain the conditional probability for observing X and Y under each S_j as follows

$$f(X, Y|S_0) = \sum_{x_0=1}^{20} \sum_{y_0=1}^{20} b_{x_0} E[P_{x_0 y_0}(v|\lambda_0) f(X|\lambda; x_0) f(Y|\lambda; y_0)]$$

$$f(X, Y|S_1) = \sum_{x_0=1}^{20} \sum_{y_0=1}^{20} b_{x_0} E[P_{x_0 y_0}(v|\lambda_0)] \times E[f(X|\lambda; x_0) f(Y|\lambda; y_0)]$$

$$f(X, Y|S_2) = \sum_{x_0=1}^{20} \sum_{y_0=1}^{20} b_{x_0} E[f(X|\lambda_1; x_0)] \times E[P_{x_0 y_0}(v|\lambda_0) f(Y|\lambda; y_0)]$$

$$f(X, Y|S_3) = \sum_{x_0=1}^{20} \sum_{y_0=1}^{20} b_{x_0} E[P_{x_0 y_0}(v|\lambda_0) f(X|\lambda; x_0)] \times E[f(Y|\lambda_2; y_0)]$$

$$f(X, Y|S_4) = \sum_{x_0=1}^{20} \sum_{y_0=1}^{20} b_{x_0} E[P_{x_0 y_0}(v|\lambda_0)] \times E[f(X|\lambda_1; x_0)] \times E[f(Y|\lambda_2; y_0)]$$

$$(3.24)$$

Therefore, the joint probability of X and Y can be generally expressed as follows

$$p(X, Y) = \sum_{j=0}^{m-1} \pi_j f(X, Y|S_j) \qquad (3.25)$$

where $m = 5$. Similar to above, maximization of the likelihood $L = \prod_k p(X^{(k)}, Y^{(k)})$ can be achieved by either Newton–Raphson or EM algorithm.

Relationship between coefficients of functional divergence Under this unifying framework, the coefficients $\pi_0, \ldots, \pi_4$ provide a complete description of functional divergence and constraint after the gene duplication. Their relationships with the coefficients of type-I and type-II functional divergences (θ_I and θ_{II}) are as follows (Table 3.3). Since type-II functional divergence results in no altered functional constraints between two clusters, it can be interpreted as the functional divergence pattern $S_1 = (F_2, F_0, F_0)$: cluster 0 is under F_2, but clusters 1 and 2 are under F_0. Therefore, the coefficient of type-II functional divergence can be defined as $\theta_{II} = P(S_1) = P(F_1, F_0, F_0) = \pi_1$. On the other hand, type-I functional divergence means that at least either cluster 1 or cluster 2 should be under F_1, regardless of the status of cluster 0. According to Table 3.3, the coefficient of type-I functional divergence is given by $\theta_I = \pi_2 + \pi_3 + \pi_4$. In general, we can define *the coefficient of overall functional divergence* as $\pi_f = 1 - P(S_0) = 1 - \pi_0$. Obviously, we have

$$\theta_I + \theta_{II} = \pi_f = 1 - \pi_0 \qquad (3.26)$$

In this regard, π_0 is called *the coefficient of functional constraint* between duplicate genes.

Predicting critical amino acid sites Based on the whole-tree likelihood for functional divergence, we can develop a site-specific profile for type-I as well as type-II functional divergence. In the case of two clusters, the posterior probability of each (nondegenerate) combined state S_i (Table 3.3) can be computed as

$$P(S_i|X, Y) = \frac{\pi_i f(X, Y|S_i)}{\sum_{j=0}^{m-1} \pi_j f(X, Y|S_j)}, \qquad i = 0, 1, \ldots, 4 \qquad (3.27)$$

where $\pi_i = P(S_i)$, and $f(X, Y|S_j)$ is given by Eq. (3.24). Thus, one can easily show that site-specific profiles for type-I and type-II functional divergence are given by

$$P(\text{type I}|X, Y) = P(S_2|X, Y) + P(S_3|X, Y) + P(S_4|X, Y)$$

$$P(\text{type II}|X, Y) = P(S_1|X, Y) \qquad (3.28)$$

respectively.

Conclusive remarks Though the unifying model for type-I and type-II functional divergences is statistically elegant, it may be difficult to implement in practice for two reasons. First, the computational time could be expensive, mainly because of the complexity of algorithm, particularly when the number of subtrees for subfamilies is large. Second, due to the stochastic error, the ML estimation could not be converged when the sequence length is short. Nevertheless, we have found that separate analyses of these two types of functional divergence have been practically effective, as illustrated in the next chapter. Besides, the analysis can be also applied to the divergence after speciation (Gribaldo *et al.* 2003; Cheng *et al.* 2009; Penn *et al.* 2008).

4

Functional Divergence after Gene Duplication: Applications and Others

As most amino acid substitutions are not related to the functional divergence but only represent the neutral evolution, it becomes crucial how to statistically distinguish between these two possibilities (Abhiman *et al.* 2005a; 2005b). In the previous chapter, we introduced the statistical methods (Gu 1999; 2001b; 2006) to solve this problem, based on the principle that functional divergence between duplicate genes is highly correlated with the change of evolutionary rate. Here we discuss several case studies for applying these methods in biological problems (Wang and Gu 2001; Jordan *et al.* 2001; Gu *et al.* 2002a; Gu and Gu 2003b; Zheng *et al.* 2007; Zhou *et al.* 2007). Besides, other related methods are briefly discussed (Knudson and Miyamoto 2001; Gaucher *et al.* 2001; Lopez *et al.* 2002; Bielawski and Yang 2004; Gribaldo *et al.* 2003; Nam *et al.* 2005; Abhiman *et al.* 2006; Xu *et al.* 2009).

4.1 DIVERGE-based analysis

We have developed the software DIVERGE, short for DetectIng Variability in Evolutionary Rates among Genes, which is available at http://www.xungulab.com. DIVERGE is a GUI-based, user-friendly software package to provide an integrated analytical tool for functional divergence analysis of protein families, which can be run under both Window and LINUX operate systems (Fig. 4.1).

4.1.1 Functional-structural basis of shifted evolutionary rates between caspases

A cascade of cysteine aspartyl proteases (caspases) is the key component in the apoptotic machinery (or programmed cell death). There are 14 members of the caspase gene family in mammals, which can be further classified into two major subfamilies, CED-3 (including caspase-2, -3, -6, -7, -8, -9, -10, and -14) and ICE (including caspase-1, -4, -5, -11, -12, and -13). CED-3-type caspases are essential for most apoptotic pathways, while the major function of the ICE-type caspases is to mediate immune response. Phylogenetic analysis showed that these major caspase subfamilies are clustered separately (Fig. 4.2).

Using DIVERGE, Wang and Gu (2001) analyzed the caspase gene family to explore the structural-functional basis for the type-I functional divergence of protein sequences

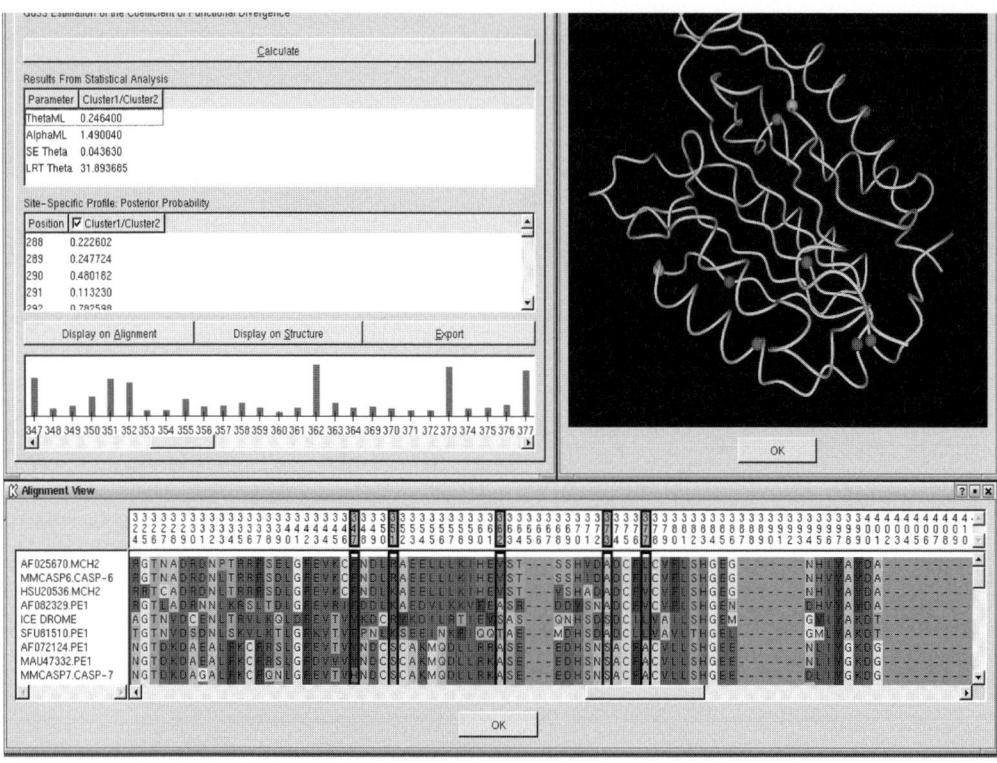

Fig. 4.1 Illustration of the software DIVERGE interface. Figure modified from Gu and Vander Velden (2002).

between CED-3 and ICE caspase subfamilies. Based on the inferred tree of caspases (Fig. 4.2), we found that type-I functional divergence is statistically significant between two major subfamilies, CED-3 and ICE ($\theta_I = 0.29, p < 0.001$). This means that, after the gene duplication, some amino acid sites may have been involved in the functional divergence between CED-3 and ICE. We further carried out the posterior profile analysis (Fig. 4.3) and predicted 29 crucial amino acid residues that are responsible for functional divergence between them at the cutoff of more than 70 per cent (posterior probability), which were onto the 3-D structure of caspases. The resolved X-ray crystal structures of human caspase-1 and -3 (Wilson *et al.* 1994; Rotonda *et al.* 1996) were used to illustrate the structural features of ICE and CED-3 subfamilies, respectively (Fig. 4.4).

From the literature, Wang and Gu (2001) found some experimental evidence for four predicted residues that are involved in the functional-structural divergence between CED-3 and ICE subfamilies (Fig. 4.3 to Fig. 4.5):

1. Residue 161(348):   (In the literature, this site is numbered as W348, according to the protein sequence of human caspase-1) is critical for CED-3 caspase substrate specificity by interacting with a unique surface loop in 3-D structures $[P(F_1|X) = 0.999]$

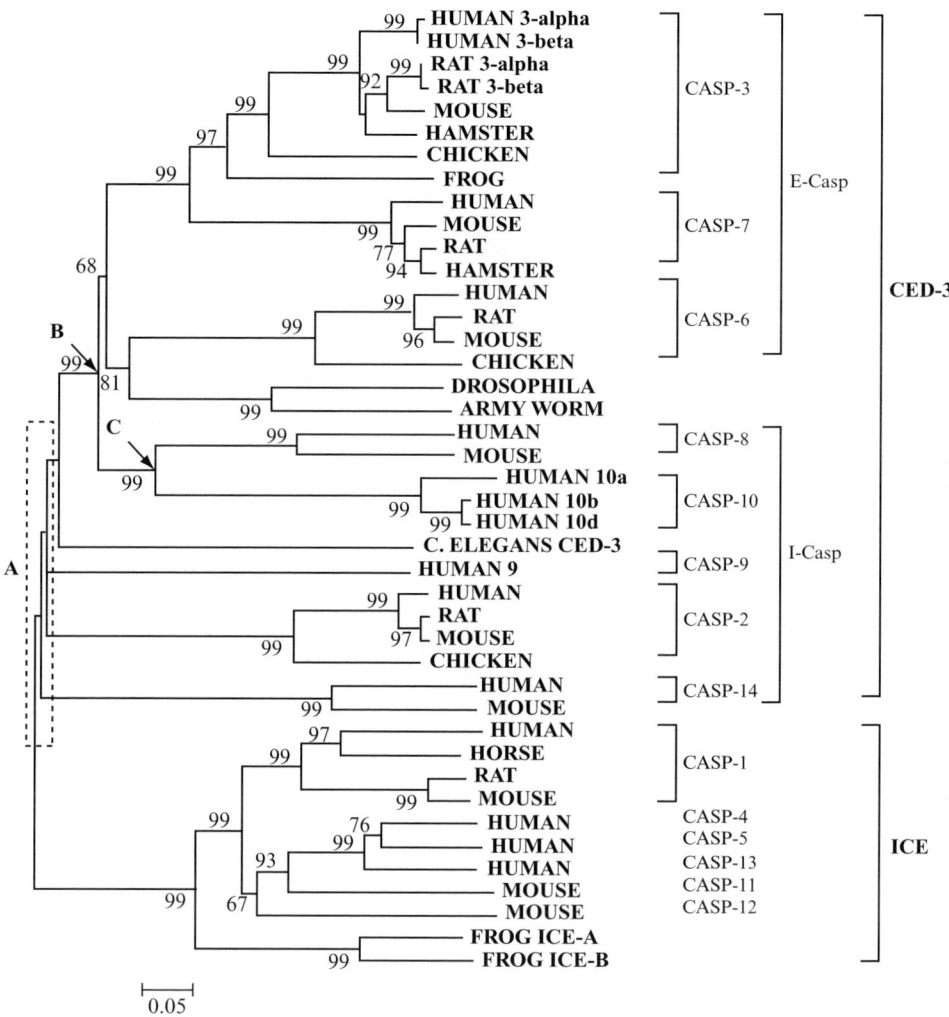

Fig. 4.2 Phylogenetic tree of the caspase gene family, inferred by the neighbor-joining method on the basis of amino acid sequences with the Poisson correction. Bootstrap values of more than 50 per cent are presented. Initiator caspases (I-casps) are involved in the upstream regulatory events, and effector caspases (E-casps) directly lead to cell disassembly. Modified from Wang and Gu (2001).

(Rotonda *et al.* 1996). At this position, all 22 sequences from the CED-3 subfamily contain an invariant tryptophan (W), whereas a variety of residues are present in the ICE subfamily (Fig. 4.5). Crystal structural analysis (Fig. 4.4) reveals that W348 is a key determinant for the caspase-3 (CED-3)-type specificity. First, W348 forms a narrow pocket with a surface loop that is highly conserved in the CED-3 subfamily. The steric constriction due to this pocket determines the preference of caspase-3 to

(A)

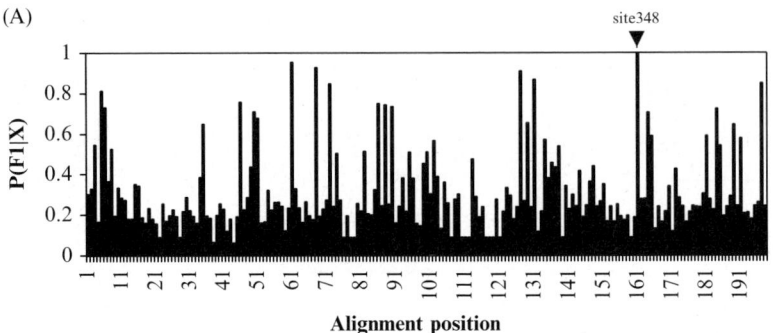

(B)

	CED-3	ICE
Sequence conservation	An Invariant Trp (W)	Highly variable
Structural features	Form a narrow pocket with an extra loop; Form a H-bond network with a group of aa's.	No extra loop; A shallow depression found.
Substrate specificity	Hydrophilic side chains	Hydrophobic side chains

Fig. 4.3 (A) Site-specific profile for predicting critical amino acid residues responsible for the (type-I) functional divergence between CED-3 and ICE subfamilies, measured by the posterior probability of being functional divergence related at each site. The arrows point to four amino acid residues at which functional divergence between CED-3 and ICE has been verified by experimentation; see (B) for details. Figure modified from Wang and Gu (2001).

the substrates with small hydrophilic side chains. Second, W348 along with a group of residues forms a hydrogen bond network, which affects the interaction with the substrate. In contrast, the surface loop shared with CED-3 caspases seems to be deleted in all ICE-type caspases, as shown in the boxed region in the multiple alignment (Fig. 4.5). Hence, the relaxed evolutionary constraint observed at this position in the ICE subfamily is likely to be caused by the 3-D structural difference.

2. Residues 86 $[P(F_1|X) = 0.75]$ and 88 $[P(F_1|X) = 0.74]$: They are responsible for 3-D difference with an unknown functional role. Indeed, in human caspase-1 (ICE), these two residues appear to lie in a small loop that is not found in the CED-3 subfamily.

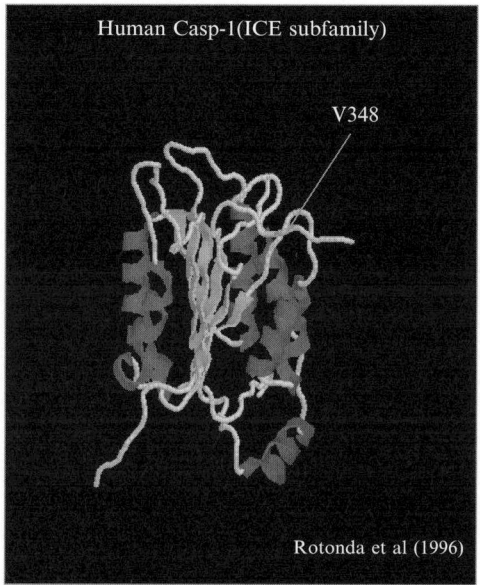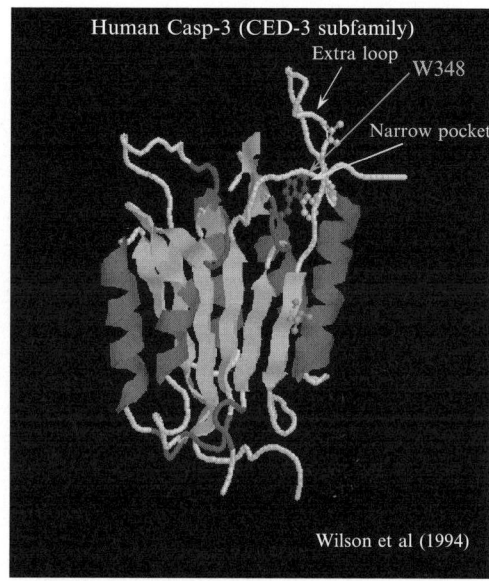

Fig. 4.4 Protein structures of human caspase-1 (ICE-type) and human caspase-3 (CED-3 type). Protein structure data from Rotonda *et al.* (1996) and Wilson *et al.* (1994), respectively.

3. Residue 131 $[P(F_1|X) = 0.866]$: It is the proteolytic site specific to the ICE subfamily. All caspases are synthesized as inactive proenzymes that need to be processed to mature forms (Nicholson *et al.* 1995). However, distinct cleavage sites within the precursors are found for two subfamilies. D131 is known as a cleavage site in human caspase-1 (ICE type; Thornberry *et al.* 1992). All ICE-type caspases preserve an Asp (D) at this position, except for mouse caspase-12 (Asn, E). However, human caspase-3 (CED-3 type) utilizes two other Asn sites for cleavage (Rotonda *et al.* 1996) so that the functional role of position 131 in CED-3 caspases is no longer important. Therefore, the altered evolutionary constraints at this position can be well explained by the different utilization of cleavage sites for the precursor processing between CED-3 and ICE subfamilies.

4.1.2 Pseudokinase domain in Jak protein kinase is functional

Jak (Janus kinase) is a nonreceptor tyrosine kinase, which plays important roles in signal transduction pathways. The unique feature of Jak is that, in addition to a fully functional tyrosine kinase domain (JH1), Jak possesses a pseudokinase domain (JH2). Although JH2 lost its catalytic function, experimental evidence has shown that this domain may have acquired some new but unknown functions. This apparent functional

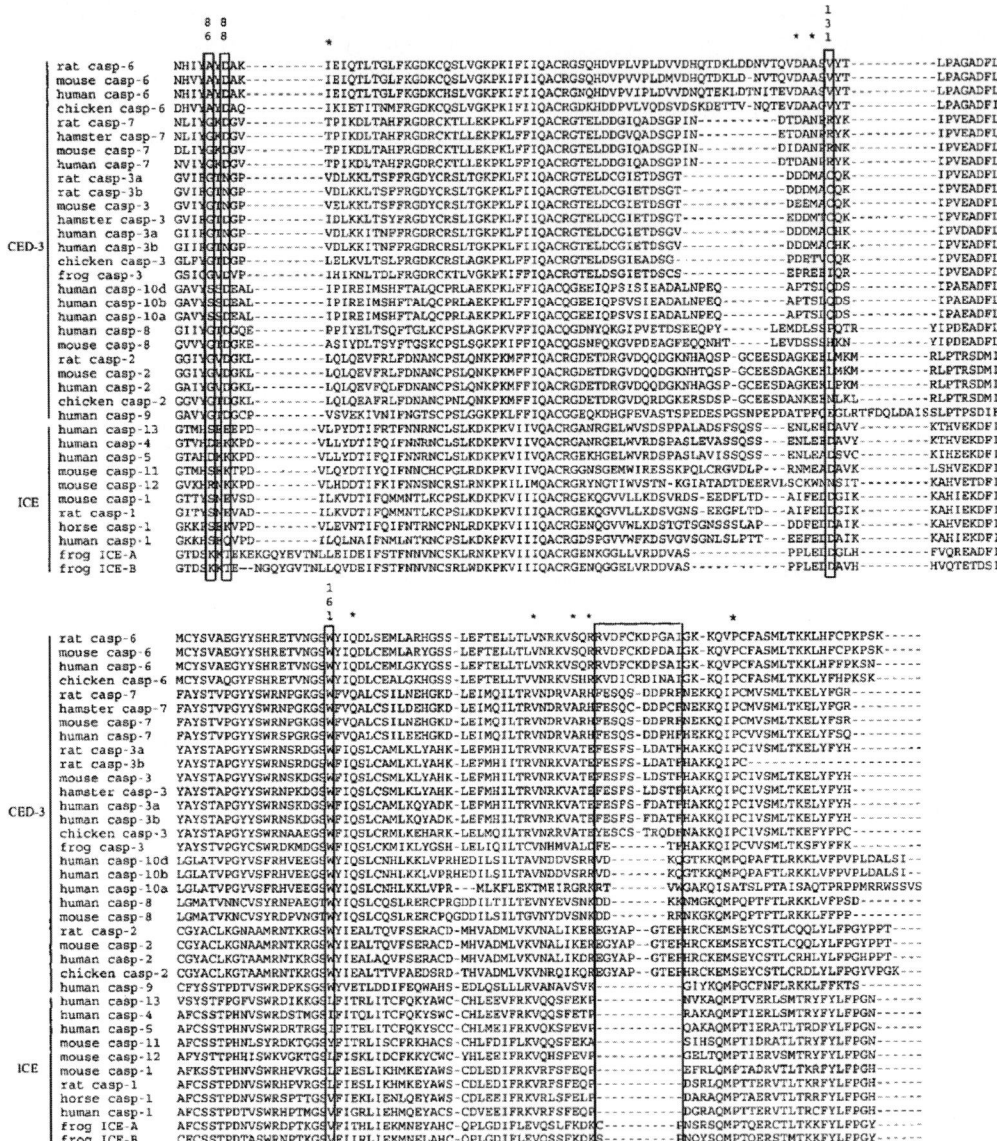

Fig. 4.5 Alignment of predicted regions of caspases. Four predicted sites with experimental evidence are highlighted. The sites with asterisks are predicted residues. The boxed region in the C-terminus is the critical region for CED-3 substrate specificity: Most CED-3 type capases form a surface loop, whereas a shallow depression is found in ICE-type caspases. Modified from Wang and Gu (2001).

divergence after the (internal) domain duplication may result in dramatic changes of selective constraints (type-I functional divergence) at some sites.

We (Gu *et al.* 2002a) conducted a data analysis to test this hypothesis. We first reconstructed a neighbor-joining (NJ) tree, including Jaks and two closely related protein tyrosine kinases, FGFR and EGFR. The inferred phylogeny shows that the tandem kinase (JH1) and pseudokinase (JH2) domains are evolutionarily distinct (Fig. 4.6). Indeed, the tandem kinase domains (JH1) in Jaks appear to be more closely related to the functional kinase domains of FGFRs and EGFRs, while the pseudokinase domains (JH2) of Jaks form a unique clade. It seems that the JH2 domain was generated before the emergence of most member genes of the protein tyrosine kinase supergene family.

As the pseudokinase domains (JH2) no longer exhibit the catalytic activity but may have acquired some new functions, it is interesting to test whether this functional divergence resulted in shifted selective constraints (different evolutionary rates) at some sites between the tandem kinase (JH1) and the pseudokinase (JH2) domains. To this end, we estimated the coefficient of type-I functional divergence between JH1 and JH2 domains $\theta_I = 0.412 \pm 0.049$, providing strong statistical evidence supporting the hypothesis of altered selective constraints between the tandem kinase (JH1) and pseudokinase (JH2) domains in Jak proteins.

A site-specific profile based on the posterior probability is used to identify critical amino acid sites that are responsible for functional divergence between the tandem kinase (JH1) and the pseudokinase (JH2) domains. Among 212 amino acid sites, 21 of them show a very high probability of being functional divergence related ($P(F_1|X) > 0.9$). These 21 sites can be definitively grouped into two categories: (I) conserved in the tandem kinase (JH1) domain but variable in the pseudokinase (JH2) domain and (II) conserved in the pseudokinase domain but variable in the tandem kinase domain (Fig. 4.7).

Category I: Of the 12 sites belonging to this category, site 137 has been demonstrated to be a determining site for the function of the tandem kinase domain (JH1), corresponding to the second tyrosine (highlighted Y) of a conserved (E/D)YY motif in the Jak2 protein. This motif, which is located in the activation loop of Jak2, regulates the kinase activity by autophosphorylation. In Tyk2, these two consecutive tyrosines (YY), have also been identified as phosphorylation sites (Gauzzi *et al.* 1996). Interestingly, the multiple alignments clearly show that site 137 is invariant in the tandem kinase (JH1) domains. In contrast, the same position in the pseudokinase domains (JH2) has a variety of amino acids with very different chemical properties. For example, some JH2 domains have amino acids with nonpolar side chains such as glycine, alanine, and proline, and some of them have uncharged polar amino acids such as serine and threonine (Fig. 4.6). This observation can be explained as a relaxed selective constraint that was caused by loss of function in phosphorylation in the JH2 domains.

Category II: Nine predicted sites belong to this category (Fig. 4.7). Among them, site 103 is predicted to be highly functional divergence related. Experimental data

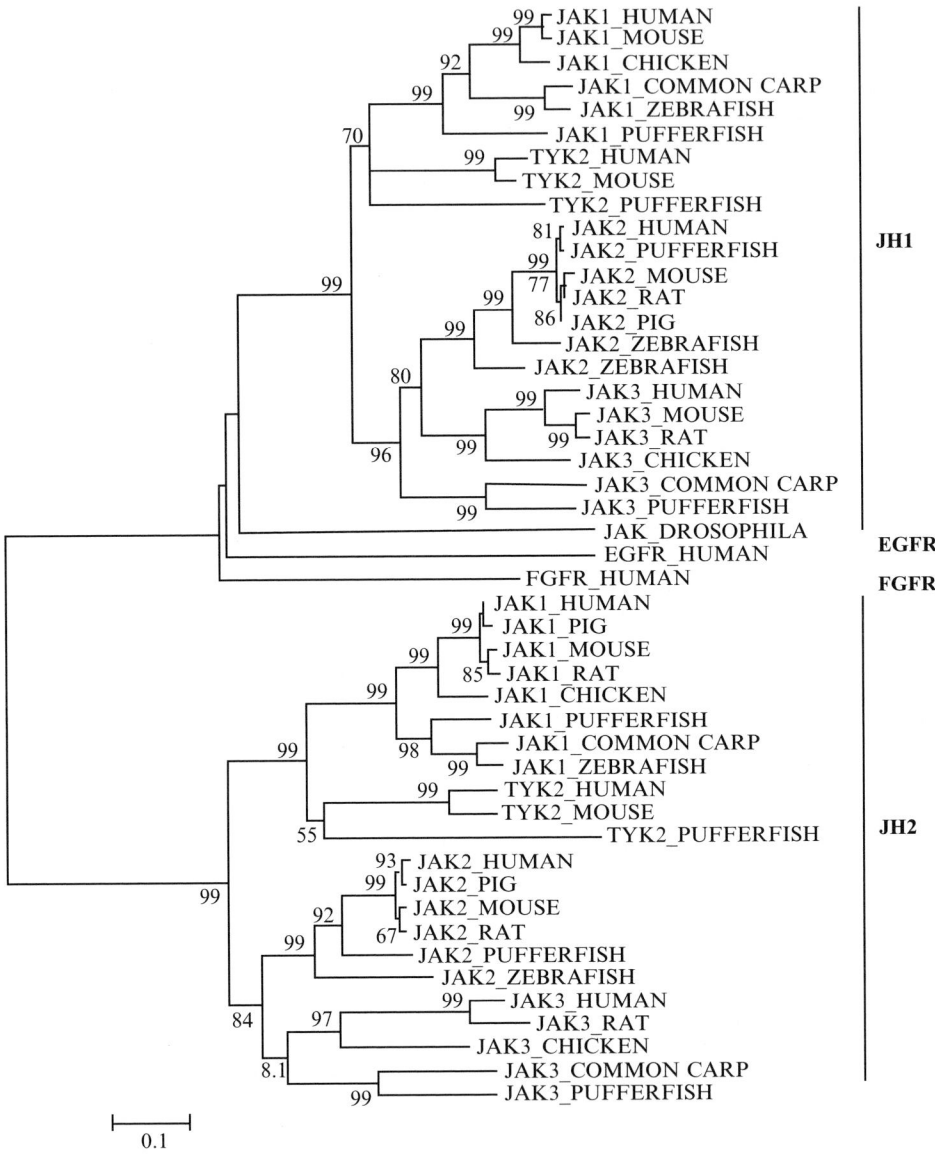

Fig. 4.6 The NJ tree of protein kinase domains from JAKs (JH1 and JH2), FGFRs and EGFRs based on the sequence alignment. The statistical reliability of the inferred phylogeny was assessed by bootstrapping values. From J. Gu *et al.* (2002).

(A)

		11112 222234735690 013606079263	
	position (k)		
	HUMAN	DPDGALPYYSRC	
	MOUSE	DPDGALPYYSRC	
	CHICKEN	DPDGALPYYSRC	Jak1
	CARP	DPDGALPYYSRC	
	ZEBRAFISH	DPDGALPYYSRC	
	PUFFERFISH	DPDGALPYYSRC	
	HUMAN	DPDGALPYSSRC	
	MOUSE	DPDGALPYSSRC	
	RAT	DPDGALPYSSRC	Jak2
	PIG	DPDGALPYSSRC	
JH1	PUFFERFISH	DPDGALPYSSRC	
	ZEBRAFISH	DPDGALPYSSRC	
	HUMAN	DPDGALPYSSRC	
	MOUSE	DPDGALPYSSRC	
	RAT	DPDGALPYSSRC	
	CHICKEN	DPDGALPYSSRC	Jak3
	CARP	DPDGALPYSSRC	
	PUFFERFISH	DPDGALPYSSRC	
	HUMAN	DPHGALPYSSRC	
	MOUSE	DPDGALPYSSRC	Tyk2
	PUFFERFISH	DPDGALPYSSRC	
	HUMAN	MDEKIVESSARK	
	PIG	MDEKIVESSARK	
	MOUSE	LDEKIVETSARK	
	RAT	LDEKIVETSARK	
	CHICKEN	LNNELVESSAMK	Jak1
	CARP	KLYEIIQSSAQE	
	ZEBRAFISH	KPYEVIQSTAQD	
	PUFFERFISH	RVSEVVQTSAQT	
	HUMAN	REQELLKPNTQA	
	MOUSE	REQKLLKPNTQT	
JH2	RAT	REQELLKPNTQT	Jak2
	PIG	REQELLKPTTQT	
	PUFFERFISH	KELQVLKPSAQI	
	ZEBRAFISH	REEKVLRPSAQT	
	HUMAN	HEEKLVHSSAQT	
	RAT	REEDLVYSNAQT	
	CHICKEN	RDEQVLRAAAQS	
	CARP	TDVTLIKGECNT	Jak3
	PUFFERFISH	SNGRFFEGTSQT	
	HUMAN	RVREVVESSMRP	
	MOUSE	RVSQVVESGTQP	Tyk2
	PUFFERFISH	QVSDVLKTRPRK	

(B)

	1111 344480029 814773510	
position (k)		
HUMAN	KERYQGRST	
MOUSE	KERYQGRST	
CHICKEN	KERYQGRSV	Jak1
CARP	WHRYTARNV	
ZEBRAFISH	WHRYTGRNV	
PUFFERFISH	SDKFTGKNL	
HUMAN	EEKQKGKNL	
MOUSE	EEKQKGKNL	
RAT	EEKQKGKNL	Jak2
PIG	EEKQKGKNL	
PUFFERFISH	EEKQKGKNL	
ZEBRAFISH	EEKQKAKSL	
HUMAN	QQKHRGRSL	
MOUSE	QQKHRGRSL	
RAT	QQKHRGRSL	Jak3
CHICKEN	EQHQTGQSL	
CARP	QQSHRQMSF	
PUFFERFISH	KKSHRQLSI	
HUMAN	KDRYQHHNL	
MOUSE	QERYQHQNL	Tyk2
PUFFERFISH	INKDQHKRL	
HUMAN	FAMSWEKRF	
PIG	FAMSWEKRF	
MOUSE	FAMSWEKRF	
RAT	FAMSWEKRF	
CHICKEN	FAMSWEKRF	
CARP	FAMSWEKRF	
ZEBRAFISH	FAMSWEKRF	
PUFFERFISH	FAMSWEKRF	
HUMAN	FAMSWEKRF	
MOUSE	FAMSWEKRF	
RAT	FAMSWEKRF	
PIG	FAMSWEKRF	
PUFFERFISH	FAMSWEKRF	
ZEBRAFISH	FAMSWEKRF	
HUMAN	FAMSWEKRF	
RAT	FAMSWEKRF	
CHICKEN	FAMSWEKRF	
CARP	FAMSWEKRF	
PUFFERFISH	FAMSWEKRF	
HUMAN	FAMSWEKRF	
MOUSE	FAMSWEKRF	
PUFFERFISH	FVMSWEKRF	

Fig. 4.7 Type-1 functional divergence related amino acid sites. (A) Category I: conserved in tandem kinase domains (JH1) and variable in pseudokinase domains (JH2). (B) Category II: conserved in pseudokinase domains (JH2) and variable in tandem kniase domain (JH1). From J. Gu *et al.* (2002).

show that a glutamic acid (E)-to-lysine (K) replacement occurring at this site in the pseudokinase (JH2) domain hyperactivated the Jak-Stat pathway in Drosophila and mammalian species (Luo *et al.* 1997). It seems likely that after the internal domain duplication, the tandem kinase domain (JH1) largely maintained the original catalytic function, while the pseudokinase domain (JH2) may have achieved some unidentified new functions, resulting in a set of JH2-specific conserved sites.

4.1.3 Pattern of type-II functional divergence

Testing the significance of type-II functional divergence   We (Gu 2006) have analyzed three gene families, COX, G-protein alpha subunits, and caspases. All these gene families show significant type-I functional divergence. It has been estimated that the coefficient of type-II functional divergence between COX1 and COX2 duplicate genes $\theta_{II} = 0.159 \pm 0.036$, which is statistically significant ($p < 0.001$). We also analyzed the duplicated isoforms (G_q and G_S) of G-protein alpha subunits, and found a similar pattern of type-II functional divergence. In contrast to Wang and Gu's (2001) finding for type-I functional divergence of caspase gene family, we found no evidence for type-II functional divergence between two major CED-3 and ICE subfamilies. This raises the question whether the type-I functional divergence is more pervasive than type-II functional divergence in gene family evolution.

Type-II functional divergence between COX-1 and COX-2   For illustration, Fig. 4.8 shows the site-specific posterior ratio profile of type-II functional divergence between COX1 and COX2. For 583 aligned sites, 492 (84 per cent) sites have received the ratio-score < 1, indicating that most sites are predicted to be unrelated to the type-II functional divergence. Moreover, we identified 28 radical cluster-specific sites that receive the highest posterior ratio score, i.e. $R(F_1|F_0)_{max} = 7.17$. In other words, if we select these radical cluster-specific sites as candidates for type-II functional divergence, the posterior probability for them is $P_{II} = 7.17/(1 + 7.17) = 87.8$ per cent, indicating that the prediction error (false-positive rate) is 12.2 per cent. Actually, it is impressive that among 111 radical amino acid substitutions in the early-stage after the gene duplication, about $29/111 \approx 26$ per cent are potentially related to type-II functional divergence between COX1 and COX2.

Effect of radical substitutions in the early-stage of duplication: For the COX gene family, we found radical substitutions for type-II functional divergence in the early-stage is about 2.7-fold increasing ($a_R/\pi_R = 2.7$ in Table 4.1). Consequently, an amino acid residue with a radical change between COX1 and COX2 may have a higher score than a conserved change for being type-II functional divergence-related. As shown by Table 4.2, the sites most likely exhibiting type-II behavior are the radical cluster-specific sites, while the conserved cluster-specific sites are less likely, as indicated by a low posterior probability (~ 0.35). This case study clearly shows the important role of statistical analysis, otherwise one cannot objectively justify whether one-less radical cluster-specific sites (i.e. there is one amino acid substitution in the late-stage) is more likely to be functional divergence related than conserved cluster-specific sites.

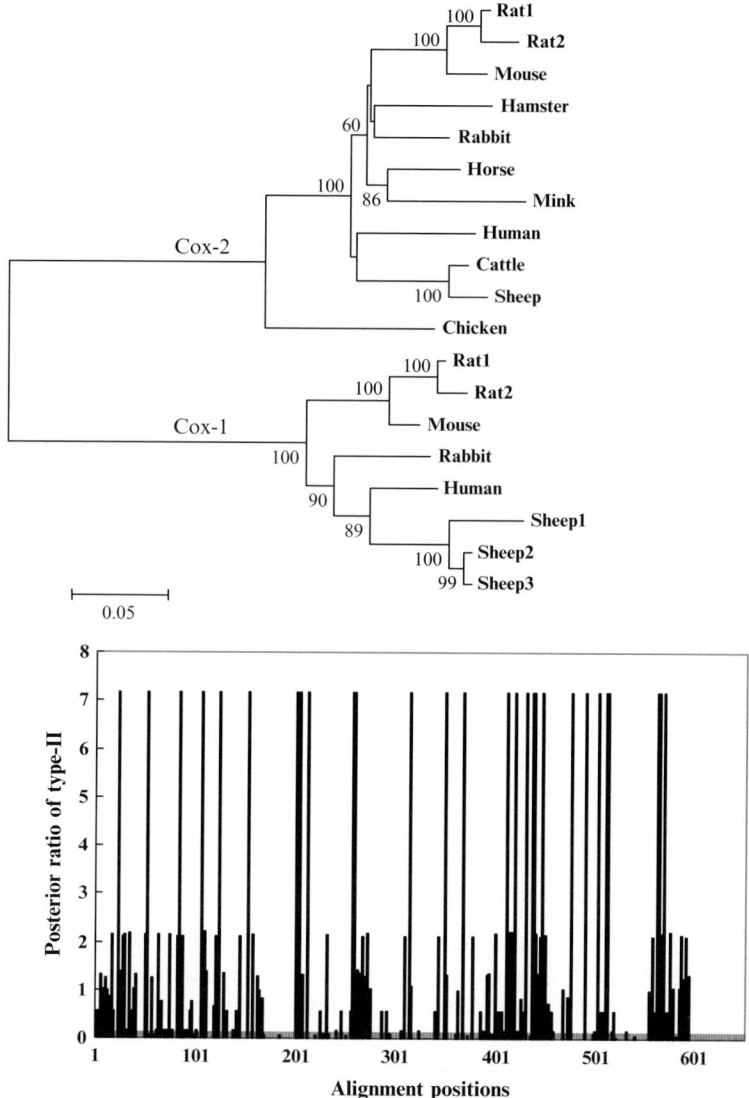

Fig. 4.8 (A) The phylogenetic tree of COX gene family, which was inferred by the neighbor-joining method, using amino acid sequences with Poisson distance. Bootstrapping values of more than 50 per cent are presented. Modified from Gu (2006). (B) Site-specific profile for type-II functional divergence between COX-1 and COX-2, measured by the posterior ratio. Horizontal lines (1)–(4) indicate cluster-specific patterns in Table 4.1. From Gu (2006).

Table 4.1 Functional ranking of several cluster-specific patterns in the COX gene family. From Gu (2006) (a.a for amino acid).

	Between clusters (Early-stage)	Within clusters (Late-stage)	*Num. of Sites*	*Ratio score*	Posterior Prob.
(1)	Radical change (Radical cluster-specific)	No a.a change	28	7.17	0.88
(2)	Radical change	One a.a change	30	2.11–2.22	0.68–0.69
(3)	Radical change	Two a.a changes	20	1.25–1.41	0.56–0.59
	. . .	. . .	. . .	. . .	. . .
(4)	Conserved change (Conserved cluster-specific)	No a.a change	31	0.55	0.35

Pattern (1): Radical cluster-specific sites. Patterns (2)–(3): Imperfect radical cluster-specific sites. Pattern (4): Conserved cluster-specific sites.

4.2 Functional distance analysis

Given the premise that shifting rates reflect changes in protein function, the coefficient of type-I functional divergence θ_I can also be interpreted as a predictor of the overall degree of functional divergence between protein genes. However, the two-cluster analysis cannot tell whether two clusters have the equal functional divergence. To solve this problem we developed the functional distance analysis.

4.2.1 Distance of functional divergence

Consider a gene family with several gene clusters. According to the notation system of Gu (2001b), let θ_i be the probability of a site being the state F_1 in cluster i that has experienced a shifted evolutionary rate from the ancestral gene. Hence, $1 - \theta_i$ is the probability of a site being the state F_0 (no rate shift). Consider two clusters i and j. Under the assumption of independence, we have $(1 - \theta_i)(1 - \theta_j)$ for the probability of a site that has no shifted rate in both clusters. In this case, it means that this site has not experienced type-I functional divergence, that is,

$$1 - \theta_{ij} = (1 - \theta_i)(1 - \theta_j) \tag{4.1}$$

where θ_{ij} is the coefficient of type-I functional divergence. Further, we define the (pairwise) distance of functional divergence as follows

$$d_{ij} = -\ln(1 - \theta_{ij}) \tag{4.2}$$

and the branch length of functional divergence is defined as

$$b_i = -\ln(1 - \theta_i) \tag{4.3}$$

Table 4.2 Summary of amino acids changes in 22 radical cluster-specific positions associated with the divergence of COX1 and COX2. From Gu (2006).

Position	COX1	COX2	Property change		
22	Y	S	H	vs	P0
51	P	E	P0	vs	−
82	W	G	H	vs	P0
103	V	S	H	vs	P0
121	I	K	H	vs	+
149	T	V	P0	vs	H
197	S	D	P0	vs	−
251	E	K	−	vs	+
253	A	T	H	vs	P0
306	T	E	P0	vs	−
340	F	H	H	vs	+
358	R	Q	+	vs	P0
401	Y	H	H	vs	+
409	A	S	H	vs	P0
419	G	A	P0	vs	H
425	D	P	−	vs	P0
427	H	A	+	vs	H
435	V	S	H	vs	P0
463	Q	E	P0	vs	−
499	S	A	P0	vs	H
548	K	Q	+	vs	P0
555	T	V	P0	vs	H

H: hydrophobic; P0: hydrophilic with neutral charge; +: charge positive; and −: charge negative.

Apparently the distance of functional divergence is additive, that is,

$$d_{ij} = b_i + b_j \tag{4.4}$$

After gene duplication, one gene copy maintains the original function, while the other copy is free to accumulate amino acid changes as a result of functional redundancy which may acquire new functions by positive selection or genetic drift (Ohno 1970; Atchley *et al.* 1994). Thus, one can predict an unequal functional divergence after the gene duplication. The branch length of functional divergence may provide insights on the degree of functional divergence of one gene copy after the gene duplication. Indeed, a large functional branch length (b_F) indicates substantial shifted functional constraints in the cluster, while $b_F \approx 0$ indicates that the evolutionary rate of each site in this duplicate gene is almost identical to the ancestral gene. In other words, a duplicate gene cluster with $b_F \approx 0$ may contain a larger component of ancestral function compared to other gene clusters. To this end, we have to consider three duplicate clusters.

4.2.2 Three-cluster analysis

Estimation of functional branch lengths   Consider three duplicate clusters 1, 2, and 3 that are phylogenetically separate. The pairwise coefficients of type-1 functional divergence are denoted by θ_{12}, θ_{13}, and θ_{23}, respectively. From Eq. (4.1) to Eq. (4.4), we have the corresponding functional distances $d_{ij} = -\ln(1 - \theta_{ij})$, functional branch lengths $b_i = -\ln(1 - \theta_i)$, as well as the additivity $d_{ij} = b_i + b_j$ ($i \neq j = 1, 2, 3$). Then one can easily show that functional branch lengths can be estimated by

$$b_1 = (d_{12} + d_{13} - d_{23})/2$$
$$b_2 = (d_{12} + d_{23} - d_{13})/2$$
$$b_3 = (d_{13} + d_{23} - d_{12})/2 \tag{4.5}$$

The total branch length $B = b_1 + b_2 + b_3$, a measure for the overall functional divergence of the gene family, can be calculated as

$$B = (d_{12} + d_{13} + d_{23})/2$$

Another related measure is the coefficient of overall functional divergence defined by $\pi = 1 - e^B$. In the case of three clusters, it is given by

$$\pi = 1 - \sqrt{(1 - \theta_{12})(1 - \theta_{13})(1 - \theta_{23})}$$

Statistical testing   It has been realized (e.g. Conant and Wagner 2003a) that the sequence divergence after gene duplication could be asymmetric. It is therefore interesting to test whether functional divergence between duplicate genes follows the same pattern. Let $\delta_{12} = d_{13} - d_{23}$. One can simply verify that $\delta = d_1 - d_2$ by the additivity. Therefore, one can construct a null hypothesis

$$\delta_{12} = 0$$

Rejection of this null hypothesis statistically indicates asymmetric functional divergence between duplicate clusters 1 and 2. A simple method to tset this null hypothesis is to compute the sampling variance of the estimate $\hat{\delta}$, which is given by

$$Var(\hat{\delta}) = Var(\hat{d}_{13}) + Var(\hat{d}_{23}) - 2Cov(\hat{d}_{13}, \hat{d}_{23}) \qquad (4.6)$$

We propose a simple method for computing the covariance between $\hat{d}_{13}$ and $\hat{d}_{23}$.

Calculation of the covariance $Cov(\hat{d}_{13}, \hat{d}_{23})$

Since the functional distance between clusters i and j is $d_{ij} = -\ln(1 - \theta_{ij})$, the sampling variance can be approximately given by

$$Var(d_{ij}) = \frac{Var(\theta_{ij})}{(1 - \theta_{ij})^2} \qquad (4.7)$$

Since $Var(\theta_{ij})$ can be obtained from the analysis of DIVERGE; also see Chapter 3, it is straightforward to compute $Var(\hat{d}_{13})$ and $Var(\hat{d}_{23})$. However, the covariance between functional distances d_{13} and d_{23} is not easy. We solve this problem as follows. Since b_3 is the common functional branch length shared by functional distances d_{13} and d_{23}, we have

$$Cov(d_{13}, d_{23}) = Var(b_3) \qquad (4.8)$$

Note that b_3 can be estimated by $b_3 = (d_{13} + d_{23} - d_{12})/2$. According to the definition of functional distance, $b_3 = -\ln(1 - \theta_3)$. Therefore we have

$$Var(b_3) = \frac{Var(\theta_3)}{(1 - \theta_3)^2}$$

On the other hand, through Fisher's z-transformation, Gu (1999) has shown that

$$Var(\theta_3) = \left(\frac{1 - r^2}{r_M}\right)^2 / (N - 3)$$

where N is the sequence length, $r = (1 - \theta_1)r_M$, and $r_M = 1 - D_3/V_3$; D_3 and V_3 is the mean and variance of the number of substitutions occurred in cluster 3. Therefore, putting this together we obtain

$$Cov(d_{13}, d_{23}) = \left(\frac{1 - r_M^2(1 - \theta_3)^2}{r_M(1 - \theta_3)}\right)^2 / (N - 3) \qquad (4.9)$$

4.2.3 Examples: vertebrate developmental gene families

We have analyzed ten vertebrate developmental gene families, each of which has three members through two rounds (2R) of gene duplications after the emergence of early vertebrates (Wang and Gu 2000). We obtained functional distances and functional branching lengths for each gene family. As indicated above, in principle, virtually zero functional branch length of a duplicate cluster indicates the ancestral function, while a

Three-cluster vertebrate gene families used in this study.

Gene family	Cluster 1	Cluster 2	Cluster 3
ADRA1(Adrenergic receptor α1)	ADRA1A	ADRA1B	ADRA1C
ADRA2(Adrenergic receptor α2)	ADRA2A	ADRA2C	ADRA2B
ADRB(Adrenergic receptor β)	ADRB1	ADRB2	ADRB3
ALDO(Aldolase)	AldoA	AldoC	AldoB
CDX(Caudal)	CDX1	CDX2	CDX4
HH(Hedgehog)	SHH	IHH	DHH
Jun	c-Jun	Jun-D	Jun-B
Myb	c-myb	A-myb	B-myb
NOS(Nitric oxidase synthase)	NOS1	E-NOS	I-NOS
SHR(Stimulating hormone receptor)	LSHR	TSHR	FSHR

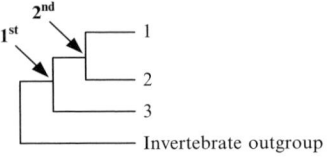

Functional branch lengths

Gene family		b_F		
		Cluster 1	Cluster 2	Cluster 3
Pattern I	ALDO	0.489	0.364	-0.22
	NOS	0.431	1.302	-0.085
	HH	-0.0125	0.117	0.072
Pattern II	ADRA1	-0.01	1.38	0.25
	ADRA2	0.533	0.056	0.267
	ADRB	-0.172	0.348	0.731
	CDX	-0.118	0.122	0.119
	Jun	1.063	-0.394	1.402
	Myb	0.74	0.034	0.478
	SHR	-0.066	0.386	0.609

Fig. 4.9 Summary of functional distance analysis for vertebrate developmental gene families. See the text for the detailed discussion.

long functional branch length indicates diversified function. Interestingly, we observed two major patterns of functional divergence after two rounds of gene duplications.

Pattern I: one copy after the first round of gene duplication maintains the original function

Three gene families (ALDO, HH, NOS) share the common pattern: Cluster 3 shows virtually zero functional branch length (Fig. 4.9). It is reasonable to speculate that after the first round of gene duplication, one of the copies (the ancestor of Cluster 3) preserves the ancestral function. This may have freed the other copy (the ancestor of Clusters 1 and 2) from selective constraint, allowing it to diverge and take on new roles in specific tissues or organs. Interestingly, these three families show two different sub-patterns of functional divergence after the second round of gene duplication:

Sub-pattern I_A In the ALDO gene family, Clusters 1 (AldoA) and 2(AldoC) both have long functional branch lengths (0.489 and 0.364, respectively). Experimental evidence has shown that AldoA and AldoC are tissue-specific: AldoA is only expressed in the fibroblast, while AldoC only functions in the brain. Similarly, in NOS gene family, Clusters 1 (NOS1) and 2 (E-NOS) show considerably large b_F values (0.431 and 1.302, respectively), which is consistent with their strong tissue specificity: NOS1 is expressed in neural system and uniquely participates in neurotransmission, and E-NOS is specifically expressed in endothelial tissues. In contrast, I-NOS, which has virtually zero (-0.085) functional branch length, is generally found expressed in several tissues such as heart and muscles. One may speculate that the tissue-specific functions of Clusters 1 and 2 in ALDO and NOS gene families may be developed under the shield of Cluster 3 which receives a strong selection pressure.

Sub-pattern I_B   In HH gene family, not only Cluster 3 (DHH) but also Cluster 1 (SHH) show virtually zero b_F value, while the b_F value (0.117) of Cluster 2 (IHH) is significantly greater than 0. This is consistent with the result of functional assay: vertebrate SHH genes typically show conserved expression in the notochord and floor plate, while IHH shows different expression patterns in different tissues in vertebrates. After the second round of gene duplication, one of the copies (SHH, in this case) had to maintain the ancestral function, whereas IHH might be the one that acquired some new functions and had escaped from the purifying selection.

Pattern II: one copy after the second round of gene duplication maintains the original function

In contrast to the three gene families showing Pattern I, the rest of the seven gene families show Pattern II: the distance branch length of Cluster 3 is significantly greater than 0, indicating that Cluster 3 may represent some novel function after two rounds of gene duplication; moreover, one gene copy after the second round of gene duplication (Cluster 1 or Cluster 2) shows virtually zero b_F value, whereas the other copy (Cluster 2 or Cluster 1) has long functional branch length. Extensive evidence has shown that certain level of functional divergence had occurred at least after the second round of gene duplications, although whether the first round had detectable impacts on the functional divergence remains unclear.

For example, in the myb gene family, dramatic functional innovations are likely to occur after both first and second rounds of gene duplications. The myb gene family consists of three member genes A, B, and c-myb that encode nuclear proteins functioning as transcription activators. The first round of gene duplication gave rise to B-myb, whereas the second produced A-myb and c-myb. Interestingly, A-myb and c-myb exert a negative regulatory effect on their transcription activation function, while B-myb functions as a positive regulator. This evidence supports the result of our functional distance analysis ($b_F = 0.478$ for B-Myb), implying that the shift of $+/-$ regulation may have taken place after the first round of gene duplication (Fig. 4.9). Moreover, gene knockout experiments show that A-myb and c-myb are functionally distinct as well. Loss of c-myb function results in embryonic lethality. In contrast, A-myb null mice are viable but exhibit growth abnormalities. This result is consistent with the inference from our functional distance analysis. A-myb with b_F value 0.034 may represent the ancestral function, whereas c-myb ($b_F = 0.740$) may carry on the novel function that is indispensable in development. Intuitively, these altered functional constraints suggest c-myb may be temporarily escaped from the evolutionary pressure by the presence of A-myb. Hence, the functional evolution of myb member genes indicate that the first round of gene duplication generated two regulation types $(+/-)$, and the second gave rise to the distinct level of negative regulation.

Analogous to the case of Myb gene family, the pattern of $+/-$ regulation changes after gene duplication is also observed in the Jun gene family. The Jun gene family encodes a component of the transcription factor Ap-1 (fos/Jun) which plays multiple roles in the functional development of hematopoietic cells and regulation of apoptosis (programmed cell death). The large b_F value (1.402) of Cluster 3 (Jun-B) implies along the lineage leading to it (i.e. after the first round of gene duplication) that there might be substantial functional innovation. This implication is supported by the experimental evidence: Jun-B exhibits strong negative regulation of apoptosis, whereas Jun-D and c-jun both regulate apoptosis positively. The altered functional constraints also appear in Clusters 1 ($b_F = -0.394$ for Jun-D) and Cluster 2 ($b_F = 1.063$ for c-Jun). Indeed, functional assay shows that in contrast to Jun-D, which only employs positive regulation, C-Jun has dual functional roles: positive regulation of apoptosis when it is solely present, and a strong negative role with Jun-B. One may speculate that the first round of gene duplication may have a contribution to the differentiation of $+/-$ negative control of apoptosis, whereas the second round may play a role in the fine regulation.

Another example is the SHR (stimulating hormone receptor) gene family. Although the detailed functional role of Cluster 3 (FSHR) is still unclear, the expression patterns of Clusters 1 (LSHR) and 2 (TSHR) may provide an insight into the underlying mechanism of the functional divergence after the second round of gene duplication. LSHR has $b_F = -0.066$ which indicates its low level functional divergence, while TSHR shows $b_F = 0.386$, which indicates high level of functional diversification. This significant difference in the functional branch length of LSHR and TSHR may reflect their tissue specificities: in contrast to LSHR, which is widely expressed in both gonadal cells and thyroid, TSHR is only uniquely expressed in thyroid. The sequence analysis suggests that tissue specific splicing have occurred after the second round of gene duplication.

In summary, in Pattern II, Cluster 3 shows significant level of functional divergence after the first round of gene duplication; and moreover, the two gene copies after gene duplications exhibit altered functional constraints, one of which seems to have inherited the ancestral function, and the other has developed some new functions.

4.3   Other methods for type-I functional divergence

4.3.1   Knudsen–Miyamoto method

Knudsen and Miyamoto (2001) developed a likelihood ratio test (LRT) for detecting significant rate shifts at specific sites in proteins. In this method, amino acid sites are analyzed individually to test whether a site from two phylogenetically related groups of sequences is evolving differently. The null hypothesis, H_0, states that a given position evolves with different rates in the two subfamilies. The likelihood under this model is calculated by using the method of Felsenstein (1981). The rate matrix used is the JTT matrix of Jones *et al.* (1992). The two rates of evolution are varied to obtain the maximum-likelihood (ML) value under this model, L_0. In contrast, hypothesis one (H_1) states that the position evolves at the same rate in the two subfamilies. Again, calculations are done according to Felsenstein (1981) and with the JTT matrix, but with a single rate used for the two subfamilies. The optimal rate is found, giving the ML value under this model, L_1.

Using an LRT statistic, we can evaluate H_1. The test statistic can be written as:

$$U = -2 \log \frac{L_1}{L_0} \qquad (4.10)$$

Because H_1 is a special case of H_0 (the hypotheses are nested), the LRT statistic U will never be negative. There are two degrees of freedom under H_0, whereas there is only one under H_1. This could indicate that under H_1, the distribution of U is approximately $\chi^2(1)$ with one degree of freedom. Though often a significance value of $P < 0.05$ is chosen in these tests, the problem is that multiple tests are being performed. At any rate, Knudsen and Miyamoto (2001) recommend that this cutoff be used to estimate the expected number of sites with $P < 0.05$ by chance alone.

4.3.2 Gaucher–Miyamoto–Benner method

Gaucher *et al.* (2001) demonstrated a function-structure analysis for divergent evolution of protein sequences between two monophyletic groups. Given the multiple alignment of protein sequences, the key step of their method is to estimate the substitution rate at each site in each group by, for instance, the Bayesian estimation under the model of rate heterogeneity among sites (Yang *et al.* 1996).

With the software package PAML, Gaucher *et al.* (2001) estimated the site-specific substitution rates of two homologous elongation factors Tu (EF-Tu) and 1a (EF-1a), and found that the distribution of rate differences over sites between EF-Tu and EF-1a was leptokurtotic, relative to the expectations of a normal distribution. Nearly 50 per cent of sites had essentially the same rate in the two groups. However, 17 sites were evolving > 2 SD (standard deviation) faster in EF-Tu than EF-1a, whereas 19 were changing > 2 SD faster in EF-1a than in EF-Tu. These sites, representing 10 per cent of alignment sites, are suggestive of functional divergence process in the EF-Tu/EF-1a family.

By integrating experimental three-dimensional structural data with these rate differences, Gaucher *et al.* (2001) reduced this initial pool of 36 sites to a subset that are most likely involved in the functional shifts between EF-Tu and EF-1a. For example, 10 sites in and around the region binding tRNAs are evolving > 2 SD faster in either EF-Tu or EF-1a. These rate changes can be correlated with a difference in biochemical function between EF-Tu and EF-1a.

4.3.3 Codon-based methods

Previous methods discussed for detecting the site-specific shift of evolutionary rate are based on amino acid sequences, which can be applied for distantly related genes. In the case of closely related genes, the codon-based approaches may be useful in studying functional divergence following gene duplication or speciation.

Forsberg and Christiansen (2003) applied Gu's (2001b) model of functional divergence to the codon model of Goldman and Yang (1994), which allows site-specific changes in selection pressure in two different parts of a phylogeny. The central parameter in the codon model is the nonsynonymous-to-synonymous rate ratio (ω), a measure of selection pressure on the protein, with $\omega = 1$, < 1, and > 1 indicating neutral evolution, purifying selection, and positive selection, respectively. Forsberg and Christiansen (2003) applied their codon model to influenza A virus nucleo-protein sequences to study the changes in selection pressure after a shift from avian to human hosts. Lacking any prior assumptions concerning the direction of such change, they assumed that ω ratio varies among sites according to a discrete distribution with three classes, while if such change occurs, the ω parameters in the two hosts become independent of each other, similar to Gu (2001). They used an LRT to test the hypothesis of $p_d > 0$ (the proportion of sites with altered selection preasures) and an empirical Bayes procedure to predict which sites had experienced a shift in selection pressures.

Later, a more complete codon-based maximum-likelihood model was proposed by Bielawski and Yang (2004). The model allows variation in ω among sites, with a fraction of sites evolving under divergent selective pressures. Divergent selection is indicated by different ω values between clades, such as between paralogous clades of a gene family. They applied the codon model to duplication followed by functional divergence of (i) the ϵ and γ globin genes and (ii) the eosinophil cationic protein (ECP) and eosinophil-derived neurotoxin (EDN) genes. In both cases likelihood ratio tests suggested the presence of sites evolving under divergent selective pressures.

4.3.4 Heterotachy model

Because of functional constraints, substitution rates vary among sites (amino acid residues) of a protein but are usually assumed to be constant at a given site during evolution (Yang 1993; Gu *et al.* 1995). However, many studies related to site-specific rate shift have convincingly demonstrated that the evolutionary rate of a given site is not always constant throughout time. Lopez *et al.* (2002) called within-site rate variations *heterotachy* (for 'different spread' in Greek), in contrast to *homotachy*.

Moreover, Lopez *et al.* (2002) suggested testing heterotachy with a chi-square test (Lopez *et al.* 1999). For two monophyletic groups, let x or y be the number of changes at a site in each group, and X or Y be the total number of changes (steps) in each group, respectively. If the substitution rate is constant at a site, the expectation of the ratio x/y is equal to the ratio of evolutionary time of each group (T_x/T_y). Since the expectation of step-ratio X/Y is also equal to T_x/T_y. A 2×2 table can be used to test this null hypothesis. This method thus allows for determining how many sites significantly reject a homotachous behavior.

4.3.5 The alpha shift measure (ASM) method

To automatically detect subfamily function divergence, Abhiman *et al.* (2006) introduced a method for large-scale prediction of functional divergence within protein families. It is called the alpha shift measure (ASM) as it is based on detecting a shift in the shape parameter (α) of the substitution rate gamma distribution. For each protein subfamily pair, they estimated the shape parameter α for both subfamilies, denoted by α_1 and α_2, respectively, as well as for the combined alignment of the two subfamilies, denoted by α_{12}, using the ML-based GZ method (Gu and Zhang 1997) and others. The ASM was then computed for each protein subfamily pair as:

$$ASM = \alpha_{12} - (\alpha_1 + \alpha_2)/2 \tag{4.11}$$

The measure ASM can be used to estimate function shift between two subfamilies, based on the following argument. If the α value from the combined sequence alignment is much larger than those from the individual subfamilies, this indicates a shift in function between them (Gu 1999; Gaucher *et al.* 2001). The rationale is that the lower α parameters of the subfamilies indicate that they have diversified to become more specific than the ancestral family (e.g. substrate specificity of enzymes). When $\alpha_1 = \alpha_2 = \alpha$ and the total evolutionary time is the same between two subfamilies, from the Appendix of Gu (1999) one can verify that $\alpha_{12} = \alpha/(1 - \theta/2) \geq \alpha$, where θ is the coefficient of type-I functional divergence. In this case, we apparently have the following relationship

$$ASM = \alpha \frac{\theta}{2 - \theta} \tag{4.12}$$

Hence, ASM=0 only when $\theta = 0$, while ASM reaches it maximum value when $\theta = 1$. It should be noted that this relationship holds strictly when the moment of method is used to estimate α; for other estimation methods, it only holds approximately.

4.3.6 Nam *et al.*'s method for detecting functional divergence of protein domains

One widely-used design to identify the functional difference is to compare a sequence with known functional domains with a new sequence by using domain swapping or site-directed mutagenesis. To facilitate such experimental studies, Nam *et al.* (2005) developed a statistical method for identifying protein domains that are likely to be functionally differentiated. In this method, two protein sequences are compared (A and B) and an outgroup sequence (C) will be used. To identify the protein regions that show a significant rate difference, Nam *et al.* (2005) used a sliding window analysis. Let n be the total number of amino acid sites and w be the window size (the number of amino acids considered for one window). This window analysis may be done by sliding the window by one amino acid position consecutively or by skipping s amino acid positions each time.

For each window, the number of amino acid substitutions a and b for branches toward genes A and B, respectively, can be estimated as follows

$$a = (d_{AB} + d_{AC} - d_{BC})/2$$
$$b = (d_{AB} + d_{BC} - d_{AC})/2 \qquad (4.13)$$

where d_{AB}, d_{AC}, and d_{BC} are the evolutionary distances between sequences, respectively. We are now interested in testing the significance level of the difference $D = a - b$. Let Z be $Z = D/\sqrt{V(D)}$, which is approximately normally distributed as long as the window size $w \geq 30$. In this case, the significance level can be determined. It is known that Z-values obtained for consecutive windows are highly correlated. Nevertheless, for the purpose to identify protein regions that should be subjected to experimental tests, any consecutive windows showing significant Z-values can be considered biologically important. This method has been applied to MIKC-type MADS box proteins that control flower development in plants. Nam *et al.* (2005) examined 23 pairs of sequences of floral MADS-box proteins from petunia and found that the rate differences for 14 pairs are significant. The significant rate differences were observed mostly in the K-domain, which is important for dimerization between MADS-box proteins. These regions may be chosen for further experimental studies.

5

Phylogenomic Expression Analysis between Duplicate Genes

Microarray technology can simultaneously monitor the expression levels of thousands of genes across many experimental conditions or treatments (Brown and Botstein 1999; Eisen 1998; Eisen and Fraser 2003; Khaitovich *et al.* 2006b), providing us with unique opportunities to investigate the evolutionary pattern of gene regulation (e.g. Wagner 2000a, Gu *et al.* 2002d; Enard *et al.* 2002; Gu and Gu 2003a; Rifkin *et al.* 2005; Caceres *et al.* 2003; Makova and Li 2003). In this chapter, we focus on how to model the expression evolution of a gene family with three goals: (*i*) Statistical methods such as likelihood ratio test can be applied for exploring the evolutionary pattern of gene expression. (*ii*) Evolutionary tracing of expression changes can be predicted by the Bayesian method. And (*iii*) the statistical model can be utilized to study the expression-motif association. Several statistical models have been developed (Gu 2004; Gu *et al.* 2005b; Oakley *et al.* 2005; Khaitovich *et al.* 2005b; Guo *et al.* 2007), most of which viewed gene expression data as continuous characters so that the modeling was based on the random-walk (Brownian) model (Edward and Cavalli-Sforza 1964; Lynch and Hill 1986). In the following we will discuss these models and their applications.

5.1 Brownian-related stochastic model

Gu (2004) has developed a statistical framework based on the Brownian model. When the phylogenetic tree of a gene family is given, usually inferred by the sequence data, the pattern of expression profiles among member genes can be modeled as a stochastic process driven by underlying evolutionary mechanisms.

5.1.1 Expression likelihood under phylogeny

In the microarray data, the expression level X of a gene is usually measured by the log-transformed signal intensity, after the normalization and bias-correction. For the (two-way) cDNA microarray, X measures the relative mRNA abundance to a pre-specified condition (control), while for Affymetrix array, X is a good predictor for the absolute mRNA abundance (Kerr and Churchill 2001; Quackenbush 2001).

Basic Brownian model (B-model)   The Brownian model assumes that expression divergence is mainly driven by small and additive genetic drifts (random effects)

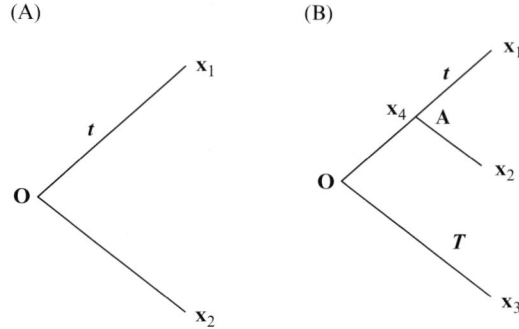

Fig. 5.1 Schematic tree of a gene family: (A) A gene family with duplicate genes that diverged from their ancestor t time units ago. At a given biological condition, the expression levels of two duplicates are denoted by x_1 and x_2, respectively. (B) The gene family tree with three duplicate genes. Gene 1 and gene 2 were duplicated t time units ago, while the gene 3 and the ancestral gene A were duplicated T time units ago. Three duplicate genes have expression levels x_1, x_2, and x_3, respectively. The expression level of the ancestor of duplicates 1 and 2 is denoted by x_4.

during the course of evolution. Consequently, given the initial expression level x_0, the expression level $X = x$ after t evolutionary time units follows a normal distribution with mean x_0 and variance $\sigma^2 t$. Formally, it can be written as

$$B(x|x_0; \sigma^2 t) = \frac{1}{\sqrt{2\pi t}\sigma} e^{-\frac{(x-x_0)^2}{2\sigma^2 t}} \tag{5.1}$$

We start from a simple two-member gene family, see Fig. 5.1(A). Let x_1 and x_2 be the expression levels of two genes, respectively. Our purpose is to derive $P(x_1, x_2)$, the joint density of x_1 and x_2. Given the initial expression value (x_0) of gene at the root O (representing a gene duplication event), the change of x_1 follows a Brownian model $B(x_1|x_0; \sigma^2 t)$, and the change of x_2 follows $B(x_2|x_0; \sigma^2 t)$. If the expression divergence is independent between lineages, refereed as the E_0-assumption, we obtain

$$P(x_1, x_2|x_0) = B(x_1|x_0; \sigma^2 t)B(x_2|x_0; \sigma^2 t)$$

The initial expression x_0 at the root O is usually unknown, but one can solve this problem by treating it as a random variable following a normal distribution, that is,

$$\pi(x_0) = \frac{1}{\sqrt{2\pi}\rho} e^{-\frac{(x_0-\mu)^2}{2\rho^2}} \tag{5.2}$$

Note that the variance ρ^2 can also be interpreted as the ancestral component of expression profiles. Therefore, the joint density can be derived according to $P(x_1, x_2) = \int_{-\infty}^{\infty} P(x_1, x_2|x_0)\pi(x_0)dx_0$, resulting in

$$P(x_1, x_2) = \int_{-\infty}^{\infty} \left[B(x_1|x_0; \sigma^2 t)B(x_2|x_0; \sigma^2 t) \right] \pi(x_0)dx_0 \tag{5.3}$$

In the three-gene case (Fig. 5.1(B)), the joint density of expressions x_1, x_2, and x_3 can be derived in the same manner. Denote the expression level at the ancestral node A by x_4. Let T and t be the evolutionary times of nodes O (the root) and A, respectively. Thus, given the initial value (x_0) at O, the change of x_4 follows $B(x_4|x_0; \sigma^2(T-t))$ and the change of x_3 follows $B(x_3|x_0; \sigma^2 T)$. Similarly, given the ancestral level x_4, the changes of x_1 and x_2 follow $B(x_1|x_4; \sigma^2 t)$ and $B(x_2|x_4; \sigma^2 t)$, respectively. According to the Markov property, we obtain the joint density

$$P(x_1, x_2, x_3, x_4|x_0) = B(x_3|x_0; \sigma^2 T)B(x_1|x_4; \sigma^2 t)B(x_2|x_4; \sigma^2 t)(x_4|x_0; \sigma^2(T-t))$$

Since the ancestral expression x_4 is unobserved, it should be integrated out, i.e.

$$P(x_1, x_2, x_3|x_0) = \int_{-\infty}^{\infty} P(x_1, x_2, x_3, x_4|x_0)dx_4$$

Together with Eq. (5.2), we have the joint density

$$P(x_1, x_2, x_3) = \int_{-\infty}^{\infty} \left[\int_{-\infty}^{\infty} B(x_3|x_0; \sigma^2 T)B(x_1|x_4; \sigma^2 t) \right.$$
$$\left. \times \ B(x_2|x_4; \sigma^2 t)(x_4|x_0; \sigma^2(T-t))dx_4 \right] \pi(x_0)dx_0 \qquad (5.4)$$

It is interesting to compare the likelihood function between expression divergence and nucleotide substitution. Since expression data are continuous, one has to use a diffusion model, of which the simplest one is the Brownian model. In contrast, as a nucleotide has 4-letter states (A, T, C, and G), the Markov chain model is more appropriate. Nevertheless, their likelihood structures are similar, as both are constructed along the phylogeny based on the Markov property. For instance, the structure of Eq. (5.4) is similar to the likelihood function of three DNA sequences (Felsenstein 1981), if the Brownian process of expression divergence $B(.)$ is replaced by the transition probability of nucleotide substitutions, $\pi(x_0)$ at the root is by the nucleotide frequencies, and the integral sign is by the sum over four-states.

Analytical results for the general expression likelihood function The joint density for two or three genes can be extended to any n-genes when a rooted phylogeny is given. After assuming independent evolution between lineages (E_0-assumption), Gu (2004) has shown that the joint expression density of $\mathbf{x} = (x_1, \ldots, x_n)$ under a given phylogeny follows a multi-variate normal distribution, with the mean $\boldsymbol{\mu}$ and the covariance matrix $\mathbf{V}$, that is,

$$P(\mathbf{x}) = \frac{1}{(\sqrt{2\pi})^n |\mathbf{V}|^{1/2}} \exp\left\{ -\frac{(\mathbf{x} - \boldsymbol{\mu})'\mathbf{V}^{-1}(\mathbf{x} - \boldsymbol{\mu})}{2} \right\} \qquad (5.5)$$

This result can be proved by using the principle of mathematical induction, also see Hansen and Martins (1996).

The covariance matrix $\mathbf{V}$ is phylogeny dependent, which can be presented in terms of *expression branch lengths*. For branch k along the phylogeny, under the B-model we define the expression branch length $E_k = \sigma^2 t_k$, where t_k is the evolutionary time.

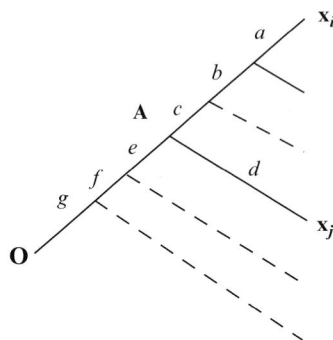

Fig. 5.2 Schematic illustration of a gene family tree, where y is the expression level at node A, the common ancestor of genes i and j. Thus, $k \in x_i = (a, b, c, e, f, g)$ for all branches from the root O to gene i; $k \in (x_i, x_j) = (e, f, g)$ for all branches from the root O to A; and also $k \in (x_i, y) = (e, f, g)$ for all branches from the root O to A.

Then, the covariance matrix $\mathbf{V}$ can be written as follows

$$V_{ij} = \begin{cases} \rho^2 + \sum_{k \in x_i} E_k & \text{if } i = j \\ \rho^2 + \sum_{k \in (x_i, x_j)} E_k & \text{if } i \neq j \end{cases} \tag{5.6}$$

where the subscript notation $k \in x_i$ runs over all branches in the lineage from the root O to gene x_i, and $k \in (x_i, x_j)$ runs over all branches shared by x_i and x_j since the root O (Fig. 5.2). For instance, in the case of two genes ($n = 2$), the covariance matrix is given by

$$\mathbf{V} = \begin{pmatrix} \rho^2 + E_1 & \rho^2 \\ \rho^2 & \rho^2 + E_2 \end{pmatrix}$$

while for three genes ($n = 3$) it is given by

$$\mathbf{V} = \begin{pmatrix} \rho^2 + E_1 + E_4 & \rho^2 + E_4 & \rho^2 \\ \rho^2 + E_4 & \rho^2 + E_2 + E_4 & \rho^2 \\ \rho^2 & \rho^2 & \rho^2 + E_3 \end{pmatrix}$$

Hence, the unknown parameters of the likelihood function can be generally presented as $(\rho^2, E_k\text{'s})$, where ρ^2 is the common expression component of the gene family expression, and these expression branch lengths measure the pattern of expression divergence along the phylogeny. Apparently, ρ^2 and the total expression branch length along the tree, $E_T = \sum_k E_k$, measure the degrees of expression conservation and divergence, respectively.

Advanced models and biological interpretations It has been shown (Gu 2004) that the general likelihood function of Eq. (5.5) holds under several advanced models, but the biological interpretation of expression branch length E (the subscript is omitted for simplicity) is model-dependent. Since the B-model assumes that the expression divergence is mainly driven by small and additive genetic drifts (random effects), it can be considered as the 'neutral-evolution' model of gene expression. Under the mutation drift model, σ^2 can be interpreted as the mutational variance (Lynch and Hill 1986, Felsenstein 1988). In this case, the rate of expression divergence σ^2 equals to the mutational variance (Lynch and Hill 1986), and the expression branch length is given by $E = \sigma^2 t$, a simple product of the rate and evolutionary time. Moreover, under the lineage-specific (L)-model, the evolutionary rate σ^2 may differ among branches along the tree.

The directional trend (D) model allows drifts of gene expression levels. That is, given the initial value x_0 at $t = 0$, the change of expression level during the evolution follows a linear function of t, i.e. $x_0 + \lambda t$; λ is called the coefficient of directional selection (Felseinstein 1988). After treating the unobserved λ as a random variable with mean 0 and variance ω^2, Gu (2004) has shown that the expression branch length is given by $E = \sigma^2 t + \omega^2 t^2$.

In contrast to the D-model that the change of gene expression is continuous with time t (the gradual evolution), the S-model demonstrates a dramatic shift (positively or negatively) in gene expression that may happen shortly after the gene duplication, and then remains relatively static, i.e. the fashion of 'punctuated-equilibrium' (Hansen and Martins 1996). Hence, after the duplication event, the expression level has independent dramatic shifts z and z' units in two copies. After further assuming that two shift variables (z and z') are random variables with mean 0 and variances s^2 and s'^2, respectively, we have the following expression branch lengths $E = s^2 + \sigma^2 t$ and $E' = s'^2 + \sigma^2 t$.

In summary, the expression branch length under the L, D, and S models can be generally written as a quadratic form of evolutionary time t

$$E = s^2 + \sigma^2 t + \omega^2 t^2 \tag{5.7}$$

which provides an approach to testing these models from genomic data. Apparently, the B and L models predict a linear t-function of E, the D-model predicts a quadratic t-function of E, and the S-model predicts an E independent of the time t.

Likelihood implementation A typical expression dataset of a gene family can be obtained from microarrays, where the k-th column represents the expression profile across the gene family in the k-th microarray experiment, and the i-th row represents the expression profile of gene i across microarray experiments. For N microarray experiments, let $\mathbf{x}_k = (x_{1,k}, \ldots, x_{n,k})$ be the expression pattern of an n-member gene family at the k-th experiment. When the phylogeny is given, the likelihood for gene expressions can be written as

$$L(\mathbf{V}, \boldsymbol{\mu}|\text{data}) = \prod_{k=1}^{N} P(\mathbf{x}_k; \boldsymbol{\mu}, \mathbf{V}) \tag{5.8}$$

The maximum-likelihood estimates of the parameters can be obtained by the Newton–Raphson iteration method, and the sampling variance of each estimate is approximated by the inverse of the information matrix.

It should be noted that microarray data usually include multiple measurements under different tissues, developmental time-courses, or environmental/experimental treatments. Apparently, some of them, e.g. two adjacent time-points, may be highly correlated. Therefore, the implementation of multi-array likelihood under the *i.i.d* assumption (independently, identically distributed) is only approximate. We will come back to this issue later.

5.1.2 Method of expression distance

Definition of additive expression distance For two duplicate genes 1 and 2 whose expression levels are x_1 and x_2, respectively, we define *the expression distance* (E_{12}) between them as the expectation of the square of expression differences, that is, $E_{12} = \mathbf{E}[(x_1 - x_2)^2]$, where the operator $\mathbf{E}[.]$ means taking expectation. After assuming that the means of x_1 and x_2 are the same, one can verify $E_{12} = V_{11} + V_{22} - 2V_{12}$, where V_{11} and V_{22} are the variances of x_1 and x_2, respectively, and V_{12} is their covariance. From Eq. (5.6) for $n = 2$, we have $V_{11} = \rho^2 + E_1$, $V_{22} = \rho^2 + E_2$, and $V_{12} = \rho^2$, directly leading to

$$E_{12} = E_1 + E_2$$

That is, the expression distance E_{12} is the sum of expression branch lengths, or E_{12} is additive. Further, the additivity of expression distance in the general (n) case has been proved by Gu (2004), that is,

$$E_{ij} = \sum_{k \in C_{ij}} E_k \tag{5.9}$$

or, the expression distance between any two genes in a phylogeny is the sum of expression branch lengths connecting between them denoted by E_{ij}.

The expression distance provides an efficient approach to study the pattern of expression divergence of gene gamily. For duplicate genes 1 and 2, let x_{1k} and x_{2k} be the expression levels, respectively, in the k-th microarray experiment, $k = 1, \ldots, m$. Let $\bar{x}_1$ and $\bar{x}_2$ be the mean expression of genes 1 and 2, respectively, over experiments. Then, the evolutionary expression distance between genes 1 and 2 can be estimated as follows

$$\hat{E}_{12} = \sum_{k=1}^{m} [(x_{1k} - \bar{x}_1) - (x_{2k} - \bar{x}_2)]^2 / (m - 1) \tag{5.10}$$

An important feature is that the expression distance is only associated with an unrooted phylogeny, in which two branches connected to the root (O) must be merged into one. Since the expression distance does not include ρ^2, the common variance component of the gene family, one may need to estimate ρ^2 separately.

Phylogeny expression mapping Given a set of microarray expression data from multiple conditions, one can use Eq. (5.10) to estimate the expression distance matrix **E** for the gene family. Since the expression distances are additive, expression branches E_k can be estimated from the expression distance matrix, when the phylogenetic tree is known or can be reliably inferred, say, from the sequence data.

While there are substantial methods in the literature, we choose the method of ordinary least squares (LS), which is widely used in molecular phylogenetic analysis. Given the topology with n genes, we use Rzhetsky and Nei's (1992, 1993) methods, for instance, to obtain the LS estimates of expression branch lengths. Let $\mathbf{d}_E = (E_{12}, E_{13}, \ldots, E_{n,n-1})'$ be the vector of $n(n-1)/2$ (pairwise) expression distances, and $\mathbf{b}_E = (E_1, \ldots, E_m)'$ be the vector of m expression lengths. The connection matrix $\mathbf{A}$ is defined as follows. The first row is for the expression distance E_{12}: $A_{1k} = 1$ if the k-th branch is in the path connecting genes 1 and 2, otherwise $A_{1k} = 0$; and so forth. Then, by the conventional linear model approach, the estimates of expression branch lengths are given by $\hat{\mathbf{b}}_E = (\mathbf{A}'\mathbf{A})^{-1}\mathbf{A}'\mathbf{d}_E$. When the number of genes is large, we may use the fast algorithm developed by Rzhetsky and Nei (1993) without using the matrix algebra. Finally, the sampling variance of any expression distance E_{ij} or any expression branch length E_k, can be approximately calculated by the bootstrapping procedure.

The index for expression conservation When the expression branch lengths are estimated by the LS method, the total branch length along the tree can be calculated by $E_T = \sum_k E_k$, where the subscript k runs for all branches. Moreover, the total variance of expression (V_T) of a gene family is defined by the sum of independent variance components along the tree, that is, $V_T = \rho^2 + E_T$. Since ρ^2 is the common variance component shared by the gene family, the expression conservation of the gene family can be measured by the following index Q

$$Q = \frac{\rho^2}{\rho^2 + E_T} \tag{5.11}$$

which is a measure for the relative expression divergence of a gene family. In order to calculate Q, one has to estimate ρ^2, which is root-dependent. If the root of the gene family tree is uncertain, one may obtain the upbound and low-bound of ρ^2 by comparing all possible rooting along the phylogeny.

5.2 Ancestral gene expression inference

Ancestral state reconstruction along a phylogenetic tree is at the center of comparative methods in evolutionary biology, for both morphological and molecular characters (Harvey and Pagel 1991; Golding and Dean 1998; Yang *et al.* 1995; Schluter *et al.* 1997). The massive microarray data make it possible to reconstruct ancestral expression pattern that is useful to trace the evolutionary changes of gene regulation. Similar to the ancestral sequence inference, Gu (2004) has developed the 'empirical' Bayesian approach for the inference of ancestral expression profiles.

Single-node ancestral inference This method provides a fast Bayesian procedure to infer ancestral expression profiles because each time it deals with one ancestral node and then runs over the tree. Let $\mathbf{x} = (x_1, \ldots, x_n)$ be the observed expression pattern and y be the expression level at the ancestral node of interest. According to the Bayes rule, the posterior density $P(y|x_1, \ldots, x_n)$ is computed as follows

$$P(y|x_1, \ldots, x_n) = \frac{P(x_1, \ldots, x_n, y)}{P(x_1, \ldots, x_n)}$$

From Eq. 5.5, we know $P(x_1, \ldots, x_n)$ is an n-variate normal density. Next we demonstrate how to derive $P(x_1, \ldots, x_n, y)$ in the case of three-gene family (Fig. 5.1b), where $y = x_4$. According to the Markov property, one can write

$$P(x_1, x_2, x_3, x_4|x_0) = B(x_3|x_0)B(x_1|x_4)B(x_2|x_4)B(x_4)$$

and

$$P(x_1, x_2, x_3, x_4) = \int_\infty^\infty B(x_3|x_0)B(x_1|x_4)B(x_2|x_4)B(x_4)\pi(x_0)dx_0$$

Similar to the derivation of $P(x_1, x_2, x_3)$, we show that $P(x_1, x_2, x_3, x_4)$ is a 4-variate normal density.

In the general case, let $M = n + 1$ and regard the ancestral level y as an additional variable x_{n+1}. It has been shown that $P(x_1, \ldots, x_n, y)$ is an $(n + 1)$-variate normal density, denoted by $N(x_1, \ldots, x_n, y; \boldsymbol{\mu}, \mathbf{V_M})$. The extended variance-covariance matrix $\mathbf{V_M}$ has the following structure: If $1 \le i, j \le n$, the ij-th element of $\mathbf{V_M}$ is equal to that of $\mathbf{V}$ in Eq. 5.6. For any $i, n + 1$-th element, $i = 1, \ldots, n + 1$, it is given by

$$V_{i,n+1} = \begin{cases} \rho^2 + \sum_{k \in y} E_k & \text{if } i = n + 1 \\ \rho^2 + \sum_{k \in (x_i, y)} E_k & \text{if } i \neq n + 1 \end{cases}$$

and $V_{n+1,i} = V_{i,n+1}$, where the subscript notation $k \in y$ runs over all branches in the lineage from the root O to the ancestral node y, and $k \in (x_i, y)$ runs over all branches shared by x_i and y since the root O (Fig. 5.2). For simplicity, we assume the mean vector $\boldsymbol{\mu} = (\mu, \ldots, \mu)'$.

Let c_{ij} be the ij-th element of $\mathbf{C} = \mathbf{V}_M^{-1}$. Gu (2004) has shown that the posterior density $P(y|x_1, \ldots, x_n)$ is a normal density, after some algebras we obtain

$$P(y|x_1, \ldots, x_n) = \frac{1}{\sqrt{2\pi}\sigma_{y|x}} \exp\left\{ -\frac{1}{2\sigma_{y|x}^2} \left[y - \mu + \sum_{i=1}^n \frac{c_{i,n+1}}{c_{n+1,n+1}}(x_i - \mu) \right]^2 \right\} \quad (5.12)$$

where $\sigma_{y|x}^2 = 1/c_{n+1,n+1}$ is the (posterior) variance of y. That is, the posterior mean of y conditional of $\mathbf{x} = (x_1, \ldots, x_n)'$ is given by

$$E[y|x_1, \ldots, x_n] = \beta_0 + \sum_{i=1}^n \beta_i x_i \quad (5.13)$$

where $\beta_i = -c_{i,n+1}/c_{n+1,n+1}$ and $\beta_0 = \mu(1 - \sum_{i=1}^n \beta_i)$. Apparently, the posterior mean prediction for the ancestral gene expression is a linear function of current gene expressions.

Joint ancestral inference   To explore the joint evolutionary pattern of expression changes after gene duplications, the single-node method may not be sufficient. Therefore we develop an approach for joint ancestral expression inference. For a gene family with n member gene, there are m ancestral nodes when the phylogenetic tree is given.

Let $\mathbf{x} = (x_1, \ldots, x_n)'$ and $\mathbf{y} = (y_1, \ldots, y_m)'$ be the vectors of current and ancestral expression levels, respectively; and $M = n + m$. The (extended) $M \times M$ variance-covariance matrix for $(\mathbf{y}', \mathbf{x}')$ is denoted by $\mathbf{V}_M$. We have shown that $P(\mathbf{y}, \mathbf{x})$ is an M-dimensional multi-normal density. It follows that the joint posterior density of ancestral nodes $\mathbf{y}$

$$P(\mathbf{y}|\mathbf{x}) = \frac{P(\mathbf{y}, \mathbf{x})}{P(\mathbf{x})} = \frac{N(\mathbf{y}, \mathbf{x}; \boldsymbol{\mu}, \mathbf{V}_M)}{N(\mathbf{x}; \boldsymbol{\mu}, \mathbf{V})}$$

is also $m \times m$ multi-normal, that is, $P(\mathbf{y}|\mathbf{x}) = N(\mathbf{y}; \boldsymbol{\mu}_{y|x}, \boldsymbol{\Sigma}_{y|x})$, where $\boldsymbol{\mu}_{y|x} = (\mu_{y_1|x}, \ldots, \mu_{y_m|x})'$ is the posterior mean vector of the ancestral nodes, and $\boldsymbol{\Sigma}_{y|x}$ is the $m \times m$ posterior variance-covariance matrix of $y_1, \ldots, y_m$.

To obtain useful analytical results for numerical calculation, we partition the matrix $\mathbf{V}_M$ as follows

$$\mathbf{V}_M = \begin{bmatrix} \mathbf{A} & \mathbf{H} \\ \mathbf{H}' & \mathbf{V} \end{bmatrix}$$

where $\mathbf{H}$ and $\mathbf{A}$ are $m \times n$ and $m \times m$ matrices, respectively. The matrix $\mathbf{H}$ is the ancestral-current expression covariance and $\mathbf{A}$ is the variance-covariance matrix among ancestral nodes. Thus, the inverse of the matrix $\mathbf{V}_M$ can be written as

$$\boldsymbol{\Lambda}_M = \begin{bmatrix} \mathbf{A} & \mathbf{H} \\ \mathbf{H}' & \mathbf{V} \end{bmatrix}^{-1} = \begin{bmatrix} \boldsymbol{\Lambda}_{yy} & \boldsymbol{\Lambda}_{yx} \\ \boldsymbol{\Lambda}'_{yx} & \boldsymbol{\Lambda}_{xx} \end{bmatrix}$$

where $\boldsymbol{\Lambda}_{xx}$, $\boldsymbol{\Lambda}_{xy}$ and $\boldsymbol{\Lambda}_{yy}$ are $n \times n$, $m \times n$ and $m \times m$ matrices, respectively. It has been shown that

$$\boldsymbol{\Sigma}_{y|x} = \boldsymbol{\Lambda}_{yy}^{-1}$$

$$\boldsymbol{\mu}_{y|x} = \boldsymbol{\mu} - \boldsymbol{\Lambda}'_{yx} \boldsymbol{\Lambda}_{yy}^{-1} (\mathbf{x} - \boldsymbol{\mu}) \tag{5.14}$$

5.3 Oakley *et al.*'s model

Oakley *et al.* (2005) introduced the phylogenetic comparative method to investigate the pattern of gene-expression evolution. Based on the simple Brownian model, they first considered three classes about how the expression branch length are specified along the phylogeny (Fig. 5.3). The first specific model type (genetic distance) assumes that genetic distance, calculated from gene-sequence data, predicts the expression branch length. The second model type (equal models) (Mooers and Schluter 1999) assumes that the expression branch lengths are equal for every branch. In this case, more change in expression is expected to occur with more duplication events. The third specific model ('free model') estimated each unconstrained expression branch length, without assuming a constant rate of expression change.

Each of the three specific model types was implemented in three different general classes for a total of nine different models (Fig. 5.3). The first type is the 'pure-phylogenetic' model, which assumes that the divergence of gene expression increases along the phylogeny of a gene family. Second, the non-phylogenetic class of models

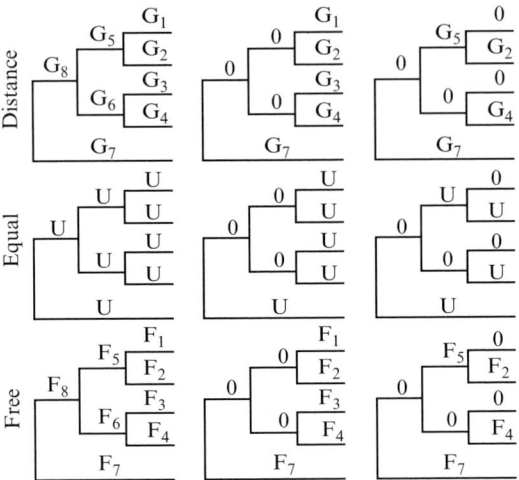

Fig. 5.3 Nine different maximum-likelihood models of gene expression evolution. From Oakley *et al.* (2005). These models predict that change in expression increases monotonically with the 'time' available for change. Time available for change is estimated in different ways for different models, as indicated by different letters above branches of a hypothetical gene tree that would be estimated from sequences of a gene family. Branches labeled 'Gi' assume expression change is equal to genetic distance of that branch, those labeled 'U' assume a unit (equal) amount of change, and those labeled 'Fi' are estimated from the expression data itself (free). Branches labeled '0' assume no change has occurred. Columns represent three different classes of models. The pure phylogenetic class assumes expression change occurs on every branch of the phylogeny, the nonphylogenetic class assumes expression change occurs only along terminal branches, and the punctuated class assumes expression change occurs on only one of every pair of descendent branches.

assumes that no change occurred in gene expression in the internal branches of a phylogeny, i.e. virtually a star-like tree pattern of expression divergence. Finally, the punctuated model class assumes that at every branching point one daughter gene changes expression, whereas the other does not.

According to the Brownian framework, Oakley *et al.* (2005) used the likelihood function (L) under the phylogenetic tree as follows

$$L = \prod_{i,i'} \frac{1}{\sqrt{2\pi\beta t_{i'}}} \exp\left\{-\frac{(x_i - x_{i'})^2}{2\beta t_{i'}}\right\}$$

Here, x_i is the expression value at node i and $x_{i'}$ is that of a descendent node (i') of i; β is the rate of expression change, and $t_{i'}$ is the time available for evolutionary change between the nodes i and i' (along a single branch). The product is taken over all branches on the tree. Oakley *et al.* (2005) constructed nine different models for gene-expression evolution, and used the AIC to compare maximum-likelihood models (L) with different numbers of parameters (P):

$$AIC = -2\ln L + 2P$$

Although not strictly a statistical significance test, the model with the lowest AIC value is considered the best fit.

Oakley *et al.* (2005) analyzed the 10 large yeast gene families, and found that expression of duplicated genes has evolved according to a nonphylogenetic model, where closely related genes are no more likely than more distantly related genes to share common expression patterns. After having discussed several possible explanations, the authors supported the notion of rapid evolution of gene expression after the gene duplication.

5.4 Expression divergence under stabilizing selection: the Ornstein–Uhlenback (OU) model

Expression divergence between duplicate genes is under some constraints. As a result, the extent of expression divergence cannot increase with time without any restriction. To describe the effect of constraint on the expression divergence, we adopted the stabilizing selection model (Lynch and Hill 1986) of quantitative traits. For a gene expressed, the stabilizing selection on the expression level x follows a Gaussian fitness function

$$f(x) = e^{-w_e(x-\theta_e)^2} \tag{5.15}$$

where θ_e is the optimal value of expression level, w_e is the coefficient for stabilizing selection on the gene expression; a large w_e means a strong selection pressure, and *vice versa*.

Under the stabilizing model of Eq. (5.15), we have shown that the expression divergence follows an Ornstein–Uhlenback (OU) process (Hansen and Martins 1996). The stochastic OU process is characterized by the infinitestimal mean $-\beta_0(x - \theta_e)$ and variance $\epsilon^2/2N_e$, where ϵ^2 is the mutational variance, N_e is the effective population size, and $\beta_0 = w_e\epsilon^2$ measures the direct force against the deviation from the optimum. Given the initial expression value x_0, the OU model claims that $x(t)$ follows a normal distribution with the mean and variance given by

$$E[x(t)|x_0] = e^{-\beta t}x_0 + (1 - e^{-\beta t})\theta_e$$

$$V[x(t)|x_0] = \frac{\epsilon^2(1 - e^{-2\beta t})}{2\beta} \tag{5.16}$$

respectively, where $\beta = 2N_e\beta_0$ is the decay-rate of expression divergence.

For two duplicate genes that have diverged t time units ago (Fig. 5.1), the expression distance can be derived similar to Gu (2004), also see above. Here we only consider a simple case of two duplicates for illustration. Let x_1 and x_2 be the expression levels of duplicates 1 and 2, respectively. Assuming that the initial expression is at the optimum ($x_0 = \theta_e$), and then diverged independently after the gene duplication, one can show that the expression variances are given by $V(x_1) = V(x_2) = \epsilon^2(1 - e^{-2\beta t})/2\beta + Var(\theta_e)$, and the covariance by $Cov(x_1, x_2) = Var(\theta_e)$, where $Var(\theta_e)$ is the variance of expression optima over various conditions. Therefore, the expression distance between duplicates, $E_{12} = E[(x_1 - x_2)^2] = V(x_1) + V(x_2) - 2Cov(x_1, x_2)$, is given by

$$E_{12} = \frac{\epsilon^2(1 - e^{-2\beta t})}{\beta} = \frac{(1 - e^{-2\beta t})}{W} \tag{5.17}$$

where $W = \beta/\epsilon^2$ is the strength of stabilizing selection on the expression divergence. It should be noticed that the expression distance under the stabilizing model has several important features: (*i*) E increases linearly with the evolutionary time t when t is small, but approaches a saturated level of expression divergence when $t \to \infty$. (*ii*) The rate of approaching the saturated level depends on the decay-rate β, a large value indicates a quick approaching, and *vice versa*. (*iii*) In particular, it is reduced to the Brownian (*B*) model when $\beta = 0$ (no constraint), in which the expression divergence is linear to the time t, i.e. $E_{12} = 2\epsilon^2 t$. And (*iv*) as expected, E_{12} is inversely related to the strength of stabilizing selection W, which also determines the saturated level of expression divergence, i.e. $E_{12} = 1/W$ when $t \to \infty$.

5.5 Likelihood and distance methods under experimental correlations

In microarray experiments, the overall expression profiles with similar types of conditions or treatments are more similar to each other, e.g. two adjacent sampling points in a time-course assay are usually highly correlated. Because of these experimental correlations, the *i.i.d* assumption seems to be unrealistic. That is, the sample of expression profiles of a gene family is not only phylogenetically but also experimentally dependent.

As the first-order approximation, we (Gu 2004) modeled the experimental correlation of the microarray data as the overall correlations among microarray experiments. Let $\mathbf{D}$ be the $m \times m$ matrix of experimental correlations, that is, the diagonal element is 1, while the off-diagonal element is the coefficient of correlation between any two microarray experiments. For the microarray chip that includes in total C genes, $\mathbf{D}$ can be estimated by the standard approach. Then one can show that the expression profiles (normalized by zero mean) of the gene family $\mathbf{X} = (\mathbf{x}_1, \ldots, \mathbf{x}_m)'$ follow a multi-variate normal distribution, with the large $[n \times m] \times [n \times m]$ variance-covariance matrix $\mathbf{V} \otimes \mathbf{D}$. The likelihood function is therefore given by

$$L(\mathbf{V} \otimes \mathbf{D}, \boldsymbol{\mu}|\text{data}) = P(\mathbf{X}; \boldsymbol{\mu}, \mathbf{V} \otimes \mathbf{D})$$

Apparently, it is reduced to Eq. (5.8) when $\mathbf{D} = \mathbf{I}$. Consider two duplicate genes with (centralized) expression profiles $\mathbf{X} = (\mathbf{x}_1, \mathbf{x}_2)'$. In this case one may define the expression distance as follows

$$E_{12} = [\mathbf{x}'_1 \mathbf{D}^{-1} \mathbf{x}_1 + \mathbf{x}'_2 \mathbf{D}^{-1} \mathbf{x}_2 - 2\mathbf{x}'_1 \mathbf{D}^{-1} \mathbf{x}_2]/(m-1)$$

$$= \sum_{i=1}^{m} (x_{1i} - x_{2i})^2/(m-1) + R_{12} \tag{5.18}$$

where the correct term for experimental correlations $R_{12} = 0$ only when $\mathbf{D} = \mathbf{I}$; in this case it is reduced to Eq. (5.10). Preliminary analysis has shown that for dense-sampled time-courses, or dosage-dependent treatments, the effect of experimental correlation should not be neglected.

5.6   Yeast GlnS gene family: an example

In yeast, the GlnS family (Glutamyl- and glutaminyl-tRNA synthetases) has three member genes (YGL245w, YOR168w and YOL033w). Phylogenetic analysis has shown that YGL245w and YOR168w are more closely related (Fig. 5.4). The cell-cycle yeast microarray data were used (Eisen *et al.* 1998).

Under the general E_0-model, the likelihood under the *i.i.d* assumption results in the ML estimates of $\hat{\rho}^2 = 0.053 \pm 0.014$, $\hat{E}_1 = 0.100 \pm 0.023$, $\hat{E}_2 = 0.062 \pm 0.020$, $\hat{E}_3 = 0.099 \pm 0.031$, and $\hat{E}_4 = 0.079 \pm 0.020$. The maximum log-likelihood value is -146.19. Next we consider the likelihood function considering the experimental correlations. We first compute the matrix $\mathbf{D}$ of microarray experiments. Using the ML estimates under *i.i.d* as initial values, we obtain $\hat{\rho}^2 = 0.055$, $\hat{E}_1 = 0.112$, $\hat{E}_2 = 0.068$, $\hat{E}_3 = 0.104$, and

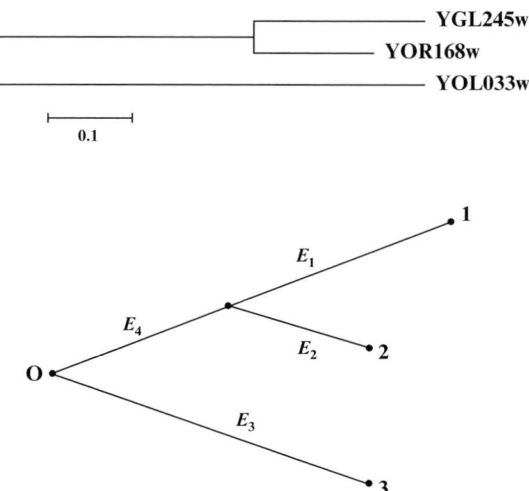

Fig. 5.4 The simplified phylogenetic tree of GlnS gene family, inferred from the multi-alignment of amino acid sequences including eukaryotes and prokaryotes. The neighbor-joining method was used. Modified from Gu (2004).

$\hat{E}_4 = 0.061$. It seems that the likelihood under the *i.i.d.* assumption is useful for fast and large-scale analysis.

Using the molecular clock approach, Gu (2004) have roughly dated the relative time of first gene duplication (between YGL245w/YOR168w and YOL033w) is 2.2 (to the E.coli/yeast split time), and the second one is 1.27. Thus, under the *B*-model, we have obtained the ML estimates (under *i.i.d.*) of $\hat{\rho}^2 = 0.057 \pm 0.015$, and $\hat{\sigma}^2 = 0.047 \pm 0.005$. The maximum log-likelihood value under the *B*-model is -154.13. Apparently, the likelihood ratio test shows the *B*-model, or the 'expression clock', is rejected at the significance level of 0.001; $\chi^2_{[3]} = 2(154.13 - 146.19) = 15.88$. As shown in Fig. 5.5, the expression profile of the ancestral ancestor of YGL245w-YOR168w has been inferred by the Bayesian method. Therefore, one can infer lineage-specific changes after gene duplication (the derived characters) from the ancestral expression pattern.

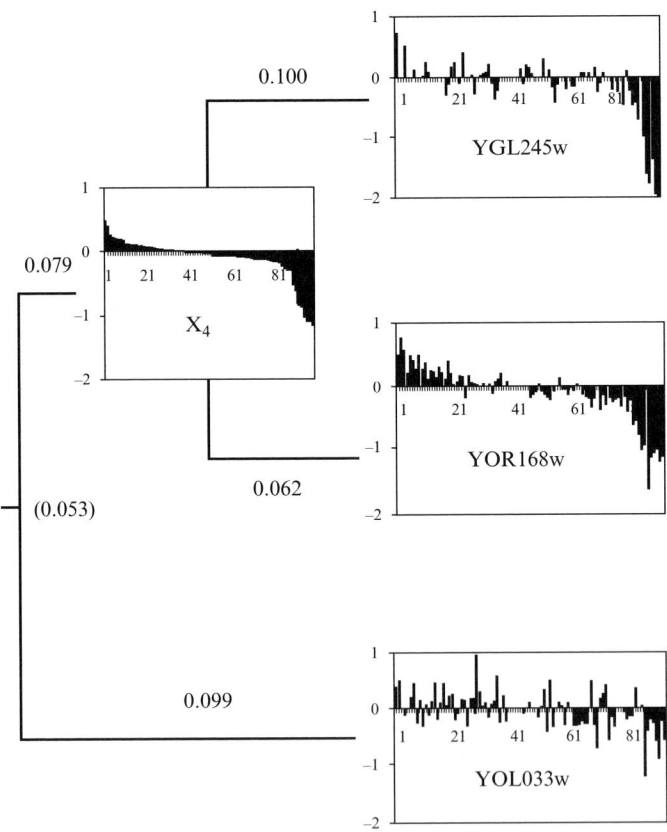

Fig. 5.5 Phylogenetic expression analysis of the yeast GlnS family. The expression branch length is indicated on each branch. The inferred ancestral expression profile for the common ancestor of YGL245W and YOR165W is presented. Modified from Gu (2004).

5.7 Estimating expression divergence based on massively parallel sequencing technology

Several new sequencing instruments, called 'next-generation' or 'massively parallel', are transforming the field of genetics and genomics; see Mardis (2008) and Jacquier (2009) for reviews. In short, they have three features: the ability to process millions of sequence reads in parallel at a time; relatively little input needed to produce a next-generation sequence-ready library; and shorter read length (35–250 bp, depending on the platform) than capillary sequencers (650–800 bp). Aside from these common features, commercially available sequencers may differ considerably, as summarized by Mardis (2008).

Since the basic measure for gene expression is to count the sequence reads, the sampling can be modeled as a simple Poisson process. Roughly speaking, the abundance of sequence reads indicates the relative level of gene expression from the biological sample. Though the Poisson-based model has been used in the study of ESTs (expressed sequence tags)(Audic and Claverie 1997; Ewing and Claverie 2000; Stekel *et al.* 2000; Ge and Epstein 2004), raw data treatment and normalization for the next-generation sequence reads remain a challenge, which depends on the technology platforms and procedures (Balwierz *et al.* 2009). Our interest here is to develop a statistical framework to conduct evolutionary analysis of transcriptome (coding genes or RNAs) based on the next-generation sequencing technology. In particular, we consider how to measure the expression divergence between homologous genes by this new data type, which can provide valuable cross-validations in these studies based on microarrays (Gu *et al.* 2004). For simplicity, we assume that raw sequence reads have been appropriately normalized, represented by tags per million (TPM) (Balwierz *et al.* 2009).

5.7.1 The Poisson-lognormal model

Both the sequence read count and the microarray intensity measure the level of gene expression. It is known that the distributions of both measures are highly skewed among genes. To address this issue, log-transformation has become the standard procedure in the microarray data normalization. However, the log-based transformation may not be applicable for the EST data that include zero count. Instead, several studies (Audic and Claverie 1997; Ewing and Claverie 2000; Stekel *et al.* 2000; Ge and Epstein 2004; Balwierz *et al.* 2009) used the Poisson-based model to describe the discrete sampling process that fits the previous EST data or the current next-generation sequence read data well.

For a given biological sample, let us denote $p(x)$ the probability to observe x sequence reads of the gene picked up randomly, which closely follows a Poisson distribution:

$$p(x|\lambda) = \frac{\lambda^x}{x!}e^{-\lambda} \tag{5.19}$$

$x = 0, 1, 2, \ldots$, where λ is the expected number of sequence reads of a gene per one million reads of the sample (tag per million, TPM). The Poisson model describes the sampling error during the experimental process. Based on this model one can study the expression divergence between duplicate genes as follows. Assume that the Poisson parameter (λ) is a random variable that can be modeled by the Poisson-lognormal regression; the simplest version is as follows

$$\ln \lambda = \mu + \gamma \tag{5.20}$$

where μ is the ground mean, and the genetic effect γ is a random variable that follows a normal distribution. In fact, γ describes the variation of sequence read abundance among samples, representing the underlying regulatory profile.

5.7.2   *U*-distance for expression divergence

Our goal is to estimate the expression divergence between duplicate genes from the sequence read abundance. Consider two duplicates X and Y that diverged t time units ago. The sequence read sampling process of each gene follows a Poisson model, with the expected numbers, λ_X and λ_Y, for genes X and Y, respectively. Applying the model of expression divergence developed by Gu (2004) to the Poisson-based model, we claim that λ_X and λ_Y can be further decomposed as follows

$$\ln \lambda_X = \mu + \gamma_X = \mu + \alpha + \beta_X$$
$$\ln \lambda_Y = \mu + \gamma_Y = \mu + \alpha + \beta_Y \tag{5.21}$$

where the genetics effects $\gamma_X = \alpha + \beta_X$ and $\gamma_Y = \alpha + \beta_Y$ indicate the evolutionary relationship of the regulatory machinery between duplicates: α is for the common genetic component shared by duplicates X and Y, while β_X and β_Y are gene-specific genetic effects for regulatory divergence after the gene duplication.

   According to Gu (2004), α follows a normal distribution with mean 0 and variance ρ^2. Under the B-model (the neutral expression evolution), one can show β_X and β_Y follows independent Brownian processes, characterized by the Gaussian distributions $N(0, \sigma_X^2 t)$ and $N(0, \sigma_Y^2 t)$, respectively. Since σ_X^2 and σ_Y^2 are the neutral rates of expression divergence, the expression distance between genes X and Y is defined as

$$U_{XY} = (\sigma_X^2 + \sigma_Y^2)t \tag{5.22}$$

   In the following we develop a statistical method to estimate the expression distance (U_{XY}) between two duplicate genes (X and Y) based on the sequence read abundance. Let $E[X]$ ($E[Y]$) and $E[X^2]$ ($E[Y^2]$) be the first and second moments of sequence read abundance of gene X (or Y), and $E[XY]$ be the cross-moment of X and Y. Our approach is to establish the relationship between U_{XY} and these statistical measures of read abundance. By the Poisson model, one can show the following conditional expectations: $E[X|\lambda_X] = \lambda_X$ and $E[X^2|\lambda_X] = \lambda_X + \lambda_X^2$, as well as $E[Y|\lambda_Y] = \lambda_Y$ and $E[Y^2|\lambda_Y] = \lambda_Y + \lambda_Y^2$. Putting these together, we obtain

$$E[\lambda_X] = E[X], \ E[\lambda_Y] = E[Y], \ E[\lambda_X \lambda_Y] = E[XY]$$

and

$$E[\lambda_X^2] = E[X^2] - E[X], \ E[\lambda_Y^2] = E[Y^2] - E[Y]$$

From Eq. (5.21), $\ln \lambda_X$ or $\ln \lambda_Y$ follows a normal distribution, with mean μ and variance $V_X = \rho^2 + \sigma_X^2 t$ or $V_y = \rho^2 + \sigma_Y^2 t$, respectively. As both λ_X and λ_Y follow a log-normal distribution, the first and second moments of λ_j ($j = X$ or Y) as

$$E[\lambda_j] = e^{\mu + (\rho^2 + \sigma_j^2 t)/2}, \;\; E[\lambda_j^2] = e^{2\mu + 2\rho^2 + 2\sigma_j^2 t}$$

respectively. Further we notice the cross-product moment

$$E[\lambda_X \lambda_Y] = E[e^{2\mu + 2\alpha + \beta_X + \beta_Y}] = e^{2\mu} \times e^{2\rho^2} \times e^{\sigma_X^2 t/2} \times e^{\sigma_Y^2 t/2}$$

because of the independence between α, β_X and β_Y. Thus, we establish the following relationship

$$\frac{E^2[\lambda_X \lambda_Y]}{E[\lambda_X^2] E[\lambda_Y^2]} = e^{-(\sigma_X^2 + \sigma_Y^2)t} \tag{5.23}$$

Suppose each gene has (normalized per million, TPM) sequence read counts for each of m tissue samples, as denoted by $x_1, \ldots, x_m$ and $y_1, \ldots, y_m$, respectively. Obviously, the ML estimates of $E[X]$, $E[X^2]$, $E[Y]$, $E[Y^2]$, and $E[XY]$ are given by $\sum_{i=1}^{m} x_i/m$, $\sum_{i=1}^{m} x_i^2/(m-1)$, $\sum_{i=1}^{m} y_i/m$, $\sum_{i=1}^{m} y_i^2/(m-1)$, and $\sum_{i=1}^{m} x_i y_i/(m-1)$, respectively. Further, we define three quantities J_{XX}, J_{YY}, and J_{XY} as follows

$$J_{XX} = \sum_{i=1}^{m} x_i^2/(m-1) - \sum_{i=1}^{m} x_i/m$$

$$J_{YY} = \sum_{i=1}^{m} y_i^2/(m-1) - \sum_{i=1}^{m} y_i/m$$

$$J_{XY} = \sum_{i=1}^{m} x_i y_i/(m-1) \tag{5.24}$$

respectively. Under the Poisson-lognormal model, all three quantities are expected to be positive, and satisfy the condition $J_{XY}^2 \le J_{XX} J_{YY}$. Apparently, J_{XX}, J_{YY}, and J_{XY} are the estimates of $E[\lambda_X^2]$, $E[\lambda_Y^2]$, and $E[\lambda_{XY}]$, respectively, under the assumption of independent sampling. Therefore, the expression distance between homologous genes X and Y can be estimated by

$$\hat{U}_{XY} = -\ln \frac{J_{XY}^2}{J_{XX} J_{YY}} \tag{5.25}$$

In practice, we set $\hat{U}_{XY} = 0$ if $J_{XY}^2 \ge J_{XX} J_{YY}$.

6

Expression between Duplicate Genes: Genome-Wide Analysis

Expression divergence between duplicate genes has long been a subject of great interest to geneticists and evolutionists. Ohno (1970) proposed that expression divergence is the first step in the functional divergence between duplicate genes and thereby increases the chance of retention of duplicate genes in a genome. Further, advances in molecular biology in the 1970s and 1980s provided a much better understanding of the molecular basis of expression divergence between duplicate genes. The availability of completely sequenced genomes and microarray gene expression and other functional genomic data have stimulated many studies on these questions (e.g. Wagner 2000b; Gu *et al.* 2002d; Zhang *et al.* 2004; Blanc and Wolfe 2004; Gu *et al.* 2005b; see Li *et al.* (2005) for a recent review). In this chapter we introduce these results.

6.1 Coding sequence divergence vs expression divergence

The relative roles of coding sequence and expression divergence, either after speciation or gene duplication, has been the central but controversial issue in the evolutionary study of developmental processes. Wagner (2000b) appears to be the first to consider whether coding sequence divergence and expression divergence between yeast duplicate genes are correlated at the genomic level. The repertoire of yeast duplicate genes was the result of a whole genome duplication occurred about 100 million years ago (Wolfe and Shields 1997), plus substantial more ancient (or recent) segmental duplications. From microarray data of yeast genes under five physiological conditions and the sequences of 114 pairs of duplicate genes, he found no significant correlation between protein sequence divergence and expression divergence and concluded that protein sequence divergence and expression divergence are decoupled. However, using synonymous divergence (K_S) as a proxy of divergence time between duplicate genes and more extensive data, Gu *et al.* (2002d) found a positive correlation between expression divergence and K_S (Fig. 6.1(A)). Because many gene pairs with a small K_S value were found to show expression divergence, the authors concluded that expression divergence between duplicate genes can occur rapidly. Further, they found a positive correlation between expression divergence and nonsynonymous divergence (K_A) for duplicates with $K_A < 0.30$ (Fig. 6.1(B)), indicating that expression divergence and protein sequence divergence are initially correlated. Similarly, using a phylogenetic analysis of yeast duplicate genes, Zhang *et al.* (2004) found a positive correlation between expression divergence and age of duplicate genes.

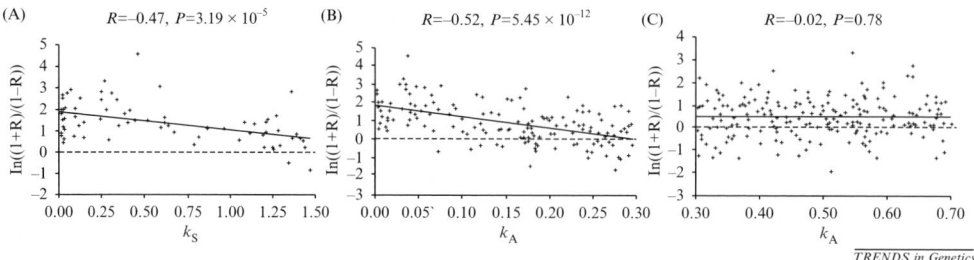

Fig. 6.1 Relationship between the correlation coefficient (R) of gene expression over all data points and K_S(orK_A) between duplicate genes. (A) A significant negative correlation between $\ln[(1 + R)/(1 - R)]$ and K_S for gene pairs. This implies a positive correlation between K_S and expression divergence because $1 - R$ can be regarded as expression divergence. (B) A significant negative correlation between $\ln[(1 + R)/(1 - R)]$ and K_A for gene pairs with $K_A >$ 0.3. (C) No correlation between $\ln[(1 + R)/(1 - R)]$ and K_A for gene pairs with $K_A > 0.3$. Figure modified from Gu *et al.* (2002d).

A similar conclusion as that in yeast was reached in an analysis of human duplicate genes. Using Affymetrix expression data in 25 human tissues, Makova and Li (2003) found a positive correlation between divergence in tissue expression and K_S (or K_A). Blanc and Wolfe (2004) studied the expression divergence in Arabidopsis duplicate pairs that are believed to be the remnants of a whole genome duplication event that occurred $\sim$ 24–30 million years ago. They observed a weak but significant negative correlation between sequence similarity and expression divergence.

6.2   Regulatory motif divergence vs expression divergence between duplicates

Compared to divergence in coding sequences, divergence in cis-regulatory sequences (motifs) between duplicate genes is likely to have a more direct effect on expression divergence. It is therefore interesting to study how cis-regulatory motifs diverge between duplicate genes. Papp *et al.* (2003) showed that for yeast young duplicates, the number of shared cis-regulatory motifs between duplicates decreases with the age of the duplicates, as measured by K_S. This conclusion, which was derived largely from computationally predicted motifs, was confirmed by Zhang *et al.* (2004), who used known yeast regulatory motifs.

Since duplicate genes are likely to share cis-regulatory motifs, one would expect a stronger co-expression pattern between duplicate genes than between two randomly selected genes. Zhang *et al.* (2004) found that this is the case for yeast genes. It is also reasonable to expect a negative correlation between expression divergence of duplicates and extent of their shared motifs. Zhang *et al.* (2004) indeed found a weak negative correlation, but estimated that only a low percentage ($\sim$2–3 per cent) of expression patterns of duplicate genes can be explained by their shared cis-regulatory motifs. That is, in addition to differences in the cis-regulatory motif structure between duplicate genes, a high percentage of expression patterns may be contributed to some unknown factors: (*a*) The analysis had not included all cis-regulatory motifs of the duplicate genes, because cis-regulatory motifs are still not completely known for most yeast genes, as pointed out by Papp *et al.* (2003). (*b*) mRNA stability and chromatin structure may have contributed to the differences in expression level between duplicates. And (*c*) unknown trans-factors in the gene network may have strong influences on the expression divergence between duplicate genes. It remains unclear how trans-factors can contribute to the expression differences between two duplicate genes.

In summary, there is a positive correlation between expression divergence and cis-regulatory motif divergence between duplicate genes, though differences in other factors such as mRNA stability and chromatin structure may also contribute to expression differences between duplicate genes.

6.3   Gene duplication and expression diversification

The theory of subfunctionalization (Force *et al.* 1999) predicts that gene duplication allows duplicates to become specialized in different tissues or developmental stages. When considering a group of closely related organisms, genes that have duplicated should have more diversified expression profiles across organisms than single-copy genes. To test this idea, Gu *et al.* (2004) used the gene expression data during the start of metamorphosis in three species of the *Drosophila melanogaster* subgroup to determine the effect of gene duplication on expression. They first used ANOVA to identify genes that were differentially expressed across the start of metamorphosis within a particular lineage and then the ones that differed between lineages in that stage. They found that the proportion of duplicated genes that showed significant changes in expression in a lineage is significantly higher than that for single-copy genes. Further, a between-genome comparison showed that duplicated genes have a much higher probability of expression divergence between species or between different strains of the same species (Table 6.1). The authors also examined differences in expression of duplicated genes between two yeast strains and found a similar pattern (Table 6.1). They therefore concluded that duplicated genes are more likely to have divergent expression profiles than single-copy genes both within and between genomes.

More recently, Huminiecki and Wolfe (2004) examined the expression profiles of orthologous genes in human and mouse with those genes having lineage specific duplications. They focused on loci where a recent lineage specific gene duplication had occurred and created paralogous pairs in either human or mouse, and examined pairs of young duplicates that also have a comparable single-copy ortholog in the other species. The presence of lineage specific duplicate genes increased the divergence between the expression profile between human and mouse in the homologous tissues. Moreover, orthologs with multiple lineage specific duplications showed even higher divergence between expression profiles. They reasoned that since orthologous genes of human and mouse are of the same age (as by definition they come from the last common ancestor), the increased divergence in expression is mainly due to the presence of recent lineage specific gene duplications.

Table 6.1 Distributions of singletons and duplicate genes in the comparisons between different strains/species of *Drosophila* and yeast. From Gu *et al.* (2004).

Difference in expression	# of Singletons	# of Duplicates
Comparison between *Drosophila* strains or species ($\chi^2 = 97.6, \mathrm{d.f.} = 1, p \approx 0$)		
Differentially expressed[a]	1201	1593
Similarly expressed[b,c]	2155	1745
Total	3356	3332
Comparison between yeast strains ($\chi^2 = 54.5, \mathrm{d.f.} = 1. p < 10^{-12}$)		
Differentially expressed	541	392
Similarly expressed	2252	925
Total	2793	1317

[a] Differentially expressed between at least two strains within a species or between two species.
[b] Similarly expressed between every pair of strains within a species and between species.
[c] If only the genes that showed expression changes during the start of metamorphosis are considered, the proportion of duplicate genes with different expression patterns between species/strains is still significantly higher than that of singletons.

6.4   Expression divergence and retention of duplicate genes

A central issue in the evolution of duplicate genes is why many duplicate genes have been retained in a genome despite the fact that the most likely fate for a redundant duplicate is nonfunctionalization. The neofunctionalization (Ohno 1970) and subfunctionalization (Force *et al.* 1999) models are two frequently considered models. The neofunctionalization model postulates that gain of new function(s) is the major factor for the retention of both copies of duplicate genes in a genome. The subfunctionalization model is also known as the duplication-degenerate-complementation (DDC) model and it assumes that the two duplicate genes undergo complementary degeneration of their cis-regulatory motifs, so that both copies are required to produce the full complement of the cis-regulatory motifs of the ancestral gene (Fig. 6.2). This model predicts (1) the total number of cis-regulatory motifs in the two genes should decrease with evolutionary time (see Fig. 6.2) and (2) genes with many paralogs should tend to have a low number of regulatory motifs because these genes should have undergone multiple rounds of gene duplication and complementary loss of motifs.

Using genomic data from yeast, Papp *et al.* (2003) tested the above two predictions of the DDC model. They found that although the number of shared motifs decreases over time, the total number of motifs remains constant among duplicates (Fig. 6.3) and that genes with numerous paralogs in the yeast genome do not have an especially low number of cis-regulatory motifs. To explain these observations, the authors suggested that either regulatory motifs with new function arise from existing ones or the loss of regulatory motifs is balanced by the gain of new regulatory motifs, keeping the total

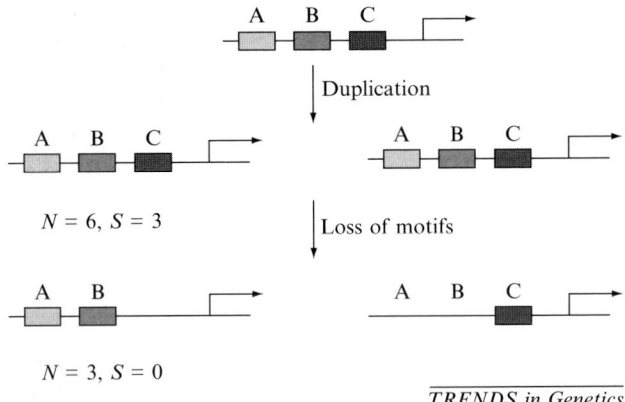

Fig. 6.2 The degenerative complementation model of evolution in regulatory regions after gene duplication. Abbreviations: A, B, C, regulatory motifs; N, the total number of motifs in the duplicates; S, the number of different motifs shared by the duplicates. Each daughter gene retains only a subset of promoter elements compared to the ancestral state. As a result, both S and N decline with age. Modified from Papp *et al.* (2003).

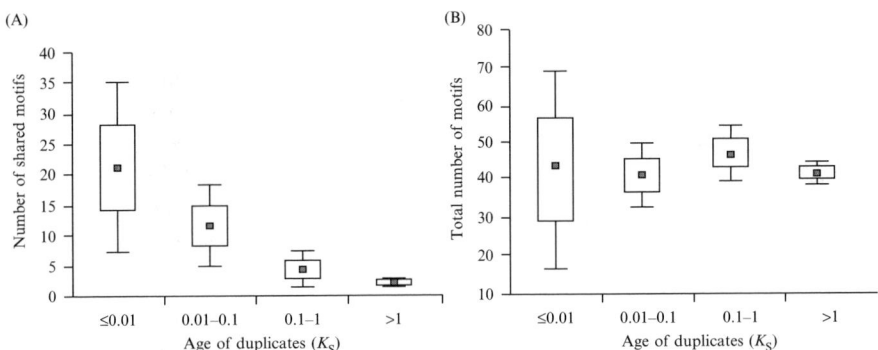

Fig. 6.3 (A) Negative association between duplicate age (the rate of evolution at synonymous sites; Ks) and the number of different shared motifs of duplicates. (B) No association between duplicate age and the total number of motifs possessed by the two copies. From Papp *et al.* (2003).

number constant. They concluded that the DDC model alone cannot fully explain duplicate gene evolution in yeast and the gain of novel function plays a substantial role in the retention of duplicate genes in the yeast genome.

The traditional view predicts a trend towards tissue specific expression following gene duplication. Huminiecki and Wolfe (2004) indeed found a general trend for paralogous genes to become more specialized in their expression patterns, showing decreased breadth and increased specificity of expression as gene family size increases.

On the other hand, they also claimed that a detailed examination of some gene families revealed examples of neofunctionalization of duplicated genes, but only one case of subfunctionalization. However, this issue remains unsettled because it is possible that neofunctionalization is a later-stage process, while subfunctionalization is an early-stage process that reduces the chance of nonfunctionalization of a duplicate (Force *et al.* 1999; Papp *et al.* 2003; Wray *et al.* 2003; He and Zhang 2005).

6.5 Evolutionary distance of expression divergence

To facilitate evolutionary analysis of expression data, we (Gu *et al.* 2005b) defined an additive expression distance (E) between duplicate genes, measured by the average of squared expression differences. For any duplicate genes 1 and 2, let x_{1k} and x_{2k} be the expression levels, respectively, in the k-th microarray experiment, $k = 1, \ldots m$. Let $\bar{x}_1$ and $\bar{x}_2$ be the means of expression, respectively. Then, we define the evolutionary expression distance between genes 1 and 2 as follows:

$$E_{12} = \sum_{k=1}^{m} \left[(x_{1k} - \bar{x}_1) - (x_{2k} - \bar{x}_2) \right]^2 / (m - 1) \qquad (6.1)$$

In other words, the evolutionary expression distance is the (centralized) squared Euclidean metric which is normalized by the sample size.

In the last chapter we showed that the expression distance defined by Eq. (6.1) satisfies the additive requirement. That is, $E_{12} = E_1 + E_2$, where E_1 and E_2 are the expression branch lengths (Fig. 6.4(A)). The additivity assures that, given the evolutionary time t between two duplicate genes, the evolutionary rate of expression divergence is given by $\lambda_E = E_{12}/2t$, which is the average rate over two lineages. For a large gene family, the additivity allows us to develop a least-square method for mapping the pairwise expression distances onto the phylogeny (Gu 2004). Thus, the mean rate of expression divergence can be estimated by $\lambda_E = E_T/T$, where E_T is the sum of expression branch lengths, and T is the total evolutionary time of the gene family.

Biological meaning of expression distance and rate   Gu (2004) (also see the last chapter) developed a statistical framework for expression evolution under the Brownian process. This simplest B-model assumes that the expression divergence of a gene family is mainly driven by small and additive genetic drifts (random effects), with a constant rate measured by σ^2, the mutational variance. In the two gene case, the B-model assures $E_1 = E_2 = \sigma^2 t$. Hence, the expression distance is given by $E_{12} = E_1 + E_2 = 2\sigma^2 t$, and the evolutionary rate of expression divergence equals to the mutational variance, i.e. $\lambda_E = E_{12}/2t = \sigma^2$. Therefore, the B-model could be considered as the 'neutral-evolution' model of gene expression; in analogy, under the classical neutral model, the evolutionary rate of DNA sequence equals to the mutation rate.

Moreover, Gu (2004) studied several evolutionary mechanisms in which selection forces may be involved. For instance, under the dramatic shift (S) model, the expres-

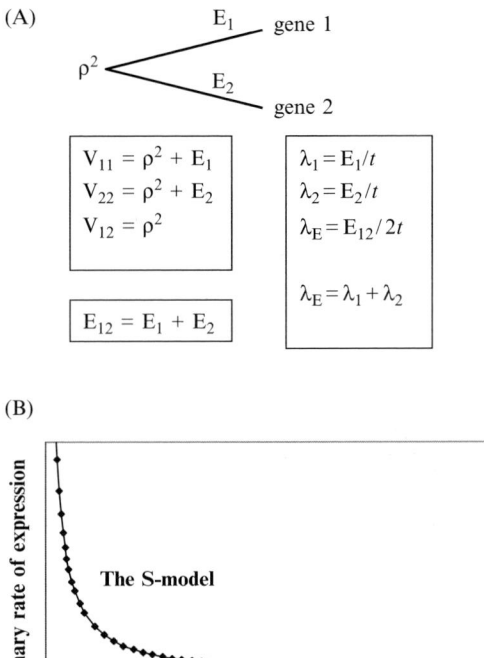

Fig. 6.4 (A) Schematic illustration for a rooted 2-gene tree. E_1 and E_2 refer to the expression distances associated with the branches after the gene duplication; ρ^2 refers to the common expression variance. The expression distance E_{12} is the sum of E_1 and E_2. (B) Schematic illustration for the time-dependency of evolutionary rate under the S-model. Modified from Gu *et al.* (2005).

sion branch lengths are $E_1 = \sigma^2 t + S_1^2$ and $E_2 = \sigma^2 t + S_2^2$, where S_1^2 and S_2^2 measure the duplication-dependent dramatic shifts in both lineages, respectively. Under the S-model, the expression distance turns out to be $E_{12} = 2\sigma^2 t + S_1^2 + S_2^2$, resulting in $\lambda_E = E_{12}/2t = \sigma^2 + S^2/t$, where $S^2 = (S_1^2 + S_2^2)/2$. Therefore, the accelerated (time-dependent) rate of expression divergence may reflect the non-neutral fashion of expression divergence after gene duplication (Fig. 6.4(B)).

Relative expression rate test   We implement the relative rate test to examine whether the expression divergence is asymmetric after the gene duplication, when an outgroup gene (gene 3) is available see Fig. 6.8A). The null hypothesis is $E_1 = E_2$, i.e. equal expression divergence after duplication (symmetric evolution). From the additivity of expression distance, i.e. $E_{13} = E_1 + E_3$, and $E_{23} = E_2 + E_3$, the relative rate test for gene expression is to compute the statistic

$$\delta_E = E_{13} - E_{23} \tag{6.2}$$

The biological interpretation of δ_E can be illustrated under the S-model, in which $\delta_E = E_{13} - E_{23} = S_1^2 - S_2^2$ is to test which lineage may have more dramatic (duplication-dependent) expression shift. Under the null hypothesis, $\delta_E = 0$, the P-value can be empirically calculated by the bootstrapping procedure. Therefore, the null hypothesis of symmetric expression evolution is rejected at the significance level α if $P < \alpha$.

Effects of microarray experimental factors Let σ_ϵ^2 be the variance component of gene expression from all (non-biological) experimental factors. Consequently, the expectation of the expression distance defined by Eq. (6.1) turns out to be

$$E_{ij}^* = E_{ij} + 2\sigma_\epsilon^2$$

That is, the expression distance, as well as the evolutionary rate of experimental divergence, tends to be overestimated. Nevertheless, the relative rate expression test in Eq. (6.2) is statistically not affected by the experimental factors, because $\delta_E^* = E_{13}^* - E_{23}^* = E_{13} - E_{23} = \delta_E$.

The ANOVA model Gu *et al.* (2005b) used the analysis of variance (ANOVA) model to account for various sources of expression variation in microarray data (Kerr and Churchill 2001). For instance, we consider the (cDNA) microarrays during the yeast sporulation. Let y_{ijkg} be the log-transformed expression intensity of gene g from array i, dye j ($j = 1$ for green and 2 for red) at the k-th treatment. The ANOVA model for y_{ijkg} can be written as follows

$$y_{ijkg} = \mu + A_i + D_j + T_k + G_g + (AG)_{ig} + (TG)_{kg} + e_{ijkg} \tag{6.3}$$

where μ is the overall mean. The error terms e_{ijkg} are independent and identically distributed with mean 0, and the variance ϵ^2. The array effects A_i account for mean expression differences of expression between arrays, the dye effects D_j for differences between the average signals from each dye. The time point effects T_k account for overall differences in the time points. The gene effects G_g capture the average levels of expression for individual genes spotted on the arrays. The array-by-gene interactions $(AG)_{ig}$ account for the effect of the spot on array i for gene g. Finally, the (normalized) interactions between genes and treatments, $(TG)_{kg}$, which capture differences from overall averages that are attributable to the specific combination of treatment-k and gene g. In short, the experimental variance σ_ϵ^2 may include the dye effect D_j, the array-by-gene interactions $(AG)_{ig}$, and the variance ϵ^2 of error terms e_{ijkg}.

6.6 Rate of expression divergence between yeast duplicate genes

Age distribution of yeast gene duplications We (Gu *et al.* 2005b) have studied 434 yeast (*S. cerevisiae*) gene families, consisting of 201 two-, 113 three-, 39 four-, 18 five-, and 63 six-or-more member gene families. For each gene family, we inferred the phylogenetic tree and estimated the age of each duplication event. The age distribution of 1369 duplication events we identified is shown in Fig. 6.5. The time scale used in

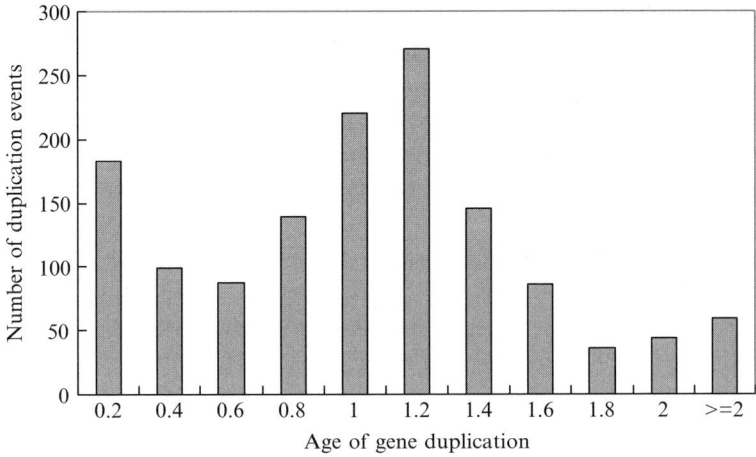

Fig. 6.5 The histogram for the estimated age distribution of yeast duplicates. The evolutionary time unit is defined by the time of bacteria/yeast split, roughly corresponding to two billion years ago. Figure modified from Gu *et al.* (2005b).

this analysis was the divergence time between prokaryotes and eukaryotes, which is about 1.4–2 billion years ago. Using amino acid distance as a proxy for duplication time gives virtually the same result. There are two peaks in the age distribution of yeast duplicates. The recent tremendous increase of duplication events could be well explained by the yeast genome duplication hypothesis (a polyploidization event) that might have occured about 100 million years ago. In addition, there exists a very ancient peak for gene duplications that occurred round the divergence time between prokaryotes and eukaryotes. The age distributions from the bacteria (*Escherichia coli K12*) and the archaea (*Thermoplasma acidophilum*) also show this very ancient component of duplicates (not shown). These observations together raise an interesting question about the role of gene duplications during the emergence of three major kingdoms.

Evolutionary rate of expression divergence after gene duplication　For each of yeast gene families with two or more duplicates, we estimated the expression distances (E_{ij}) between any two duplicate genes i and j, based on 276 yeast microarray datasets. Given the inferred phylogeny from the multi-alignment of protein sequences, the total expression branch length (E_T) was estimated by the least-square mapping of expression distances on the topology. As the total evolutionary time (T) of the gene family was obtained from the estimated duplication times (Fig. 6.5), we estimated the evolutionary rate ($\lambda_E = E_T/T$) of expression divergence for each gene family.

Overall, the mean rate among yeast gene families under study is 0.977 per time unit (the Bacteria/yeast split); the 95 per cent quantile (0.09–6.50) reveals a substantial variation of expression rates. If one further assumes the bacteria/yeast split about 2 billion years (Byr) ago, the mean rate of expression divergence after yeast duplications turns out to be 0.49×10^{-9} per year. Interestingly, the evolutionary rate

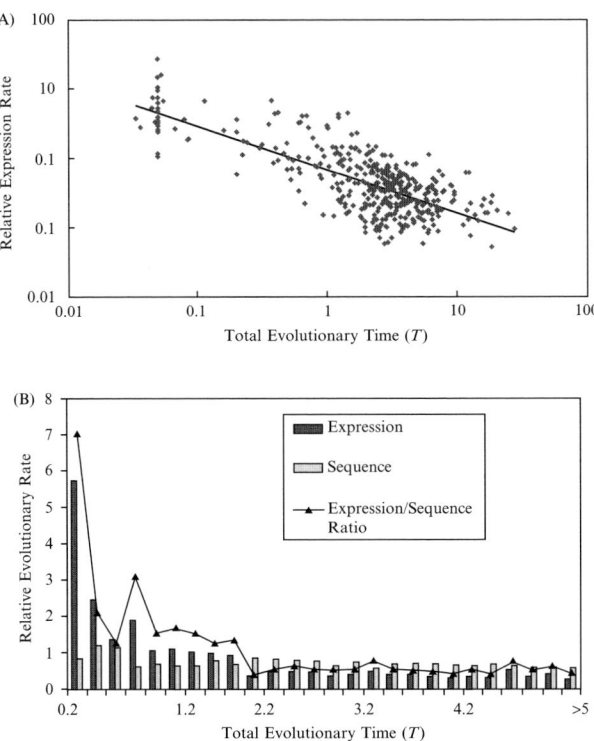

Fig. 6.6 The log-log regression (A) between the evolutionary rate (λ_E) of gene expression and the total evolutionary time (T) of the gene family. In (B), the evolutionary rate of gene expression, that of the protein sequence, as well as the expression/sequence ratio, averaged over each bin (0.2 time unit), are plotted against T. The evolutionary time unit is defined by the time of bacteria/yeast split, roughly about 2 billion years ago. Figure modified from Gu *et al.* (2005b).

(λ_E) of gene expression after duplications is time-dependent (Fig. 6.6(A)); the log-log regression shows λ_E is negatively correlated with the total evolutionary time (T) of the gene family ($R = -0.75$, $P < 10^{-8}$). Figure 6.6(B) shows that in the early-stage, the mean initial rate of expression divergence would be as high as 5.8 per time unit, or 2.9×10^{-9} per year, which is more than 20-fold higher than the baseline expression rate (0.14×10^{-9} per year). Our finding supports the notion of rapid expression divergence shortly after gene duplication, which is much more dramatic than that in the sequence evolution of duplicate genes. Indeed, only a moderate (~ 20 per cent) increase in the rate of protein sequence evolution for young gene families ($R = -0.18$, $P < 0.01$). Consequently, the relative ratio of expression rate to protein sequence rate is high (~ 7.1) for young duplicates and decreases rapidly with the evolutionary time (Fig. 6.6B).

Evolutionary rate of regulatory interaction between TF-genes   Rapid expression divergence reflects rapid evolution of regulatory network in the early stage

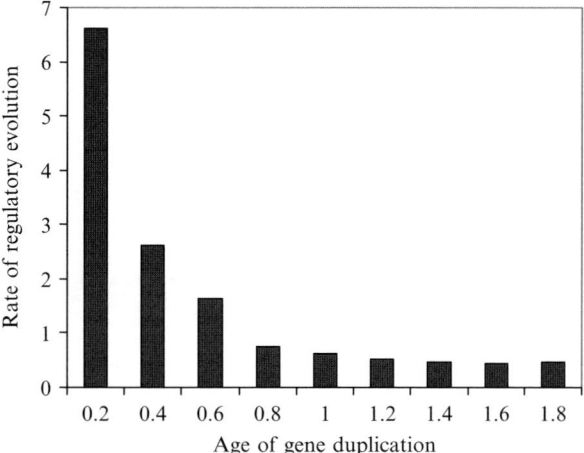

Fig. 6.7 Evolutionary rate of regulatory interaction turnover, averaged over each bin (0.2 time uint), is plotted against the age of gene duplication. The evolutionary time unit is defined by the time of bacteria/yeast split, roughly about 2 billion years ago. Figure modified from Gu *et al.* (2005b).

after gene duplications. We tested this prediction using the transcription factor (TF)-target gene interactions by the large-scale chromatin immunoprecipitation (ChIP) experiments (called regulatory interactions thereafter) (Lee *et al.* 2002). This technology detects the regulatory network without accurate identification of binding sites. We used the parsimony to infer the evolutionary events of regulatory interactions. Since the root of gene family tree is uncertain in many cases, we used the turn-over events (gain or loss) in our study. Then we estimated that the mean evolutionary rate (λ_R) of regulatory interactions $\lambda_R \approx 0.722$ per time unit, or 0.36×10^{-9} per year. Similar to the rate of expression divergence, we grouped duplicate genes with similar duplication age (with a bin of 0.2 time unit), and estimated the mean evolutionary rate for each group. As shown in Fig. 6.7, the regulatory evolution in the young duplicate group is almost 10-fold faster than that in the ancient group. The null hypothesis of equal rate among age groups is highly rejected ($P < 10^{-5}$).

6.7   Asymmetric expression evolution after gene duplications

The view of asymmetric evolution after gene duplication predicts that only one of duplicate copy has undergone rapid expression evolution shortly after the gene duplication (high-rate expression divergence), while the other copy largely kept the ancestral pattern (low-rate expression divergence). We utilized the relative expression rate test to study 111 yeast duplicate gene pairs (Fig. 6.8(A)), while an out-group duplicate gene was determined by the phylogenetic analysis. Overall, 60 gene families (54 per cent) shows the null hypothesis (equal expression divergence, $E_1 = E_2$) is rejected at the 0.05 significance level, and 47 gene families (42 per cent) at the 0.01

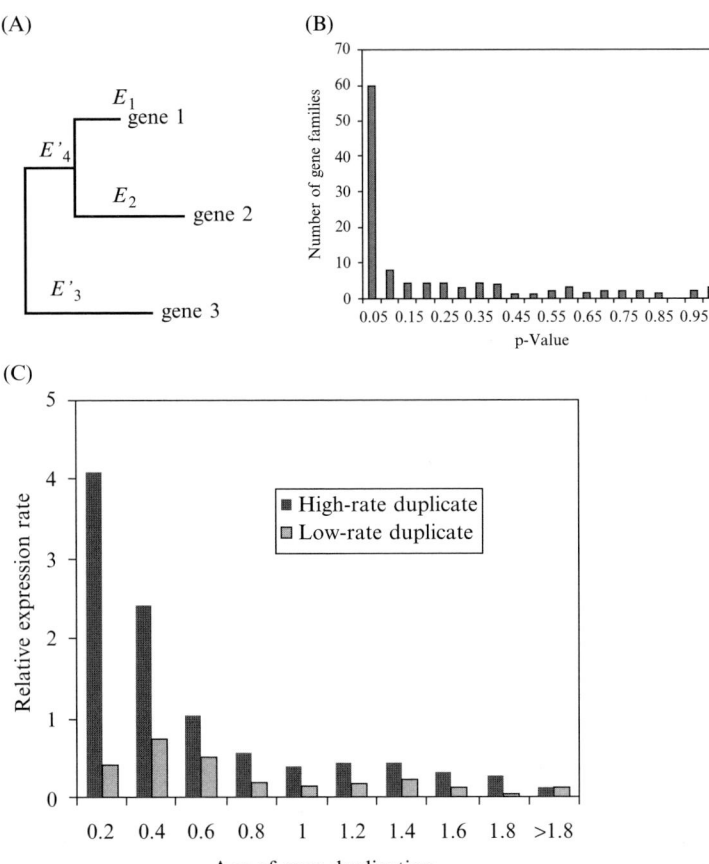

Fig. 6.8 (*A*) Schematic illustration for the relative expression rate test, whereas $E_3 = E_{3'} + E_{4'}$. (*B*) The histogram for the p-values obtained from the bootstrapping in each individual test. (*C*) Expression rates for highly-diverged copy as well as lowly-diverged copy after gene duplication, averaged over each time bin (0.2 unit), are plotted against the total evolutionary time, relative to the bacteria/yeast split. Figure modified from Gu *et al.* (2005b).

significance level. It has been recommended that a reasonable measure for type-I error under multiple tests is the false discovery rate (FDR). Given the histogram of the p-values (Fig. 6.8, (B)), we estimated FDR = 11.8 per cent at the 0.05 significance level, that is, there are about 60×11.8 per cent ≈ 7 cases that could be false positive. This result indicates that the expression evolution of duplicate gene pairs is highly asymmetric. We tentatively classified these duplicate genes into high (*H*) or low (*L*) expression-rate group, respectively. Fig. 6.8(C) shows that the mean expression rate in the *H*-group is much higher in young duplicates, whereas that in the *L*-group has no difference between young and ancient duplicates.

Expression distance corrected for the experimental noises One remaining question is whether our results could be affected by the (non-biological) experimental factors. Gu *et al.* (2005b) applied the ANOVA procedure to the yeast sporulation microarray data for obtaining a rough estimate of $\sigma_\epsilon^2 \approx 0.275$. We then recalculated the evolutionary rate of expression divergence after gene duplication. It showed that the mean initial rate of expression divergence now is down to 1.7×10^{-9} per year, which is about 41 per cent lower than the original estimate. However, it is still more than 10-fold higher than the baseline expression rate. In other words, the relatively high level of experimental noises does not alter the main result.

6.8 Concluding remarks

Although much progress has been made in our understanding of the mode and tempo of expression divergence between duplicate genes, much remains for further study. It should be noted that the rate of expression divergence may have been overestimated by the noisiness of microarray data and the effect of gene conversion between duplicate genes. Another topic that needs much further investigation is the role of expression divergence in the retention of duplicate genes in a genome. In yeast, we observed that there is at least 10-fold increase in the initial rate for both expression and regulatory evolution shortly after gene duplication, whereas only ~ 20 per cent rate increase in the early stage for protein sequences. Moreover, relative expression rate tests suggested that the expression of duplicate genes tends to evolve asymmetrically, that is, the expression of one copy evolves rapidly, whereas the other one largely maintains the ancestral expression profiles. Since studies on this topic have been made mainly on the yeast, there is a need to expand the research to other eukaryotes, as yeast is only a single cell organism with a large effective population size. It is interesting to test whether the conclusions drawn from the yeast data are also applicable to other species. For the future outlook, one exciting and novel area that remains to be investigated is to explore the expression divergence of duplication in the context of the gene network.

7

Tissue-Driven Hypothesis of Genomic Evolution

Understanding the underlying regulatory mechanism is a fundamental step to exploring the emergence of genome complexity (King and Wilson 1975). An important issue is the role of tissue-specific factors in genomic evolution. Several studies have suggested that tissue-specific constraints may generate among tissue variation of expression divergence between the human and chimpanzee (Enard *et al.* 2002; Gu and Gu 2003a; Khaitovich *et al.* 2004a, 2004b; 2005a, 2005b, 2006c), between the human and mouse (Gu and Su 2007), or between fruitflies (Rifkin *et al.* 2005). Duret and Mouchiroud (2000) showed that the rate of protein divergence was negatively associated to the tissue broadness of gene expression. In this chapter, we first discuss *the tissue-driven hypothesis* (Gu and Su 2007), based on an explicit evolutionary model for providing testable predictions. This theory claims that stabilizing selections for both expression and sequence divergences may be affected simultaneously by common tissue factors. In the second part, we address an interesting problem about the expression evolution in the human brain since the split of human and chimpanzee (Enard *et al.* 2002; Gu and Gu 2003a; 2004; Jordan *et al.* 2005; Yang *et al.* 2005; Zhang and Li 2004).

7.1 Tissue-driven hypothesis of genomic evolution

7.1.1 Expression divergence under stabilizing model

We (Gu and Su 2007) invoked the stabilizing selection model (Hansen and Martins 1996) of quantitative characters to describe the tissue-specific constraint on expression divergence. For a gene expressed in a certain tissue (ti), the stabilizing selection on the expression level x follows a Gaussian fitness function

$$f_{ti}(x) = e^{-w_{ti}(x-\theta_e)^2} \tag{7.1}$$

where θ_e is the optimal expression level, w_{ti} is the coefficient of stabilizing selection on gene expression in tissue ti; a large w_{ti} means a strong selection pressure, and *vice versa* (Fig. 7.1).

Under the stabilizing selection model of Eq. (7.1), we have shown that the expression divergence follows an Ornstein–Uhlenback (OU) process. The stochastic OU process is characterized by the infinitesimal mean $-\beta_0(x - \theta_e)$ and variance $\epsilon^2/2N_e$, where ϵ^2 is the mutational variance, N_e is the effective population size, and $\beta_0 = w_{ti}\epsilon^2$ measures

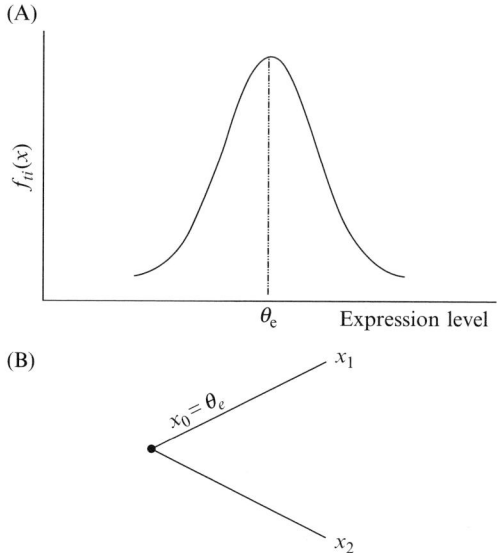

Fig. 7.1 (A) Fitness function plotting against the expression level under the stabilizing selection; θ_e is the optimal expression level. (B) Scheme of inter-species expression divergence between two orthologous genes. Here we assume that the ancestral expression level is at the optimal value $(x_0 = \theta_e)$.

the direct force against the deviation from the optimum. Given the initial expression value x_0, the OU model claims that $x(t)$ follows a normal distribution with the mean and variance

$$E[x(t)|x_0] = e^{-\beta t}x_0 + (1 - e^{-\beta t})\theta_e$$

$$V[x(t)|x_0] = \frac{\epsilon^2(1 - e^{-2\beta t})}{2\beta} \tag{7.2}$$

respectively, where $\beta = 2N_e\beta_0$ is the decay-rate of expression divergence.

For two genomes, such as the human and mouse, that have diverged t time units ago (Fig. 7.1B), the expression distance can be derived as follows. Let x_1 and x_2 be the expression levels of two orthologous genes 1 and 2, respectively. Assuming that the initial value is at the optimum $(x_0 = \theta_e)$, from Eq. (7.2) we have $E[x_1|x_0] = E[x_2|x_0] = \theta_e$. If the gene expression diverged independently, we have $Cov(x_1, x_2) = Var(\theta_e)$, that is, the covariance between x_1 and x_2 is equal to the variance of ancestral expression. Similarly, one can show $V(x_1) = V(x_2) = \epsilon^2(1 - e^{-2\beta t})/2\beta + Var(\theta_e)$. Therefore, the (Euclidean) expression distance for any gene pair g in tissue ti, defined by $E_{ti,g} = E[(x_1 - x_2)^2]$, is given by

$$E_{ti,g} = \frac{\epsilon_g^2(1 - e^{-2\beta_g t})}{\beta_g} = \frac{(1 - e^{-2\beta_g t})}{W_{ti,g}} \tag{7.3}$$

Similar to Eq. (7.1) and Eq. (7.2), ϵ_g^2 is the mutational variance, β_g is the decay-rate of expression divergence, and $W_{ti,g} = \beta_g/\epsilon_g^2$ is the strength of stabilizing selection on the expression divergence of gene pair g. Eq. (7.3) shows that $E_{ti,g}$ is inversely related to $W_{ti,g}$; when $t \to \infty$, $E_{ti,g} = 1/W_{ti,g}$.

7.1.2 Tissue-dependent rate of protein evolution

Gu (2007a) has studied the evolutionary rate of a protein sequence, based on a multi-dimensional stabilizing selection model as briefly introduced below; a detailed discussion about this subject will be in the next chapter. For a protein that only has a single function that is hypothetically represented by a variable called *molecular phenotype* (y), the stabilizing selection on y follows a simple Gaussian form (Fig. 7.2)

$$f(y) = e^{-(y-\theta_g)^2/2\sigma_g^2} \tag{7.4}$$

where a smaller σ_g^2 means a strong stabilizing selection and θ_g is the optical molecular phenotype. Thus, the coefficient of selection on y is given by $s(y) = f(y) - 1 \approx -(y - \theta_g)^2/2\sigma_g^2$. On the other hand, random (nonsynonymous) mutations in the coding region affect the molecular phenotype y according to a distribution with the

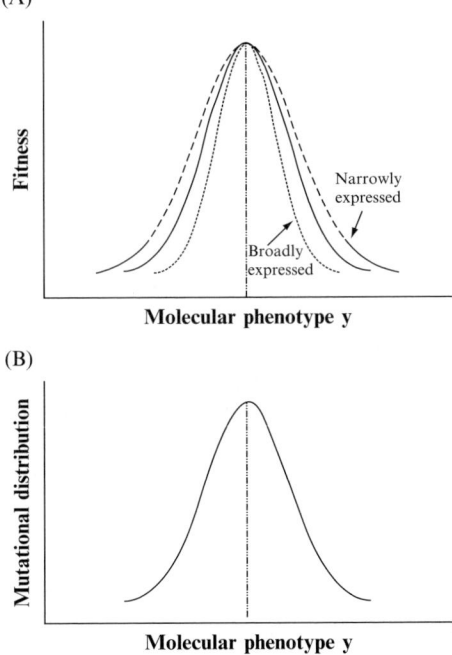

Fig. 7.2 (A) Fitness function plotting against the molecular phenotype (y) of protein function under the stabilizing selection. (B) Distribution of random mutations that affect the molecular phenotype y.

mean θ_g and the mutational variance σ_m^2. Consequently, the mean selection of coefficient is given by $\bar{s} = -E[(y - \theta_g)^2]/2\sigma_g^2 = -\sigma_m^2/2\sigma_g^2$, and the selection intensity $S_g = 4N_e\bar{s} = -2N_e\sigma_m^2/\sigma_g^2$. In the general case of multiple (K) molecular phenotypes of a protein, Gu (2007a) has shown

$$S_g = -2N_e \sum_{i=1}^{K} \frac{\sigma_{m,i}^2}{\sigma_{g,i}^2} \qquad (7.5)$$

where the subscript i assigns parameters σ_m^2 and σ_g^2 to the i-th molecular phenotype.

The stabilizing selection of molecular phenotypes may be tissue-dependent. This idea can be modeled as $\sigma_{g,i}^2 = a_{g,i}^2/Z_g$: While $a_{g,i}^2$ is molecular phenotype-dependent but tissue-independent, Z_g measures the accumulated tissue effect on the gene evolution; a larger Z_g means a greater tissue effect, and *vice versa*. Together, the mean selection intensity of the gene in Eq. (7.5) can be written as

$$S_g = S_0 \times Z_g \qquad (7.6)$$

where S_0 is the tissue-independent component. According to the theory of molecular evolution (Kimura 1983), the evolutionary rate of gene g is given by $\lambda_g \approx vS_g/[1 - e^{-S_g}]$, where v is the mutation rate. From Eq. (7.6) it indicates

$$\lambda_g = v\frac{S_0 Z_g}{1 - e^{-S_0 Z_g}} \qquad (7.7)$$

Eq. (7.7) links between tissue-effects and evolutionary rate of protein sequence. It predicts an inverse relationship between the evolutionary rate and the accumulated tissue effect Z_g.

7.1.3 Tissue-driven hypothesis

The tissue-driven hypothesis of genomic evolution postulates that the tissue factor plays an important role of functional constraint on the rate of genomic evolution, because genes influence phenotypic traits through regulated expressions in specific tissues. The phenotypic consequences of genetic variations in the regulatory and coding sequences are both affected by the common micro-environment of tissues. The tissue-driven hypothesis predicts several interesting genomic correlations, as shown below.

Tissue expression distance (E_{ti}): To measure the expression difference of a tissue between two species, we define E_{ti} as the mean expression distance over N orthologous genes in tissue ti, that is, $E_{ti} = \sum_{g=1}^{N} E_{ti,g}/N$, where $E_{ti,g}$ is given by Eq. (7.3). Gu and Su (2007) have shown that E_{ti} can be approximated by

$$E_{ti} \approx \left(1 - e^{-2\bar{\beta}t}\right)/W_{ti} \qquad (7.8)$$

where the mean tissue factor W_{ti} is the (harmonic) mean tissue factor, $\bar{\beta}$ is the mean decay-rate of expression divergence, and t is the time of speciation. Eq. (7.8) indicates that the tissue expression distance increases with time t, and decreases with the mean tissue factor W_{ti}. When $\bar{\beta}$ is close to zero (nearly neutral) or t is short (closely-related species), Eq. (7.8) is reduced to $E_{ti} \approx 2\epsilon^2 t$, converged to the Brownian model (Gu 2004,

Gu *et al.* 2005b), where ϵ^2 is the mean mutational variance over genes. In the case of distantly-related genomes when the expression divergence approaches the steady state, the time-dependent property of Eq. (7.8) vanishes, resulting in $E_{ti} \approx 1/W_{ti}$.

Tissue expression and sequence distances: the $E_{ti} - D_{ti}$ correlation. For a set (N_{ti}) of genes that are expressed in tissue ti, the mean evolutionary distance is denoted by D_{ti}. With some mild assumptions, Gu and Su (2007) showed that

$$D_{ti} \approx 2vt \times \frac{\bar{S}_{ti}}{1 - e^{-\bar{S}_{ti}}} \tag{7.9}$$

where the mean selection intensity of tissue (ti) over expressed genes is given by

$$\bar{S}_{ti} \approx \bar{S}_0 \times Z_{ti}$$

and Z_{ti} is the mean of accumulated tissue-(ti) factors over expressed genes, and $\bar{S}_0$ is the averaged tissue-independent components.

According to the tissue-driven hypothesis, two mean tissue factors W_{ti} in Eq. (7.8) and Z_{ti} in Eq. (7.9) should be positively correlated, because they represent the effects of common micro-environment of tissue (ti) on the expression divergence and protein sequence divergence, respectively. This argument predicts a positive correlation between tissue expression distance (E_{ti}) and tissue sequence distance (D_{ti}).

Inter-species and inter-duplicate tissue expression divergence: the $E_{ti} - T_{dup}$ correlation. Moreover, the tissue-driven hypothesis predicts that tissue factors may affect the expression divergence between duplicate genes. To be clear, we use $Q_{ti,g}$ for the tissue factor of expression divergence between duplicate pair g. For a set of duplicate genes, let T_{dup} be the mean evolutionary distance between duplicate pairs, called the tissue (ti) duplicate distance. Similar to the derivation of Eq. (7.8), we obtain

$$T_{dup} \approx \left(1 - e^{-2\bar{\gamma}\bar{\tau}}\right) / Q_{ti} \tag{7.10}$$

where Q_{ti} is the mean tissue factor for the inter-duplicate expression divergence in tissue ti, $\bar{\gamma}$ is the mean decay-rate of expression divergence, and $\bar{\tau}$ is the mean duplication time. Hence, positively correlated W_{ti} and Q_{ti} under the tissue-driven hypothesis leads to a testable prediction of positive correlation between E_{ti} and T_{dup}.

Tissue broadness and preference: Since any gene g may be expressed in multiple tissues, the tissue effect should be accumulated with the increased number of tissues (L_g). Therefore, the accumulated tissue effect by Eq. (7.6) can be further written as

$$Z_g = L_g \times Z_0 \tag{7.11}$$

where Z_0 is the average tissue factor of gene g, which measures the effect of tissue preference on the expression divergence. In short, the accumulated tissue effect (Z_g) can be decomposed into two factors: tissue broadness (L_g) and tissue preference (Z_0). Together, Eq. (7.9) and Eq. (7.11) indicate that the protein sequence becomes more conserved if the gene is expressed in more tissues, or in tissues with more stringent constraints.

While many studies have showed the effect of tissue broadness (e.g. Duret and Mouchiround 2000), Gu and Su (2007) studied the effect of tissue preference by grouping genes with the same tissue broadness (L_g). When L_g is the same, the larger the Z_0 value, the greater the selection intensity S_g, and so the lower evolutionary rate λ_g. This prediction can be tested, as shown later.

7.2   Testing the tissue-driven hypothesis

7.2.1   Estimation of genomic distances

Based on 29 human and mouse orthologous tissues (Fig.7.3), Gu and Su (2007) used substantial genomic data to test these genomic correlations predicted from the tissue-driven hypothesis. To this end, we have to estimate several evolutionary distances from various functional genomics datasets.

Tissue expression distance (E_{ti}):   Consider a set (N) of orthologous genes between species 1 (human) and species 2 (mouse). Let $x_{g1,ti}$ and $x_{g2,ti}$ be the (log2-transformed) expression levels of the g-th orthologous genes in tissue ti, respectively. It has been shown that the tissue (ti) expression distance defined by Eq. (7.8) can be estimated by

$$\hat{E}_{ti} = \sum_{g=1}^{N} (x_{g1,ti} - x_{g2,ti})^2 / N \tag{7.12}$$

Tissue protein distance (D_{ti}):   The mean of evolutionary distances of proteins that are expressed in tissue ti can be estimated by the conventional methods, such as the Poisson model. Obviously, D_{ti} is dependent of the cutoff used to identify expressed genes in a tissue. Gu and Su (2007) considered two expression status of a gene in tissue ti, i.e. high expression or normal expression.

Tissue duplicate distance (T_{dup}) for expression divergence: Consider a set (N_{dup}) of duplicate gene pairs. For the j-th duplicate pair, the expression levels of two duplicate genes in a given tissue (ti) are denoted by x_j and y_j, respectively. Similar to Eq. (7.12), T_{dup} can be estimated by

$$\hat{T}_{dup} = \sum_{j=1}^{N_{dup}} (x_j - y_j)^2 / N_{dup} \tag{7.13}$$

Roughly speaking, a large T_{dup} value reflects the plasticity of tissue-specific developmental constraint that allows more expression divergence between duplicate genes, and *vice versa*.

Estimation of tissue broadness and preference: The number (L_g) of tissues in which gene g is expressed, or the tissue broadness, can be statistically inferred (Gu and Su 2007). For gene g that is expressed in L_g different tissues, let E_j ($j = 1, \ldots, L_g$) be the j-th tissue expression distance between the human and mouse. Since a large E_j means less tissue constraint on the expression divergence, we propose the following index to measure the effect of tissue preference

$$t_g = \sum_{j=1}^{L_g} E_j^{-1} / L_g \tag{7.14}$$

where tissue expression distance E_j is estimated by Eq. (7.12). In particular, when the expression divergence is close to the steady-state, we have $E_j \approx 1/W_j$ so that

$$t_g \approx \sum_{j=1}^{L_g} W_j/L_g$$

That is, t_g is an estimate of the mean tissue factor of gene g. As the tissue-driven hypothesis predicts $W_j \approx Z_j$, t_g is a proxy for the effect of tissue preference $\bar{Z}_g = \sum_{j=1}^{L_g} Z_j/L_g$ on the sequence conservation, creating a negative correlation between t_g and the evolutionary distance of protein sequence (d_g).

7.2.2 Tissue expression divergence between human and mouse

Based on 8936 human-mouse orthologs, we (Gu and Su 2007) estimated the tissue expression distance E_{ti} for each of 29 tissues. Figure 7.3 shows a substantial variation of E_{ti} among tissues. Indeed, there is a 2.4-fold difference from the lowest $E_{ln} = 0.85$ (lympho-node, ln) to the highest $E_{pc} = 0.206$ (pancreas, pc).

Khaitovich *et al.* (2005a, 2005b) observed that, in primates, the brain may have more expression conservation than four other tissues (testis, heart, liver, and kidney). More extensively, Gu and Su (2007) have found an overall expression conservation in some neuro-related tissues, e.g. pituitary(pi), amygdala (ad), hypothalamus (hp), and cerebellum (cb) (Fig. 7.3). In contrast, testis (ts) may have a rapid inter-species expression divergence. One possible reason is that the overall relaxed developmental constraint in the testis may facilitate the operation of sexual selection after speciation. Moreover, some hormone-related tissues, such as the pancreas, may have more developmental plasticity to allow rapid expression divergence, likely through the interactions with environmental cues during the evolution. In short, substantial variation of E_{ti} among tissues may imply the role of tissue-specific factors in the mammalian genomic evolution.

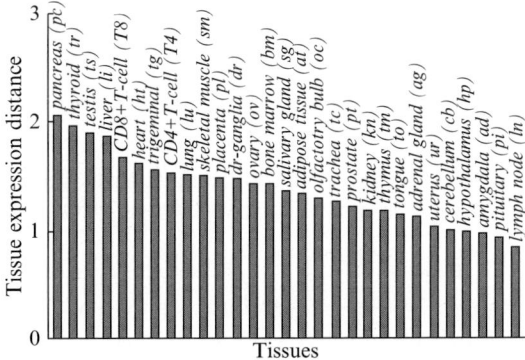

Fig. 7.3 Variation of human–mouse tissue expression distances (E_{ti}) among 29 tissues. Abbreviations for these tissues are shown in the parentheses. Modified from Gu and Su (2007).

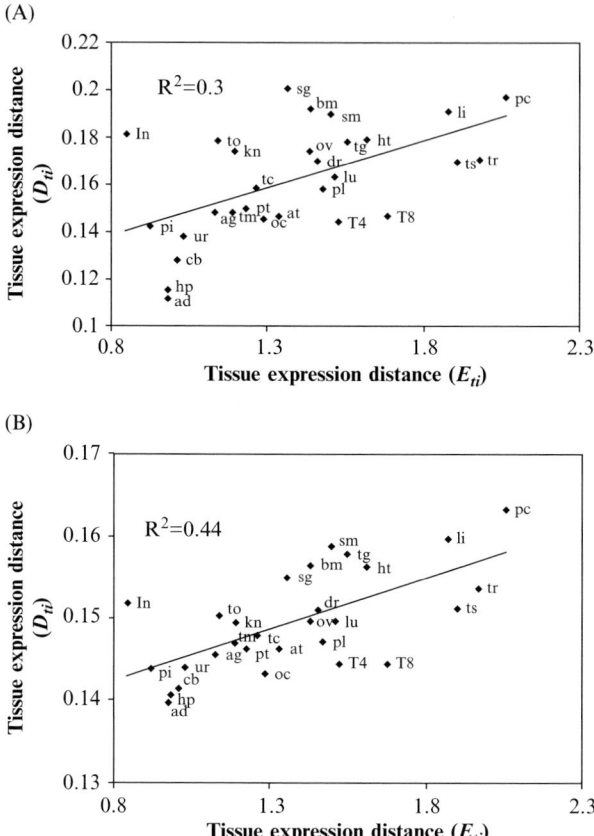

Fig. 7.4 Correlations between tissue expression distance (E_{ti}) and tissue protein distance (D_{ti}): for highly expressed proteins (A), and for normal expressed proteins (B). See Fig. 7.3 for the description of abbreviations of tissue names. In each case the correlation is statistically highly significant ($P < 0.001$). Modified from Gu and Su (2007).

7.2.3 Correlation ($E_{ti} - D_{ti}$) between tissue expression and sequence divergence

For each tissue ti, we calculated the tissue protein distance (D_{ti}) between the human and mouse. The tissue-driven hypothesis expects a positive correlation between E_{ti} and D_{ti}, driven by the same tissue-specific developmental constraint that may affect both tissue expression divergence and sequence divergence of expressed proteins. We indeed found a highly significant correlation between E_{ti} and D_{ti} based on 29 human–mouse tissues (Fig. 7.4). In the case of high expression (panel A), the (Pearson) coefficient of correlation is $R = 0.55$ ($p < 0.001$), while $R = 0.66$ ($p < 0.001$) in the case of normal expression (panel B). Hence, the significance of $E_{ti} - D_{ti}$ correlation provides statistically convincing evidence to support the tissue-driven hypothesis.

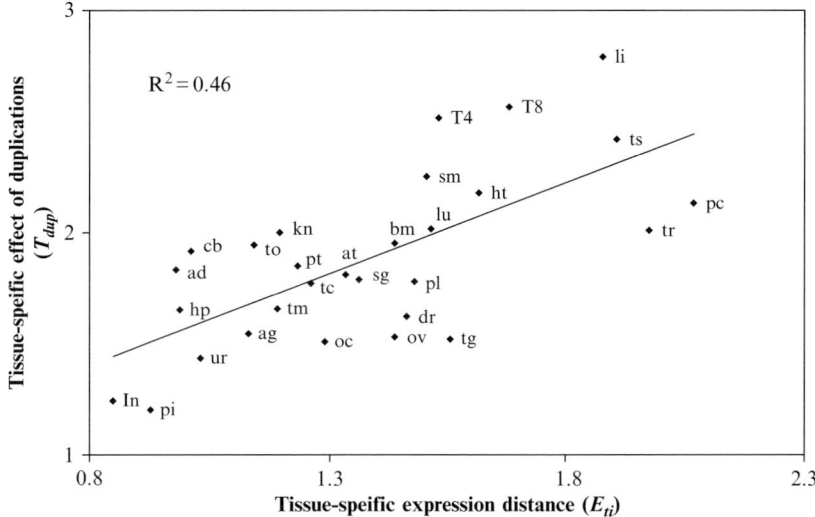

Fig. 7.5 The correlation between tissue expression distance (E_{ti}) and tissue duplicate distances (T_{dup}). Here T_{dup} is the average of human and mouse duplicates. The correlation is statistically highly significant ($P < 0.001$). Modified from Gu and Su (2007).

7.2.4 Tissue correlation ($E_{ti} - D_{ti}$) between inter-species and duplicate expression divergence

A positive $E_{ti} - T_{dup}$ correlation implies that when a tissue allows more inter-species expression divergence, it should also tolerate more extensive expression divergence between duplicated genes. Based on 1312 duplicate pairs that were duplicated before the human–mouse split, we estimated the duplicate tissue distance (T_{dup}) in each tissue ti. Figure 7.5 shows a highly significant correlation between tissue expression distance (E_{ti}) and T_{dup} ($p < 0.001$). This result supports the tissue-driven hypothesis that duplicated genes tend to have more expression divergence in a tissue with relaxed developmental constraint, and *vice versa*.

7.2.5 Evolutionary rate of protein sequence under multiple tissue constraints

We further classified these 8936 human-mouse orthologous genes into groups according to the number (L_g) of tissues in which they are expressed, i.e. $L_g = 1, 2, \ldots 29$. In each group, we calculated the effect of tissue preference (t_g) for each gene in both human and mouse. Noticeably, for each group, a negative correlation between the protein distance (d_g) and t_g is observed (Fig. 7.6(A)). Twenty-five cases are statistically significant ($p < 0.05$); in particular, 15 cases show highly statistically significant ($p < 0.0001$). For instance, Fig. 7.6(B) shows the d_g versus t_g correlation in the case of $L_g = 5$.

Given the similar tissue-broadness, the overwhelming negative $d_g - t_g$ correlation indicates that genes that are expressed in stronger constrained tissues (e.g.

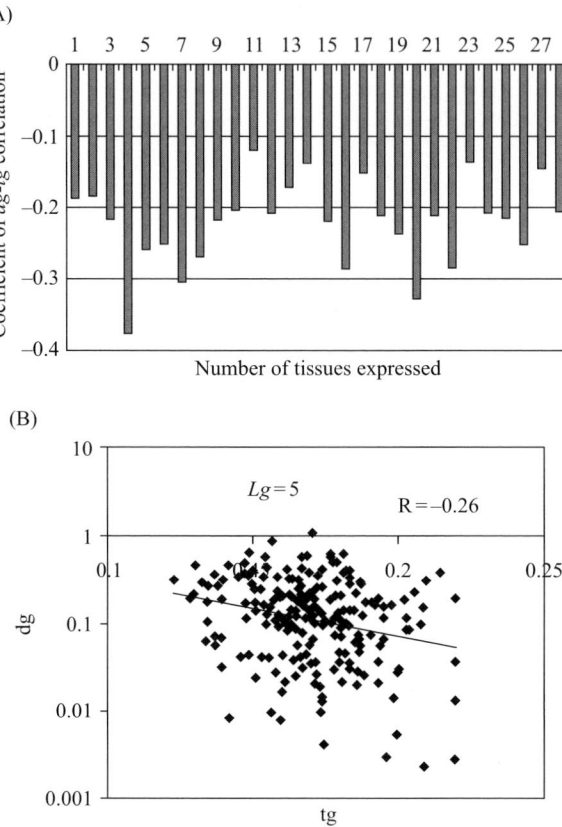

Fig. 7.6 (A) Negative coefficients of t_g–d_g correlations for gene groups with the same tissue broadness (L_g). (B) The t_g–d_g plotting in the case of $L_g = 5$. Modified from Gu and Su (2007).

neuro-related) tend to evolve more slowly at the sequence level than those expressed in weaker constrained tissues (e.g. hormone-related), as predicted by the tissue-driven hypothesis. On the other hand, if a protein is expressed in several different tissues, the evolution of protein sequence may be under multiple tissue-specific constraints. That explains why broadly expressed genes generally tend to evolve slowly at the sequence level.

7.2.6 Some comments

In spite of some studies (Yanai *et al.* 2004, 2006) that claimed that similar gene expression profiles do not imply similar tissues functions, highly significant correlations between tissue expression distance (E_{ti}) and tissue sequence distance (D_{ti}), and between E_{ti} and the duplicate tissue distance (T_{dup}), support the notion that the evolution of expression pattern and protein sequence may be under the same constraint of tissue factors. Moreover, for genes with the same expression broadness, we found that genes expressed in more stringent tissues tend to evolve slower than those in more relaxed tissues. These findings together formulated the 'tissue-driven hypothesis', providing some new insights on how the genomic evolution is shaped by the up-level physiological systems of the organism.

A basic assumption of the tissue-driven hypothesis is that the genome evolves largely under functional constraints maintained by stabilizing selections at multiple levels from cell-physiology to development. In some cases, episodic adaptive selection may happen either in the expression pattern or in the protein function (Enard *et al.* 2002; Gu and Gu 2003, also see below).

7.3 Compound-Poisson model of expression evolution

Khaitovich *et al.* (2005b) introduced a stochastic model that describes neutral changes of gene expression over evolutionary time as a compound Poisson process. Formally, let random variable $M(t)$ be the number of mutations occurring in the regulatory region in the time interval t. Moreover, the effect of mutation i on the expression level is denoted by X_i, which follows a specified distribution with zero mean. Hence, the expression level $Y(t)$ after t time units is given by $Y(t) = Y(0) + \sum_{i=1}^{M(t)} X_i$, which defines a compound Poisson process. Let Z_{12} describe a gene expression difference between the same tissues from two species, which have evolved independently on branches of length t_1 and t_2 from a common ancestor. Then, we have

$$Z_{12} = Y_1(t_1) - Y_2(t_2) = \sum_{i=1}^{M(t_1)} X_i - \sum_{j=1}^{M(t_2)} X_j \qquad (7.15)$$

since the common ancestry implies $Y_1(0) = Y_2(0)$. Though a closed formula for the density function of Z_{12} does not exist, moments can be derived using characteristic functions. Let $\mu(X)$ denote the mean and $\mu_k(X)$ the k-th (central) moment of random variable X and define its coefficients of skewness and kurtosis as $\gamma_1(X) = \mu_3(X)/[\mu_2(X)]^{3/2}$ and $\gamma_2(X) = \mu_4(X)/[\mu_2(X)]^2$, respectively. Khaitovich *et al.* (2005) suggested to estimate the expression distance $(t_1 + t_2)$ by the following relationship

$$t_1 + t_2 = \frac{\gamma_2(X)}{\gamma_2(Z_{12}) - 3} \qquad (7.16)$$

While $\gamma_2(Z_{12})$ can be calculated from the data, one has to specify the distribution of X to determine $\gamma_2(X)$. Two types of distributions were studied by Khaitovich *et al.* (2005b): A normal distribution corresponds to the symmetric case where a random mutation causes equally likely a decrease and an increase in expression. In this case $\gamma_2(X) = 3$. Alternatively, a specific form of extreme value distribution describes a situation where a mutation is more likely to reduce the expression of the gene, leading to $\gamma_2(X) = 5.4$. Khaitovich *et al.* (2005b) applied this model to analyze the expression divergence between the human and chimpanzee.

7.4 Expression shifts in the human brain

7.4.1 Evolution of the human brain

An unsolved mystery in human genetics and evolution is how humans and chimpanzees differ so considerably in many morphological, behavioral, and cognitive aspects, given only about 4 per cent difference in their genomic DNA sequences. Using microarray technology to measure gene expression in the human and chimpanzee, Enard *et al.* (2002) concluded that since the split of the human and chimpanzee, gene expression in the human brain has altered more dramatically than that in the chimpanzee. This hypothesis has attracted a lot of attention because the result supports the long-term notion that the genetic basis of human–chimpanzee differences is the alteration in gene expression rather than in coding sequence (King and Wilson 1975). Later, Gu and Gu (2003a) conducted a careful statistical analysis, and found some new interesting findings.

7.4.2 Enard *et al.*'s analysis

After defining a statistical measure for the expression distance, Enard *et al.* (2002) examined the expression of 12,600 genes (on an Affymetrix chip) in the human, chimpanzee, and orangutan, for both brain and liver. By comparing human, chimpanzee, and orangutan Affymetrix microarray expression samples, an overall expression distance for each lineage was obtained (Fig. 7.7). Enard *et al.* (2002) estimated that for the brain sample, the ratio of expression distance of human versus chimpanzee lineage is about 3.8. For the liver sample, this ratio is lower, about 1.7. Because the expression distance was interpreted as a measure for the change of gene expression, they concluded that it is the brain rather than the liver that has undergone a dramatic change in gene expression in the human lineage.

7.4.3 Gu and Gu's analysis

We (Gu and Gu 2003a) have conducted a deep analysis for the Affymetrix microarray data of Enard *et al.* (2002). Given a tissue sample (brain or liver), we adopted a statistical protocol to select genes that have significant expression differences between the human and chimpanzee under a given significance level (α). To this end, we first assigned a P-value (t-test) for each gene. For a given significance level (α), a gene was said to be expressed differently between the human and chimpanzee brains (or livers) if $P < \alpha$. For instance, in the case of brain samples, we found 1988, 1087, 670, and 131 genes at $\alpha = 0.05$, 0.02, 0.01, and 0.001, respectively. For each gene selected, we then used the orangutan as a reference to infer first, the phylogenetic location of the change of expression (i.e. did it occur in the human lineage or the chimpanzee lineage?), and second, the trend of expression change (i.e. induced or repressed). At a given significance level (α), we finally classified selected genes into four groups (Fig. 7.8)

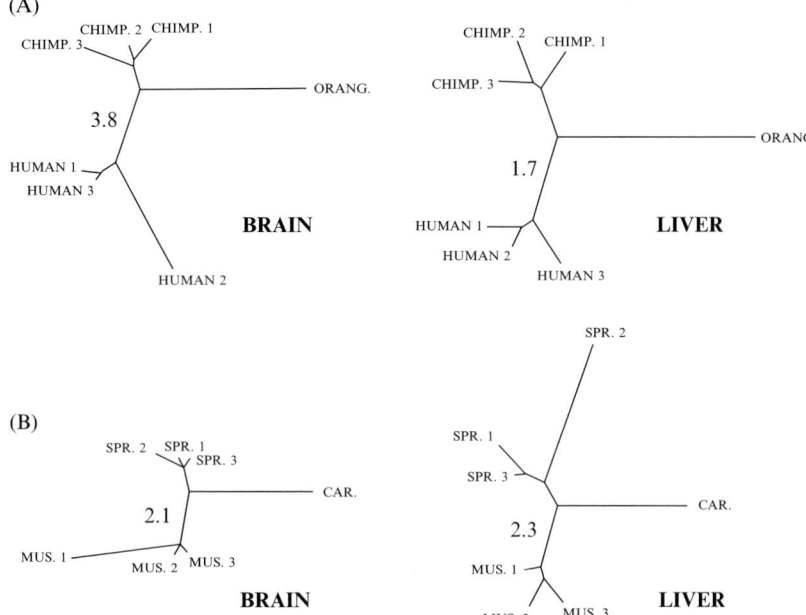

Fig. 7.7 Distance trees representing the relative extent of expression changes in brain and liver among (**A**) three primate and (**B**) three mouse species: MUS., *M. musculus*; SPR., *M. spretus*; and CAR, *M. caroli*. Numbers refer to the ratio between the changes common to humans and chimpanzees, and *M. musculus* and *M. spretus*, respectively. From Enard *et al.* (2002).

(1) Diversified expression pattern: where the gene expression level in the orangutan (O) is significantly different from the gene expression levels in both the chimpanzee and human (C and H, respectively).

(2) Chimpanzee-lineage (L_C) specific events: where O is significantly different from C but not from H, suggesting the expression change occurred in the chimpanzee lineage after the human chimpanzee split.

(3) Human-lineage (L_H) specific events: where O is significantly different from H but not C, suggesting the expression change occurred in the human lineage after the human chimpanzee split.

(4) Unclassified: where O is not significantly different from both C and H.

Figure 7.8 shows clearly that for the brain, changes in gene expression in the human lineage are statistically more frequent than in the chimpanzee lineage, as measured by the ratio of changes in the human to chimpanzee lineage, L_H/L_C. However, this is not the case for the liver. This result holds true regardless of which significance level (α) is used for the selection of differential expressed gene between humans and chimpanzees. Indeed, the L_H/L_C ratio for brain-expressed genes ranges from 2.76 to

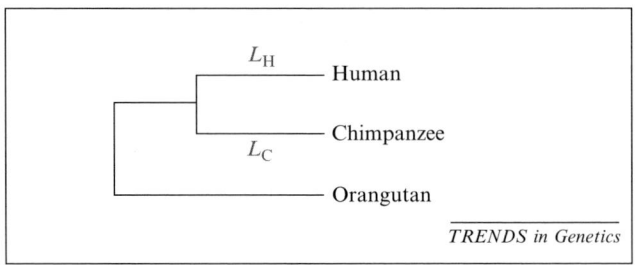

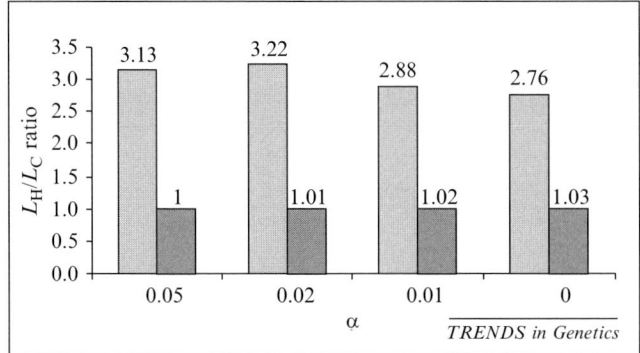

Fig. 7.8 (A) A schematic tree for the human–chimpanzee–orangutan relationship, showing the human-lineage-specific expression change (L_H) and the chimpanzee-lineage-specific expression change (L_C). (B) Ratio of human-lineage-specific expression changes to chimpanzee-lineage-specific expression changes (L_H/L_C) in both brain (green) and liver (purple) under different significance levels ($\alpha = 0.05, 0.02, 0.01$ and 0.001). In all cases, the L_H/L_C ratio of brain expressed genes is significantly greater than one, whereas L_H/L_C ratio of liver expressed genes is not. Modified from Gu and Gu (2003a).

3.22 for $\alpha = 0.05$ to 0.001; in each case the null hypothesis of $L_H/L_C = 1$ is rejected at $P < 0.001$. By contrast, the L_H/L_C ratio for liver-expressed genes is virtually equal to one in any case. Hence, our result provides statistical support for the notion of dramatic gene expression changes in the brain of the human lineage (Enard *et al.* 2002).

For gene expression changes that are chimpanzee-lineage-specific or human-lineage-specific, i.e. cases (2) and (3), we can infer the change in direction of the evolutionary event; that is, from low to high expression level (induction, denoted by I), or from high to low expression level (repression, R). In the human lineage, the induction/repression (I/R) ratio in brain ranges from 2.21 to 5.9; in each case it is greater than one with $P < 0.001$. By contrast, no evidence shows that the I/R ratio for liver-expressed genes in humans is significantly greater than one (Fig. 7.9); but it is not the case in the chimpanzee lineage.

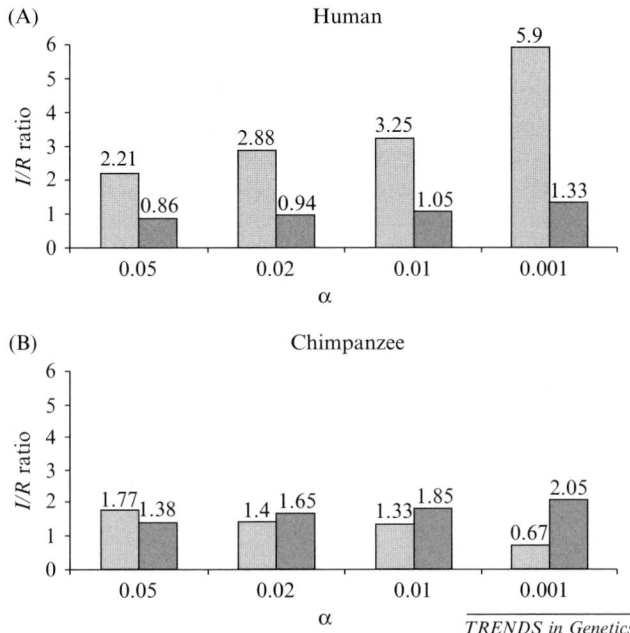

Fig. 7.9 Induction/repression (I/R) ratios for genes showing lineage-specific expression patterns. (A) In the human brain (green) and liver (purple). The I/R ratio of brain-expressed genes is statistically greater than one, whereas the I/R ratio of liver-expressed genes is not significant. (B) In the chimpanzee brain and liver. The I/R ratios for both brain- and liver-expressed genes are relatively similar, and sensitive to the significance level. Modified from Gu and Gu (2003a).

7.4.4   Concluding comments

Analyses by Enard *et al.* (2002) and Gu and Gu (2003) have provided strong evidence to show that after the split of humans and chimpanzees, the change of expression pattern in the human brain was more dramatic than that in the chimpanzee. Moreover, these changes in the human brain involved induction (increased gene expression) much more frequently than repression. This pattern is not observed in chimpanzee brain, nor in the liver of humans or chimpanzees. The enhanced expression of genes in the human brain since the split from chimpanzees could be important in the emergence of human beings, and certainly deserves further investigation.

8

Gene Pleiotropy and Evolution of Protein Sequence

The concept of gene pleiotropy may play a key role in developing an integrated view of protein evolution (Pal *et al.* 2006a). Although the capacity of a gene that may affect multiple phenotypic characters has been used to explain many biological phenomena (Fisher 1930; Wright 1968; Barton 1990; Waxman and Peck 1998; Wagner 1989; Zhang and Hill 2003; Welch and Waxman 2003; Otto 2004; MacLean *et al.* 2004; Dudley *et al.* 2005; Martin and Lenormand 2006; He and Zhang 2006), little has been known about the extent of pleiotropy at the genome level (Wagner *et al.* 2007). Based on the premise that links between gene pleiotropy and the dimensionality of organismal fitness, Gu (2007a) has provided a feasible statistical framework to estimate the degree of gene pleiotropy. In this chapter, we shall discuss this issue.

8.1 Model for protein sequence evolution

8.1.1 Fisher's model and molecular phenotypes

We refer to a class of phenotype-genotype models as Fisher-related models, including Fisher's original geometric models (1930) and various multivariate models (e.g. Lande 1980; Turelli 1985; Waxman and Peck 1998). Fisher's (1930) geometrical model was originally devised to promote a micro-mutational view of adaptation, in which adaptive walks toward a fitness peak consist mostly of small steps. There are two types of modeling. One is for a phenotypic trait affected by many genes, and the other is for a gene that affects multiple phenotypes.

In the abstract sense, multi-functions of a gene, or pleiotropy, can be represented by K distinct components in the fitness, called *molecular phenotypes*. These K-molecular phenotypes can be viewed as a K-dimensional space in Fisher's model. Random mutations of the gene generate a mutational distribution for K-molecular phenotypes. Theoretically, for molecular phenotypes denoted by $(y_1, \ldots, y_K)$, each y_i represents a nontrivial component of genetic variation in the organismal fitness, as a result of a specified (yet unknown) biological process (Fig. 8.1). At one extreme, molecular phenotypes could correspond to subcomponents of protein function, regardless of the biological processes. At another extreme, molecular phenotypes could be determined mainly by distinct physiological processes in various tissues. Since these underlying

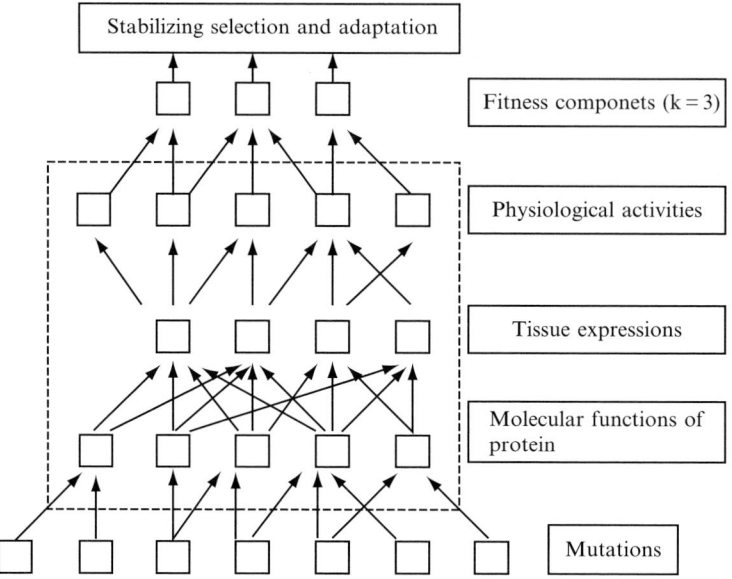

Fig. 8.1 Schematic presentation for the concept of molecular phenotypes that affect the fitness of an organism. Figure from Gu (2007a).

biological processes are usually intractable, the concept of molecular phenotypes may avoid the difficulty during the analysis.

In the following we introduce the modeling of gene pleiotropy; one may see Table 8.1 for some mathematical notations. It is common to assume a Gaussian-like fitness function for K-molecular phenotypes $\mathbf{y} = (y_1, \ldots, y_K)'$ of a gene, that is,

$$w(\mathbf{y}) = \exp\left[-\frac{(\mathbf{y} - \mu)'\boldsymbol{\Sigma}_w^{-1}(\mathbf{y} - \mu)}{2}\right] \tag{8.1}$$

where μ is the fitness optimum, $\boldsymbol{\Sigma}_w$ is a (positive definite) symmetric matrix characterizing the correlated stabilizing selections on K-molecular phenotypes. The i-th diagonal element $\sigma_{w,i}^2$ measures the strength of stabilizing selection on the i-th molecular phenotype, while the ij-th non-diagonal element $\sigma_{w,ij}$ measures the correlated stabilizing selection on y_i and y_j. Apparently, when $K = 1$, Eq. (8.1) can be reduced to a simple form of

$$w(y) = e^{-(y-\mu)^2/2\sigma_w^2}$$

Let $\mathbf{y_0}$ be the (current) population mean of molecular phenotypes of a gene. Then, the coefficient of selection for the molecular phenotype $\mathbf{y}$ can be generally defined as

$$\rho(\mathbf{y}|\mathbf{y_0}, \mu) = w(\mathbf{y})/w(\mathbf{y_0}) - 1 \tag{8.2}$$

In the following we discuss two important cases.

Table 8.1 Mathematical notations in the pleiotropy model of molecular evolution.

Notations	*Interpretations*
$\boldsymbol{y} = (y_1, \dots y_K)'$	K-molecular phenotypes of a gene, each presenting a non-trivial component of genetic variation in the organismal fitness
$\rho(\boldsymbol{y}\|\boldsymbol{y}_0, \mu)$	Coefficient of stabilizing selection of molecular phenotypes, given the population mean $\boldsymbol{y}_0$ and the optimum μ
$\rho(\boldsymbol{y})$	Coefficient of stabilizing selection when the randomly-shifted optimum μ is integrated out
$p(\boldsymbol{y})$	Multi-variate normal distribution for mutational effects on molecular phenotypes
$S(\boldsymbol{y})$	Selection intensity defined as $S(\boldsymbol{y}) = 4N_{\mathrm{e}}\rho(\boldsymbol{y})$; N_{e} is the effective population size
Σ_w	Diagonal elements of matrix Σ_w for strengths of stabilizing selections on molecular phenotypes; non-diagonal elements for correlated stabilizing selections between them.
Σ_m	Diagonal elements of matrix Σ_m for strengths of fitness-optimum shifting on molecular phenotypes; non-diagonal elements for correlated strengths between them.
Σ_μ	Diagonal elements of matrix Σ_μ for strengths of fitness-optimum shifting on molecular phenotypes; non-diagonal elements for correlated strengths between them.
$\boldsymbol{U} = \left[\Sigma_w^{-1} - \Sigma_w^{-1}\Sigma_\mu\Sigma_w^{-1}\right]^{-1}$	The matrix for describing the effect of stabilizing selection and micro-adaptation on molecular phenotypes.
$A = \Sigma_m^{-1}U$	The matrix whose dimension and eigenvalues point to the strength of stabilizing selection and micro-adaptation on evolutionary rate and functional importance.
K	Gene pleiotropy: the number of fitness components or molecular phenotypes
K_e	Effective gene pleiotropy: the number of fitness components or molecular phenotypes strongly affected by random mutations
S	Mean selection intensity of a gene
B_0	Mean selection intensity of a molecular phenotype: Baseline selection intensity

8.1.2   Stabilizing selection

Under the stabilizing model (Turelli 1985; Waxman and Peck 1998), the population mean of molecular phenotypes ($\mathbf{y_0}$) is always fixed at the optimum (μ) so that $w(\mathbf{y_0}) = 1$. Without loss of generality, we assume $\mathbf{y_0} = \mu = \mathbf{0}$. In the case of $K = 1$, one can obtain

$$\rho(y) = e^{-y^2/2\sigma_w^2} - 1 \approx -y^2/2\sigma_w^2$$

In general, from Eq. (8.2) one can show (assuming $\mathbf{y_0} = \mu = \mathbf{0}$)

$$\rho(\mathbf{y}) \approx -\mathbf{y}'\mathbf{\Sigma}_w^{-1}\mathbf{y}/2 \le 0,$$

reflecting the stabilizing (purifying) selection against deleterious mutations leading to a deviation from the optimum. One consequence of this model is that sequence evolution is dominated by the fixation of very slightly deleterious mutations (Ohta 1973; Kimura 1983). In the following the selection intensity $S(\mathbf{y})$, defined as

$$S(\mathbf{y}) = 4N_e\rho(\mathbf{y}) = -2N_e(\mathbf{y}'\mathbf{\Sigma}_w^{-1}\mathbf{y}),$$

is more frequently used, where N_e is the effective population size.

8.1.3   Micro-adaptation

Consider the case of stabilizing selection when the fitness optimum (μ) is no longer fixed during the evolution. Rather, μ can be shifted by either environmental changes or internal physiological perturbations (Hartl and Taubes 1996, 1998; Poon *et al.* 2000; Sella and Hirsh 2005; West-Eberhard 2005a, 2005b; Orr 2005). Each shift of μ results in a micro-adaptation toward new fitness optimum; see Fig. 8.2 for a schematic illustration. Hence, it is called the *S*tabilizing selection with *M*icro-adaptation (*SM*) model.

Similar to above, we assume that the population mean is at $\mathbf{y_0} = \mathbf{0}$. Since the direction and strength of the μ-shift are unobservable, we treat μ as a (K-dimensional) random variable that follows a multi-variate normal distribution $\phi(\mu)$ with mean $\mathbf{0}$ and covariance matrix $\mathbf{\Sigma}_\mu$. Gu (2007a) has shown that, under the *SM* model, the coefficient of selection of $\mathbf{y}$ is given by $\rho(\mathbf{y}) \approx -\mathbf{y}'\mathbf{U}^{-1}\mathbf{y}/2$, where the matrix $\mathbf{U} = [\mathbf{\Sigma}_w^{-1} - \mathbf{\Sigma}_\mathbf{w}^{-1}\mathbf{\Sigma}_\mu\mathbf{\Sigma}_\mathbf{w}^{-1}]^{-1}$ characterizes the correlated nature of molecular phenotypes in fitness after stabilizing selection and micro-adaptation. It follows that the selection intensity defined by $S(\mathbf{y}) = 4N_e\rho(\mathbf{y})$ is given by

$$S(\mathbf{y}) = -2N_e\mathbf{y}'\mathbf{U}^{-1}\mathbf{y} \tag{8.3}$$

Apparently, the stabilizing selection model is a special case when $\mathbf{\Sigma}_\mu = \mathbf{0}$ so that $\mathbf{U} = \mathbf{\Sigma}_w$. Moreover, unlike the stabilizing selection model that $S(\mathbf{y}) \le 0$ holds always, the selection intensity $S(\mathbf{y})$ under the *SM* model can be positive when the μ-shift becomes huge, creating a strong adaptive process.

We use $K = 1$ for illustration. Given the mean population mean $y_0 = 0$, from Eq.(8.2) one can show $\rho(y|\mu) = w(y)/w(0) - 1 = e^{\mu y/\sigma_w^2 - y^2/2\sigma_w^2} - 1$. By adopting

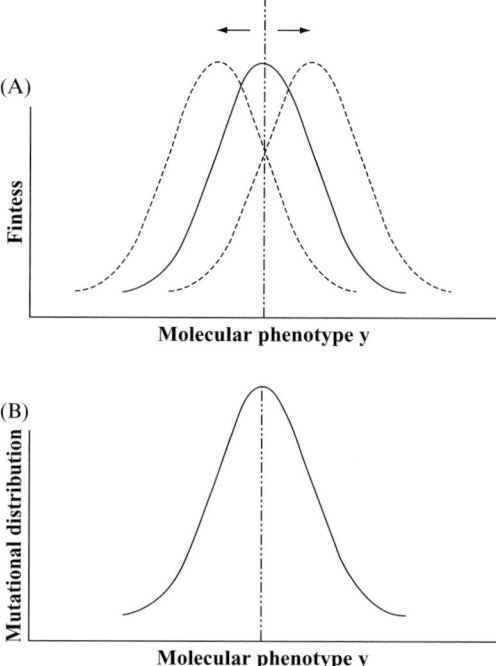

Fig. 8.2 (A) Stabilizing selection model with micro-adaptation in the case of a single molecular phenotype. (B) Distribution of mutational effects on the molecular phenotype. In both cases, we set $K = 1$ for simplicity. Figure from Gu (2007a).

Taylor expanding until the order of y^2, we have

$$\rho(y|\mu) \approx \left(\mu y / \sigma_w^2 - y^2/2\sigma_w^2\right) + \frac{1}{2}\left(\mu y/\sigma_w^2 - y^2/2\sigma_w^2\right)^2 \approx \mu y/\sigma_w^2 - \frac{y^2}{2\sigma_w^2}\left(1 - \frac{\mu^2}{\sigma_w^2}\right)$$

Hence, if the fitness optimum μ varies according to a normal distribution with mean 0 and variance σ_μ^2, we then obtain

$$\rho(y) = -\frac{y^2}{2\sigma_w^2}\left(1 - \frac{\sigma_\mu^2}{\sigma_w^2}\right)$$

8.1.4   Distribution of mutational effects

Next we consider the effects of mutations on molecular phenotypes. A random mutation in the coding region of a gene may affect molecular phenotypes in a correlated fashion. Such mutational effects can be described by a multi-variate normal distribution

$$p(\mathbf{y}) = N(\mathbf{y}; \mathbf{y_m}, \boldsymbol{\Sigma}_m) \tag{8.4}$$

where $\mathbf{y_m}$ is the mean mutational effect on molecular phenotypes, and the covariance matrix $\mathbf{\Sigma}_m$ measures the correlated mutational effects. The coefficient of selection $\rho(\mathbf{y}|\mathbf{y_0}, \mu)$ and the distribution of mutational effects $p(\mathbf{y})$ provide a connection between the organismal fitness and protein sequence evolution through molecular phenotypes $\mathbf{y}$. For mathematical convenience, we assume that the mutational distribution, $p(\mathbf{y})$ in Eq. (8.4), is centered at the population mean, namely, $\mathbf{y}_m = \mathbf{y}_0 = \mathbf{0}$.

8.1.5 S-distribution

The distribution of mutational effects on molecular phenotypes, $p(\mathbf{y})$, links between the selection and mutation. According to the theory of probability, the distribution of selection intensity $f(S)$, short for S-distribution, can be uniquely determined. However, the analytical form of $f(S)$ is available only for some special cases such as the W-model, which assumes that stabilizing selection and mutations have very similar correlation structures among molecular phenotypes, that is, $\mathbf{\Sigma}_w = \sigma_w^2 \mathbf{W}$ and $\mathbf{\Sigma}_m = \sigma_m^2 \mathbf{W}$, where matrix $\mathbf{W}$ is positively-definite and symmetric. Then, after choosing appropriate transformations, the transformed molecular phenotypes are independent and identical on stabilizing selections and mutations. In this case, it has been shown (Gu 2007a) that $f(S)$ follows a negative gamma distribution

$$f(S) = \frac{(2B_0)^{-K/2}}{\Gamma(K/2)}(-S)^{K/2-1}e^{S/2B_0} \tag{8.5}$$

for $S \leq 0$, where $B_0 = 2N_e\sigma_m^2/\sigma_w^2$ is the mean selection intensity for single molecular phenotype (the baseline selection intensity). In fact, the gamma-like distribution for the selection intensity has been used *ad hoc* (e.g. Piganeau and Eyre-Walker 2003; Nielsen and Yang 2003). The W-model provides a biological interpretation for the theoretical study (Kimura 1979) and those studies of mutational effects (Keightley 1994; Imhof and Schlotterer 2001; Shaw *et al.* 2002; Eyre-Walker 2006; Loewe *et al.* 2006).

8.2 Selection intensity and model classification

Though the S-distribution is not tractable in general, the mean selection intensity $\bar{S}$ can be derived concisely. Moreover, $\bar{S}$ can be used to classify the SM model, which is crucial for developing statistical methods to estimate K (gene pleiotropy).

8.2.1 Mean of selection intensity

From Eqs. (8.3) and (8.4), the mean of S can be calculated by

$$\bar{S} = \int S(\mathbf{y})p(\mathbf{y})d\mathbf{y} = -2N_e \int \mathbf{y}'\mathbf{U}^{-1}\mathbf{y}p(\mathbf{y})d\mathbf{y}$$

where $p(\mathbf{y})$ is the distribution of mutational effects on molecular phenotypes. Denote matrix $\mathbf{A} = \mathbf{U}^{-1}\mathbf{\Sigma}_m$, which characterizes the net effects of correlated mutations on fitness under the SM model. It has been shown (Gu 2007a) that the joint effects of all selective, micro-adaptive, and mutational covariances can be reduced to K eigenvalues of matrix $\mathbf{A}$, $\alpha_1, \ldots, \alpha_K$, leading to

$$\bar{S} = -2N_e \sum_{i=1}^{K} \alpha_i = -\sum_{i=1}^{K} B_i \tag{8.6}$$

where the i-th baseline selection intensity $B_i = 2N_e\alpha_i$ corresponds to an independent molecular phenotypic direction, on which mutation, micro-adaptation, and stabilizing selection act with an average effect $-B_i$ on the mean selection intensity ($\bar{S}$) of the gene.

Next we present the results of $\bar{S}$ that are biologically more interpretable. Note that three matrices, $\boldsymbol{\Sigma}_w$, $\boldsymbol{\Sigma}_\mu$, and $\boldsymbol{\Sigma}_m$, determine the distribution of selection intensity S under the SM model. Due to the arbitrary nature of the original K-molecular phenotypes, without loss of generality one may choose a convenient coordinate system. Here we adopt a canonical form of K molecular phenotypes in order to define independent stabilizing selections. This leads to a diagonal matrix $\boldsymbol{\Sigma}_w$; each diagonal element $\sigma_{w,i}^2$ measures the independent stabilizing selection on the i-th (canonical) molecular phenotype. Let $\sigma_{m,i}^2$ be the i-th diagonal element of matrix $\boldsymbol{\Sigma}_w$, or the mutational variance for the i-th (canonical) molecular phenotype. Gu (2007a) has shown that the canonical presentation for the mean of selection intensity is given by

$$\bar{S} = -2N_e \sum_{i=1}^{K} \frac{\sigma_{m,i}^2}{\sigma_{w,i}^2} \left(1 - \gamma_i\right) \tag{8.7}$$

where the parameter γ_i measures the net effect of micro-adaptation on the i-th (canonical) molecular phenotype. In other words, the i-th baseline selection intensity (B_i) can be written as

$$B_i = 2N_e\alpha_i = \frac{\sigma_{m,i}^2}{\sigma_{w,i}^2} \left(1 - \gamma_i\right) \tag{8.8}$$

Hence, the effect on the i-th (canonical) molecular phenotype depends on $\sigma_{m,i}^2$ that measures how much phenotypic variation could be caused by random mutations, and $\sigma_{w,i}^2$ that measures the severity of deleterious effects on the phenotypic variation. Together, the ratio $\sigma_{m,i}^2/\sigma_{w,i}^2$ determines the strength of stabilizing selection. Further, micro-adaptation driven by the random shift of the optimum of molecular phenotypes can reduce the average selection intensity for this molecular phenotype.

Derivation of $\bar{S}$ in special cases For illustration, we derive the canonical form of $\bar{S}$ in two special cases. The first one is the stabilizing selection without micro-adaptation. Since stabilizing selections on the canonical molecular phenotypes are independent, the fitness function can be simplified as

$$w(\mathbf{y}) = \exp\left[-\sum_{i=1}^{K} y_i^2/2\sigma_{w,i}^2\right] \approx 1 - \sum_{i=1}^{K} y_i^2/2\sigma_{w,i}^2$$

and the coefficient of stabilizing selection is $\rho(\mathbf{y}) = w(\mathbf{y}) - 1$. It follows that the selection intensity is given by

$$S(\mathbf{y}) = 4N_e\rho(\mathbf{y}) = -2N_e \sum_{i=1}^{K} y_i^2/\sigma_{w,i}^2$$

Since the distribution of mutational effect $p(\mathbf{y})$ has zero expectation for each (canonical) molecular phenotype, we have the second expectation $E[y_i^2] = \sigma_{m,i}^2$, resulting in

$$\bar{S} = -2N_e \sum_{i=1}^{K} E[y_i^2]/\sigma_{w,i}^2 = -2N_e \sum_{i=1}^{K} \frac{\sigma_{m,i}^2}{\sigma_{w,i}^2}$$

The second case is when the micro-adaptation is independent among the canonical molecular phenotypes. In this case, the parameter γ_i in Eq. (8.8) is given by

$$\gamma_i = \sigma_{\mu,i}^2/\sigma_{w,i}^2$$

where $\sigma_{\mu,i}^2$ is the variance of the optimum shift for the i-th molecular phenotype.

Mean effects of mutations and optimum-shifts In the above analysis, we generally assume that the population mean $\mathbf{y}_0$, the mean of optimum shifts μ, and the mean of mutational effects $\mathbf{m}_0$ of molecular phenotypes are the same. Without loss of generality, one may assume $\mathbf{y}_0 = \mu = \mathbf{m}_0 = \mathbf{0}$. Here we use a special case of $K = 1$ to show the effect if it is not the case. If the fitness optimum μ varies according to a normal distribution with mean μ_0 and variance σ_μ^2, one can show

$$\rho(y) = \mu_0 y/\sigma_w^2 - \frac{y^2}{2\sigma_w^2}\left(1 - \frac{\sigma_\mu^2 + \mu_0^2}{\sigma_w^2}\right)$$

Further, if the mutational effects follow a normal distribution with mean m_0 and variance σ_m^2, by some simple algebras we obtain

$$\bar{\rho} = \frac{\mu_0 m_0}{\sigma_w^2} - \frac{1}{2}\left(\frac{\sigma_m^2 + m_0^2}{\sigma_w^2}\right)\left(1 - \frac{\sigma_\mu^2 + \mu_0^2}{\sigma_w^2}\right)$$

Let $q = 1 - (\sigma_\mu^2 + \mu_0^2)/\sigma_w^2$. We therefore have the following results: (i) For fixed μ_0, $\bar{\rho}$ reaches the maximum at $m_0 = \mu_0/q$ when $q > 0$, or minimum when $q < 0$. (ii) For a fixed μ_0, $\bar{\rho}$ reaches the minimum at $\mu_0 = -m_0/q^*$, where $q^* = (\sigma_m^2 + m_0^2)/\sigma_w^2$. And ($iii$) $\bar{\rho} < 0$ when the condition $2\mu_0 m_0 < \sigma_w^2 q \times q^*$ holds. The implication of these results on molecular evolution remains for further investigation.

8.2.2 Model classification

As the stabilizing selection acts against nucleotide substitutions (negative selection), micro-adaptation may provide an opposite force (positive selection) to increase the evolutionary rate. The mean selection intensity given by Eqs. (8.6) and (8.7) can be used to classify the pattern of sequence conservation under the SM model.

(1) Strong stabilizing selection with weak micro-adaptation (SM_w): In this case, the magnitude of μ-shift of molecular phenotypes during the evolution is small, relative to the strength of stabilizing selection. It indicates a dominant purifying selection in the protein sequence evolution. The SM_w model can be defined as having matrix **A** positive-definite such that each eigenvalue $\alpha_i > 0$ or $B_i > 0$, resulting in $\bar{S} < 0$. In the canonical form of Eq. (8.7), it also means $\gamma_i < 1$ for any $i = 1, \ldots, K$. The pure stabilizing selection is a special case of $\gamma_i = 0$ (no micro-adaptation). Moreover, $\bar{S} < 0$ implies that the ratio of nonsynonymous to synonymous substitutions (d_N/d_S) is less than 1. Therefore, the SM_w model may describe the general pattern in protein sequence evolution.

(2) Episodic micro-adaptation under strong stabilizing selection (SM_E): Some genes may have experienced episodic adaptive processes in a few molecular phenotypes, resulting in negative eigenvalues of matrix **A**, i.e. $\alpha_i < 0$ ($B_i < 0$) for some i's, but the mean selection intensity remains negative ($\bar{S} = -\sum_{i=1}^{K} B_i < 0$). That is, matrix **A** is no longer positive-definite but a positive trace ($\sum_{i=1}^{K} \alpha_i > 0$) remains. In the canonical form, it is also means that some of γ_i are more than 1, though $\bar{S} < 0$. Apparently, the d_N/d_S ratio under the SM_E model may be increased by episodic micro-adaptations, but $d_N/d_S < 1$ holds. A neutral-like behavior of protein sequence evolution may occur when $\sum_{i=1}^{K} \alpha_i \approx 0$, result in a zero net effect on fitness by canceled episodic micro-adaptation and stabilizing selection.

(3) Strong micro-adaptation under stabilizing selection (SM_s): In a few genes, positive selection and adaptation may dominate the evolution in many molecular phenotypes. In these cases, $\sum_{i=1}^{K} \alpha_i < 0$ results in $\bar{S} > 0$ and the ratio $d_N/d_S > 1$. An extreme case occurs if $\alpha_i < 0$ or $\gamma_i > 1$ (the canonical form) for all $i = 1, \ldots, K$. In this case, matrix **A** is negative-definite, indicating the overwhelming adaptation forces for virtually all substitutions.

8.3   Evolutionary rate of protein sequences

A fundamental problem is how the degree of gene pleiotropy may affect the evolutionary rate of protein sequences, for instance, whether a highly pleiotropic gene evolves slowly. In this section we derive the analytical result for the relationship between gene pleiotropy K and the evolutionary rate (λ).

8.3.1   General formula

We first briefly introduce the standard theory of molecular evolution (Kimura 1983), which claims that the rate (λ) of protein evolution is given by

$$\lambda = v \frac{S}{1 - e^{-S}}$$

where v is the mutation rate and $S = 4N_e s$ is the selection intensity; N_e is the effective population size, and s is the coefficient of selection. Then one can easily show that $\lambda = v$ when $S = 0$ (neutral selection), $\lambda < v$ when $S < 0$ (purifying selection) and $\lambda > v$ when $S > 0$ (adaptive selection). In other words, the process of natural selection on phenotypic evolution, which further affects on the sequence evolution, was characterized by a single parameter S.

The pleiotropy model of molecular evolution (Gu 2007a) suggests that the selection intensity of a gene is defined by the means of molecular phenotype. Therefore, the evolutionary rate of a mutant affecting the molecular phenotypes ($\mathbf{y}$) can be written as

$$\lambda(\mathbf{y}) = v\frac{S(\mathbf{y})}{1 - e^{-S(\mathbf{y})}}$$

Thus, given $p(\mathbf{y})$, the distribution of mutational effects that generates a variation of $\mathbf{y}$, the (mean) evolutionary rate of a gene is therefore given by

$$\bar{\lambda} = v \int \frac{S(\mathbf{y})}{1 - e^{-S(\mathbf{y})}} p(\mathbf{y})d\mathbf{y} \tag{8.9}$$

Noticeably, molecular phenotypes $\mathbf{y}$, a theoretical representation of protein functions, are actually hidden variables in Eq. (8.9), providing a footnote for the definition of protein function in molecular evolution: They are molecular-specific phenotypic variables generated by random mutations, responding to the stabilizing selections and adaptations (Fig. 8.1).

To compute the mean evolutionary rate $\bar{\lambda}$ in Eq. (8.9), Gu (2007a) used the approximation

$$S/(1 - e^{-S}) \approx e^{-|S|}(1 + c|S|)$$

where $c \approx 0.5772$ is the Euler's constant (Fig. 8.3(A)). One may confirm that this approximation is numerically satisfactory when $S \leq 0$. Thus, under the SM_w model (stabilizing selection with weak micro-adaptation) that always assures $S(\mathbf{y}) = -2N_e\mathbf{y}'\mathbf{U}^{-1}\mathbf{y} < 0$, the mean evolutionary rate $\bar{\lambda}$ can be approximately given by

$$\bar{\lambda} = v \prod_{i=1}^{K}[1 + 2B_i]^{-1/2}\left(1 + c\sum_{i=1}^{K}\frac{B_i}{1 + 2B_i}\right) \tag{8.10}$$

where $B_i = 2N_e\alpha_i$ (Gu 2007a). That is, the evolutionary rate of a gene is determined by the mutation rate (v), gene pleiotropy (K) measured by the number of molecular phenotypes, and a number of baseline selection intensities (B_i) of molecular phenotypes. Note that under the SM_w condition, all $B_i \geq 0$.

Identical baseline selection intensities In the simplest case when all baseline selection intensities are identical, i.e. all $B_i = B_0$ for $i = 1, \ldots, K$, one can easily show

$$\bar{\lambda} = v\left(1 + 2B_0\right)^{-K/2}\left[1 + c\left(\frac{B_0 K}{1 + 2B_0}\right)\right]$$

and the mean selection intensity is given by $\bar{S} = -K \times B_0$. When $B_0 = 0$, we have $\bar{\lambda} = v$, the case of neutral evolution.

(A)

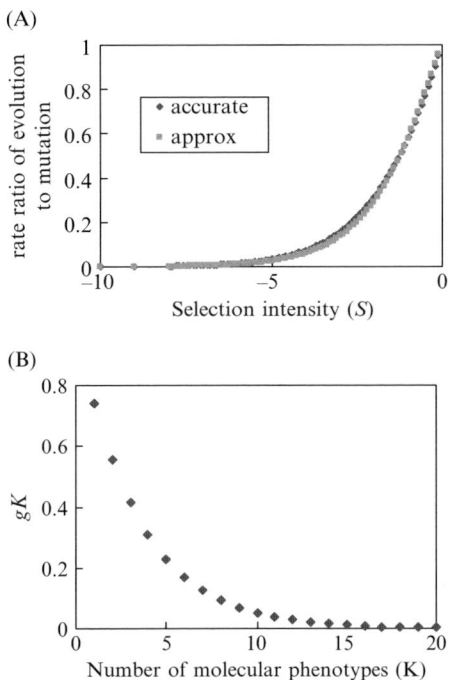

(B)

Fig. 8.3 (A) The λ/v-S (the rate ratio of evolution to mutation **vs.** the selection intensity) relationship, for the accurate formula and the approximation. (B) The $g_K - K$ plotting for estimating the effective number of molecular phenotypes. Figure from Gu (2007a).

8.3.2 K-mode and B-mode for the rate of protein evolution

There have been substantial amounts of evidence to show that the variation of mutation rates among genes could affect the evolutionary rate of protein sequences, such as C_pG mutational hotspots. Since we are focused on the functional constraints of proteins, we will not discuss this issue; readers who are interested may see Li (1997).

It becomes apparent that the rate of protein evolution is determined by two parameters related to functional importance (constraint) of a gene: Gene pleiotropy K relates to the importance in the biological network, and the baseline selection intensity B_0, or the mean of B_is in general, relates to the local biochemical property of the protein sequence, e.g. the protein stability (Dean *et al.* 2002; Koehl and Levitt 2002; DePristo *et al.* 2005; Parisi and Echave 2005), the designability or mutability (Bloom *et al.* 2005; Guo *et al.* 2004), as well as the translational efficiency (Sharp and Li 1987; Akashi and Gojobori 2002). Consequently, it reveals two fundamental modes to explain why the rate of protein evolution varies considerably among genes:

(1) The K-mode claims that gene pleiotropy (K) differs considerably among genes whereas the baseline selection intensity (B_0) remains relatively invariant. Thus,

functional importance of a gene is mainly due to the role in biological systems and networks.

(2) The B-mode claims that B_0 differs considerably among genes whereas K remains relatively constant. Hence, the biochemical-structural property of the protein sequence is the major determinant for the rate of protein sequence evolution.

In short, the theory of K and B-modes provides a novel approach to scrutinizing the inverse relationship between the evolutionary rate and the functional importance of a gene.

8.3.3 Effect of B_i-variation on the evolutionary rate

In Eq. (8.10), the baseline selection intensities, $B_1, \ldots, B_K$, vary among molecular phenotypes, but are usually unknown at the empirical level. To investigate the effect of B_i-variation on the evolutionary rate, a practical approach is to assume that each B_i is an independent realization of a random variable B that follows a common distribution $\psi(B)$ with the mean B_0 and variance $V(B)$.

Rewrite the general formula of $\bar{\lambda}$ in Eq. (8.10) as follows

$$\bar{\lambda}/v = (1 + cK/2) \prod_{i=1}^{K} [1 + 2B_i]^{-1/2} - cK/2 \sum_{i=1}^{K} \left(\prod_{j=1, j \neq i}^{K} [1 + 2B_j]^{-1/2} \right) (1 + 2B_i)^{-3/2}$$

Under the $i.i.d$ assumption of B_is, the mean rate of protein evolution is then given by

$$\bar{\lambda} = v E_0^K \left[1 + \frac{c}{2} K \left(1 - \frac{E_1}{E_0} \right) \right] \tag{8.11}$$

where two parameters are defined as

$$E_0 = \int_0^\infty (1 + 2B)^{-1/2} \psi(B) dB$$

and

$$E_1 = \int_0^\infty (1 + 2B)^{-3/2} \psi(B) dB,$$

respectively. Therefore, one may define the effective baseline selection intensity $(\tilde{B}_0)$ such that $E_0 = (1 + 2\tilde{B}_0)^{-1/2}$ holds. That is,

$$(1 + 2\tilde{B}_0)^{-1/2} = \int_0^\infty (1 + 2B)^{-1/2} \psi(B) dB \approx (1 + 2B_0)^{-1/2} \left(1 + \frac{3}{2}\eta^2 \right)$$

where the parameter η^2 is given by

$$\eta^2 = \frac{V(1 + 2B)}{(1 + 2B_0)^2} = \frac{4V(B)}{(1 + 2B_0)^2}$$

which reflects the variation of baseline selection intensities among molecular phenotypes of a gene. Apparently, $\tilde{B}_0 \geq B_0$ and $\tilde{B}_0 \to B_0$ when $\eta \to 0$.

Next we define another effective baseline selection intensity B^* such that $E_1 = (1 + 2\tilde{B}_0)^{-3/2}$ holds. Similarly, we have

$$(1 + 2B_0^*)^{-3/2} = \int_0^\infty (1 + 2B)^{-3/2} \psi(B) dB \approx (1 + 2\bar{B})^{-3/2} \left(1 + \frac{15}{2}\eta^2\right)$$

Putting all these arguments and Eq. (8.11) together, we then obtain

$$\bar{\lambda} = v \left(1 + 2\tilde{B}_0\right)^{-K/2} \left[1 + \frac{c}{2}K\left(1 - \frac{\delta_B}{1 + 2\tilde{B}_0}\right)\right] \tag{8.12}$$

where δ_B is given by

$$\delta_B = \left(\frac{1 + 2\tilde{B}_0}{1 + 2B_0^*}\right)^{3/2} \approx \frac{1 + 7.5\eta^2}{(1 + 1.5\eta^2)^3} \tag{8.13}$$

Therefore, we conclude that $0 \leq \delta_B \leq 1$ for the range from the case of identical B_is to the extreme heterogeneity among B_is. Numerical analysis has demonstrated that the effect of B-variation among molecular phenotypes on the rate of protein evolution is usually marginal.

8.4 Estimation of gene pleiotropy and selection intensity

The pleiotropy model described above involves a number of unknown parameters that are difficult to be estimated from the sequence data. Instead, Gu (2007a) developed an effective approach, focusing on two important parameters: (*i*) K, the number of molecular phenotypes that are critical for understanding the gene pleiotropy and multi-functionality, and (*ii*) $\bar{S} = -\sum_{i=1}^K B_i$, the measure for overall sequence conservation, or more specifically, the baseline selection intensity $B_0 = \sum_{i=1}^K B_i/K$.

8.4.1 The second-moment of evolutionary rate

In addition to the mean evolutionary rate in Eq. (8.10), we consider the second moment of the evolutionary rate, $\bar{\lambda}^2$. From the general formula

$$\bar{\lambda}^2 = v^2 \int \left[\frac{S(\mathbf{y})}{1 - e^{-S(\mathbf{y})}}\right]^2 p(\mathbf{y}) d\mathbf{y}$$

With a close approximation, Gu (2007a) has derived the second moment of λ as follows

$$\bar{\lambda}^2 = v^2 \prod_{i=1}^K [1 + 4B_i]^{-1/2} \left[\left(1 + c\sum_{i=1}^K \frac{B_i}{1 + 4B_i}\right)^2 + c^2 \sum_{i=1}^K \frac{2B_i^2}{(1 + 4B_i)^2}\right] \tag{8.14}$$

where $c \approx 0.5772$.

8.4.2 Effective gene pleiotropy (K_e)

In general, the mean (λ) and the second moment ($\bar{\lambda}^2$) of the evolutionary rate depends on a number of K of distinct B_is, which are difficult to estimate especially when K

is unknown. To deal with this problem, we first, by both numerical and simulation analysis, notice that Eqs. (8.10) and (8.14) can be well approximated as follows

$$\frac{\bar{\lambda}}{v} \approx \prod_{i=1}^{K} [1 + 2B_i]^{-1/2} \left(1 + cK/2\right)$$

$$\frac{\bar{\lambda^2}}{v^2} \approx \prod_{i=1}^{K} [1 + 4B_i]^{-1/2} \left(1 + cK/2 + c^2K/8 + c^2K^2/16\right) \qquad (8.15)$$

respectively, with a broad range of B_i values. Gu (2007a) has realized that the ratio of second-moment to mean of the evolutionary rate, normalized by the mutation rate, is crucial for estimating the gene pleiotropy, i.e. $g_K = [\bar{\lambda^2}/v^2]/[\bar{\lambda}/v]$, resulting in

$$g_K = 2^{-K/2} \left[\prod_{i=1}^{K} \left(1 + \frac{1}{1 + 4B_i}\right)^{1/2} \right] \left[1 + \left(\frac{c}{4}\right)^2 \frac{K(K+2)}{1 + cK/2}\right] \qquad (8.16)$$

We define the effective gene pleiotropy (K_e) as the effective number of molecular phenotypes that have experienced strong stabilizing selection, i.e. with a large baseline selection intensity. Hence, K_e is less than the true number of molecular phenotypes. Technically, one may claim that $B_i \geq a > 0$ holds for all these effective molecular phenotypes. If the low-bound cutoff a is large enough to satisfy $1 + 1/(1 + 4a) \approx 1$, we approximately obtain

$$g_{K_e} = 2^{-K_e/2} \left[1 + \left(\frac{c}{4}\right)^2 \frac{K_e(K_e+2)}{1 + cK_e/2}\right] \qquad (8.17)$$

which is only K_e-dependent ($c \approx 0.5772$). As shown in Fig. (8.3B), g_{K_e} decreases when K_e increases; $g_{K_e} = 1$ when $K_e = 0$, and $g_{K_e} \to 0$ when $K_e \to \infty$. Eq. (8.17) indicates that the effective gene pleiotropy (K_e) can be estimated if the g_K-measure can be estimated from the protein sequence. To this end, we use a widely-used measure for $\bar{\lambda}/v$, that is, the ratio of nonsynonymous to synonymous substitutions (d_N/d_S). The second moment of rate $\bar{\lambda^2}/v^2$ is related to the H-measure (Gu *et al.* 1995) for the rate variation among sites, defined by

$$H = 1 - (\bar{\lambda})^2/\bar{\lambda^2}$$

Ranging from 0 to 1, a higher value of H means a greater rate variation among sites, and *vice versa*. Therefore, by equating the first and second moments of the evolutionary rate to

$$\bar{\lambda}/v = d_N/d_S$$
$$\bar{\lambda^2}/v^2 = (d_N/d_S)^2/(1 - H)$$

respectively, we show that the estimate of g_K-measure, denoted by $\hat{g}$, is given by

$$\hat{g} = \frac{d_N}{d_S}/(1 - H) \qquad (8.18)$$

8.4.3   Estimation pipeline

Gu (2007a) has developed a simple computational procedure to estimate the effective number of molecular phenotypes (K_e), including the following steps:

(1) Infer the phylogenetic tree from a multiple alignment of homologous protein sequences.
(2) Estimate the nonsynonymous to synonymous ratio (d_N/d_S) from the closely-related coding sequences.
(3) Estimate the H-measure for rate variation among sites: Use the method of Gu and Zhang (1997) to infer the (bias-corrected) number of changes at each site, under the inferred phylogeny. Let $\bar{x}$ and $V(x)$ be the mean and variance of number of changes over sites, respectively. Assuming a Poisson process at each site, we obtain the mean the evolutionary rate $\bar{\lambda} = \bar{x}/T$, where T is the total evolutionary time along the tree. Similarly, the variance of evolutionary rate among sites is given by $V(\lambda) = [V(x) - \bar{x}]/T^2$. Then, H can be estimated by

$$\hat{H} = \frac{V(\lambda)}{V(\lambda) + (\bar{\lambda})^2} = \frac{V(x) - \bar{x}}{V(x) + \bar{x}(\bar{x} - 1)} \tag{8.19}$$

(4) Estimate K_e based on Eqs (8.17) and (8.18).
(5) The sampling variance of K_e can be approximately calculated by the delta-method. From Eq. (8.17) one may formally write $K_e = f(g)$. The delta method results in

$$Var(K_e) \approx [f(g)']^2 Var(\hat{g}) = \frac{V(\hat{g})/\hat{g}^2}{[(\ln 2/2)(1 - a_1)]^2}$$

where $a_1 = (2/\ln 2)\phi'/(1 + \phi)$; $\phi = (c/4)^2 K_e(K_e + 2)/[1 + cK_e/2]$, and $\phi' = [2(K_e + 1) + cK_e^2/2]/(1 + cK_e/2)^2$. Further, the variance $Var(\hat{g})$ can be written as

$$V(\hat{g}) \approx \hat{g}^2 \left[\frac{V(d_N/d_S)}{(d_N/d_S)^2} + \frac{V(H)}{(1 - H)^2} \right]$$

Together we have

$$Var(K_e) \approx \left[\frac{V(d_N/d_S)}{(d_N/d_S)^2} + \frac{V(H)}{(1 - H)^2} \right] / [(\ln 2/2)(1 - a_1)]^2 \tag{8.20}$$

While the variance of d_N/d_S can be estimated conventionally, we implement a bootstrapping approach to calculate the variance of H, $V(H)$.

8.4.4   Effective selection intensity

The mean selection intensity $\bar{S}$ can be written as $\bar{S} = -K \times B_0$, where B_0 is the (mean) baseline selection intensity. Replacing $\bar{\lambda}/v$ by d_N/d_S in Eq. (8.15), and K by

K_e, we have

$$\frac{d_N}{d_S} = \prod_{i=1}^{K_e}(1 + 2B_i)^{-1/2}(1 + cK_e/2)$$

which indicates that the product of $\prod_{i=1}^{K_e}(1 + 2B_i)$ is estimable. Hence, we define the effective baseline selection ($\tilde{B}_0$) such that $(1 + \tilde{B}_0)^{K_e} = \prod_{i=1}^{K_e}(1 + 2B_i)$, which can be estimated as follows

$$\tilde{B}_0 = \frac{1}{2}\left\{\left[\prod_{i=1}^{K_e}(1 + 2B_i)\right]^{1/K_e} - 1\right\} = \frac{1}{2}\left\{\left[\frac{1 + cK_e/2}{d_N/d_S}\right]^{2/K_e} - 1\right\} \qquad (8.21)$$

By the delta method, the sampling variance of $\tilde{B}_0$ is approximately given by

$$Var(\tilde{B}_0) = (1 + 2\tilde{B}_0)^2\left[aVar(d_N/d_S) + bVar(K_e)\right]$$

where $a = (K_e d_N/d_S)^2$ and $b = [(\ln d_N/d_S)^2 + (1 - h)^2]/K_e^4$; $h = 1/(1 + cK_e/2) + ln(1 + cK_e/2)$. It follows that the effective selection intensity can be estimated as $\tilde{S} = -K_e \times \tilde{B}_0$, and the approximate sampling variance is given by

$$Var(\tilde{S}) \approx K_e^2 Var(\tilde{B}_0) + \tilde{B}_0^2 Var(K_e)$$

8.4.5   Bias-corrected estimation of effectively gene pleiotropy

Since K_e provides a conserved estimate for the degree of gene pleiotropy, it is desirable to correct the underestimation bias without introducing additional assumptions. To this end, we consider the following problem: When the function g is given, what is the difference between K_e and the true K. To be concise, let $\phi(K) = (c^2/4)K(K + 2)/(1 + cK/2)$ and so $\phi(K_e)$. By equating Eq. (8.16) with Eq. (8.17), we have

$$2^{-K/2}\left[\prod_{i=1}^{K}\left(1 + \frac{1}{1 + 4B_i}\right)^{1/2}\right][1 + \phi(K)] = 2^{-K_e/2}[1 + \phi(K_e)]$$

After some simple algebras, it turns out the following relationship between K_e and the true pleiotropy K

$$K_e = K - \sum_{i=1}^{K}\log_2(1 + \frac{1}{1 + 4B_i}) + 2\log_2\frac{1 + \phi(K)}{1 + \phi(K_e)}$$

Therefore, we propose a bias-corrected estimate of effective gene pleiotropy ($\tilde{K}$) by replacing B_is with the effective baseline selection intensity $\tilde{B}_0$, leading to

$$K_e = K(1 - \eta_0) + 2\log_2\frac{1 + \phi(K)}{1 + \phi(K_e)} \qquad (8.22)$$

where η_0 is given by

$$\eta_0 = \log_2(1 + \frac{1}{1 + 4\tilde{B}_0}) \tag{8.23}$$

Numerically, one can design a straightforward iteration to obtain $\tilde{K}$. We have conducted extensive computer simulations (Huang and Gu, unpublished) and found that the following approximation may be sufficient. After neglecting the third term of the right hand of Eq. (8.22), we obtain a simple $K - K_e$ relationship $K = K_e/(1 - \eta_0)$. We define a second correction term

$$\epsilon_0 = \frac{2}{K_e} \log_2 \frac{1 + \phi[K_e/(1 - \eta_0)]}{1 + \phi(K_e)}$$

Then, replacing the third term of the right hand of Eq. (8.22) by $K_e\epsilon_0$, we obtain $K_e(1 - \epsilon_0) = \tilde{K}(1 - \eta_0)$, leading to a simple bias-corrected estimate of effective gene pleiotropy

$$\tilde{K} = \left(\frac{1 - \epsilon_0}{1 - \eta_0}\right) K_e \tag{8.24}$$

8.5 Preliminary analysis of gene pleiotropy

Based on over 300 vertebrate genes, we (Su *et al.* 2010) conducted a preliminary analysis to study the pattern of gene pleiotropy. The vertebrate gene sets includes eight genomes (human, mouse, dog, cow, chicken, Xenopus, fugu, zebrafish) (Fig. 8.4).

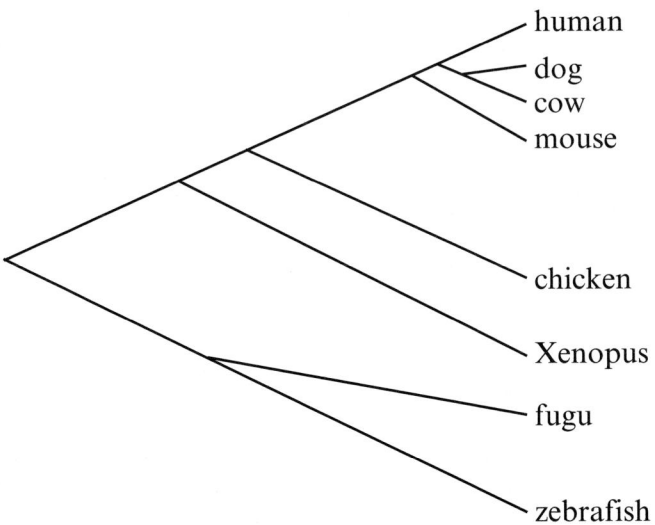

Fig. 8.4 The phylogenetic tree of vertebrates used in our data analysis.

The number of synonymous substitutions per synonymous site (d_S) and the number of nonsynonymous substitutions per nonsynonymous site (d_N) between the human and mouse orthologs were estimated by the PAML package (Yang 1997) using the likelihood method. Multiple protein sequence alignment for each homologous group was generated by Clustal W (Thompson *et al.* 1994), and the phylogenetic trees were inferred by the neighbor-joining method (Saitou and Nei 1987). Under the inferred phylogeny, we used the method of Gu and Zhang (1997) to estimate the parameter H.

8.5.1 Extent of gene pleiotropy

We analyzed 321 vertebrate proteins. For these vertebrate genes $d_N/d_S \ll 1$, with the mean 0.098, while the mean of H is 0.517. Then, for each vertebrate gene with d_N/d_S and H available, we estimated the effective gene pleiotropy (K_e). As shown in Table 8.2, most genes show certain degrees of pleiotropy ($K_e > 3$), supporting the notion that gene pleiotropy is a general feature. Moreover, the mean of effective gene pleiotropy is $K_e = 6.52$, that is, random mutations of a gene typically affect $6 \sim 7$ molecular phenotypes, or, correspondingly, fitness components.

To examine the potential effect of d_N/d_S estimation, we examined other pairwise combinations of four mammals. As shown in Table 8.3, the estimated effective gene pleiotropy varies but the scale of variation is small. In addition, we found that when the number of changes at each site is estimated by the parsimony method, the H-index would be underestimated. Consequently, the degree of gene pleiotropy tends to be overestimated (Table 8.3). At any rate, these 5–10 per cent estimation differences should not affect the general pattern about the degree of gene pleiotropy. For illustration, Tables 8.4 and 8.5 list the highly pleiotropic genes ($K_e > 10$) and lowly pleiotropic genes ($K_e < 3$), respectively. As expected that highly pleiotropic genes tend to evolve very slowly at the sequence level, while lowly-pleiotropic genes tend to evolve quickly.

8.5.2 Biological relevance

Su *et al.* (2010) conducted several analyses to test whether the effective gene pleiotropy (K_e) is biologically interpretable. Intuitively, one may view the concept of molecular phenotypes as canonical biological processes that connect between the biochemical-structural features of the encoded protein and the organismal fitness. Hence, the pleiotropic model predicts that more highly pleiotropic genes tend to be involved in more biological processes. Though a detailed documentation of biological processes remains largely lacking, the gene ontology (GO) has provided a first-order approximation to addressing this issue. With the help of the biological process (BP) category in GO, we counted the number of biological process GO terms of each gene. In spite of the fact that BP terms in the BP category of GO are far from completeness, it indeed indicates a significant portion of genes that may be involved in two or more biological processes (Table 8.2). Indeed, we found a positive correlation between the number of BP and K_e ($P < 0.01$).

Table 8.2 Summary of vertebrate gene pleiotropy analysis (From Su et al. 2010).

K_e	Num	dN/dS	dN	dS	H	S	B_0	Biological processes	Expression broadness
<3	26	0.269 ± 0.022	0.161 ± 0.018	0.609 ± 0.033	0.482 ± 0.033	-6.89 ± 0.76	3.38 ± 0.49	1.62 ± 0.28	6.10 ± 2.61
3–4	29	0.184 ± 0.010	0.112 ± 0.008	0.607 ± 0.030	0.496 ± 0.025	-7.14 ± 0.24	2.08 ± 0.06	1.76 ± 0.35	5.73 ± 2.76
4–5	47	0.134 ± 0.006	0.090 ± 0.006	0.651 ± 0.029	0.492 ± 0.022	-8.30 ± 0.23	1.83 ± 0.05	1.53 ± 0.23	2.26 ± 0.55
5–6	63	0.103 ± 0.003	0.072 ± 0.004	0.684 ± 0.031	0.483 ± 0.016	-9.03 ± 0.12	1.65 ± 0.02	1.70 ± 0.27	8.64 ± 2.05
6–7	42	0.065 ± 0.002	0.045 ± 0.004	0.678 ± 0.042	0.557 ± 0.016	-10.59 ± 0.14	1.62 ± 0.02	1.74 ± 0.31	8.36 ± 2.41
7–8	37	0.051 ± 0.002	0.034 ± 0.003	0.651 ± 0.043	0.537 ± 0.018	-11.42 ± 0.15	1.53 ± 0.02	2.08 ± 0.38	6.82 ± 2.79
8–9	27	0.036 ± 0.002	0.022 ± 0.002	0.609 ± 0.046	0.570 ± 0.023	-12.63 ± 0.20	1.50 ± 0.02	2.85 ± 0.80	10.70 ± 3.77
9–10	19	0.026 ± 0.002	0.018 ± 0.002	0.668 ± 0.062	0.559 ± 0.031	-13.75 ± 0.25	1.44 ± 0.03	2.00 ± 0.48	10.57 ± 4.03
>10	31	0.010 ± 0.001	0.006 ± 0.001	0.579 ± 0.044	0.611 ± 0.027	-18.03 ± 0.61	1.36 ± 0.02	2.26 ± 0.62	13.36 ± 3.24

Note: Effective gene pleiotropy (K_e) was estimated based on the human–mouse orthologous genes for d_N/d_S and Gu–Zhang's (1997) method for H. Biological processes were counted from the gene ontology (GO). Expression broadness is the number of mouse tissues in which a gene is expressed, based on Su et al. (2010).

Table 8.3 Effects of protein sequence analysis on the estimation of gene pleiotropy. From Su *et al.* (2010).

	d_N/d_S	H	K
(1) effect of d_N/d_S estimation			
Human-mouse	0.098	0.525	6.52
Human-cow	0.112	0.525	6.25
Human-dog	0.118	0.525	6.14
Mouse-cow	0.096	0.525	6.56
Mouse-dog	0.097	0.525	6.54
Cow-dog	0.120	0.525	6.11
(2) effect of H estimation			
Parsimony	0.098	0.468	6.75

Table 8.4 A list of highly pleiotropic genes ($K_e > 10$); BP: biological processes; EB: expression broadness). From Su *et al.* (2010).

Description	DN/DS	H	Ke	S	B_0	BP	EB	SWISSPROT (human)
Amyloid beta A4 protein precursor (APP) (ABPP)	0.0187	0.521	10.89	14.68	1.347	2	0	P05067
Gamma-soluble NSF attachment protein (SNAP-gamma)	0.0181	0.531	10.93	14.77	1.352	0	8	Q99747
transcription elongation factor B binding protein 1	0.0236	0.383	10.97	14.09	1.284	1		NP_065746
Nucleolar protein 10	0.0214	0.429	11.04	14.36	1.301	2	55	[Q9BSC4
Protein C4orf8 (Protein IT14)	0.0148	0.605	11.05	15.36	1.390	0	3	P78312
A20-binding inhibitor of NF-kappaB activation 2	0.0144	0.614	11.05	15.43	1.396	0		NP_077285
Methionine synthase reductase, mitochondrial precursor	0.0088	0.736	11.42	16.88	1.479	0	6	Q9UBK8

Nucleoporin p58/p45 (Nucleoporin-like 1)	0.0127	0.610	11.49	15.84	1.379	4	0	Q9BVL2
Protein disulfide-isomerase A6 precursor (EC 5.3.4.1)	0.0120	0.571	11.98	16.08	1.342	7		Q15084
Intraflagellar transport 140 homolog	0.0124	0.504	12.35	16.07	1.301	2	2	Q96RY7
Cathepsin B heavy chain	0.0051	0.796	12.37	18.58	1.502	14	22	P07858
Serine/threonine-protein kinase PRP4	0.0031	0.872	12.41	20.08	1.617	0	0	Q13523
Nucleoporin Nup98	0.0142	0.360	12.74	15.83	1.242	14	2	P52948
T-complex protein 1-epsilon	0.0106	0.490	12.97	16.63	1.282	1	6	P48643
adducin 1 (alpha) isoform d	0.0066	0.674	13.05	17.90	1.372	2	0	NP_789771
Sodium-dependent glutamate/aspartate transporter 3	0.0076	0.622	13.08	17.53	1.340	1	1	P43005
Rho-associated protein kinase 2 (EC 2.7.1.37)	0.0019	0.884	13.78	21.64	1.570	3	28	O75116
ATPase class I type 8A member 2) (ML-1)	0.0051	0.599	14.52	18.83	1.297	3	0	Q9NTI2
Zinc finger protein 294	0.0051	0.440	15.59	19.07	1.224	0	24	O94822
tetracycline transporter-like protein	0.0010	0.886	15.83	23.63	1.493	2	45	NP_001111
Triosephosphate isomerase	0.0010	0.777	17.77	23.61	1.329	1	60	P60174
membrane-associated ring finger (C3HC4) 6	0.0010	0.738	18.34	23.77	1.296	1		NP_005876
HN1-like protein	0.0008	0.754	18.75	24.34	1.298	2	30	Q9H910
N-glycosylase/DNA lyase	0.0011	0.544	19.98	24.03	1.203	0	2	O15527
Nucleolar complex protein 14	0.0010	0.507	20.40	24.27	1.189	1	0	P78316

Table 8.5 A list of lowly pleiotropic genes ($K_e < 3$); BP: biological processes; EB: expression broadness). From Su *et al.* (2010).

Description	DN/DS	H	K	S	B_0	BP	EB	SWISSPROT (human)
Defender against cell death 1 (DAD-1)	0.309	0.595	0.911	10.052	11.03	1	1	P61803
PolyA binding protein II (PABII)	0.605	0.189	0.977	2.270	2.32	0		Q86U42
Dehydrogenase/reductase SDR family member 4	0.269	0.626	1.108	9.808	8.85	2	4	Q9BTZ2
Leukotriene B4 receptor 1 (LTB4-R 1)	0.444	0.364	1.202	3.807	3.16	1	0	Q15722
Syntaxin-binding protein 6	0.260	0.573	1.670	6.712	4.01	2	17	Q8NFX7
RNA-binding protein Nova-1	0.363	0.378	1.810	4.406	2.43	4	58	P51513
Syntaxin-binding protein 1- like 2 (Sly1p)	0.340	0.403	1.901	4.690	2.46	2	0	Q8WVM8
Cochlin precursor (COCH-5B2).	0.277	0.484	2.103	5.596	2.66	0	0	O43405
C14orf125 protein (Fragment)	0.109	0.785	2.297	12.324	5.36	3	9	Q86XA9
Rho-GTPase-activating protein 5 (p190-B)	0.301	0.383	2.423	5.052	2.08	0	5	Q13017
Neuronal PAS domain protein 3 (Neuronal PAS3)	0.317	0.347	2.446	4.845	1.98	4		Q8IXF0
E2F-associated phosphoprotein (EAPP).	0.046	0.904	2.475	22.959	9.27	2	3	Q56P03
Cofilin-2 (Cofilin, muscle isoform).	0.235	0.485	2.654	6.060	2.28	2	6	Q9Y281
ATP-utilizing chromatin assembly and remodeling factor 1) (hACF1)	0.188	0.586	2.681	7.162	2.67	4	1	Q9NRL2
Signal recognition particle 54 kDa protein (SRP54).	0.281	0.368	2.746	5.292	1.92	2		P61011
Proteasome subunit alpha type 6	0.352	0.201	2.775	4.498	1.62	1	0	P60900

NF-kappaB inhibitor alpha	0.210	0.520	2.800	6.519	2.32	1		P25963
breast cancer metastasis-suppressor 1-like	0.176	0.598	2.804	7.394	2.63	5	0	NP_115728
MAP3K12 binding inhibitory protein 1	0.224	0.477	2.879	6.204	2.15	1	0	Q9NS73
Homeobox protein Nkx-2.1) (Homeobox protein NK-2 homolog A)	0.274	0.355	2.904	5.386	1.85	1	2	P43699
SMN-interacting protein 1	0.176	0.583	2.914	7.285	2.50	0	0	O14893
Trafficking protein particle complex subunit 6B	0.181	0.568	2.940	7.135	2.42	0	22	Q86SZ2
Nuclear protein SDK3	0.282	0.318	2.988	5.282	1.76	1	0	Q9H307

Similarly, the pleiotropic model predicts that highly pleiotropic genes tend to be expressed in multiple tissues, i.e. expression broadness. It implies that the same biochemical function of a gene (enzyme or binding activity) can be related to several different components of the organismal fitness by having expressions in multiple cell types or tissues, and found a significant correlation between the effective gene pleiotropy (K_e) and the number of tissues where the gene is highly expressed ($P <$ 0.001). Hence, the increased number of expression tissues may increase the likelihood to involve many phenotypic traits that could affect a number of fitness components. It is therefore hypothesized that a number of genomic measures that are correlated with the rate of portein evolution (Pal *et al.* 2006a; Salathe *et al.* 2006) may be the consequence of gene pleiotropy, whereas the rate-expression relationship (Pal *et al.* 2001; 2003; Wyckoff *et al.* 2005) may be complex.

8.6 Comments on gene pleiotropy

In this chapter we introduce the pleiotropy model of molecular evolution (Gu 2007a). Further, under this framework, we develop a statistical method for estimating the capacity of a gene that can significantly affect distinct components in the fitness, called the effective gene pleiotropy (K_e). This work, for the first time, provides a computational estimation of the gene pleiotropy. Moreover, modeling the concept of gene pleiotropy may fill the gap between molecular evolution and phenotypic evolution.

Needless to say, the pleiotropy model and the method for estimating K_e require a number of assumptions that need to be addressed in the future. First, the pleiotropy model involves several normal assumptions. One may refer to Martin and Lenormand (2006) for a comprehensive summary of rationales and criticisms. The normal assumption also implies the single fitness optimum. Under this condition the model actually assumes that disruptive selection with alternative optima following an adaptive process occurs rarely at the molecular level. Lande (1980) showed that when the population mean is close to the optimum, a Gaussian-like function can be a local kernal approximation for many arbitrary fitness functions. It should be noted that the normal-like effect of mutations on molecular phenotypes may be insufficient to account for lethal mutations. Besides, we shall optimize our estimation procedure to improve statistical properties of estimates K_e and B_0. In particular, we assume that gene pleiotropy is a constant in the molecular evolution, which may not hold in practice. Hence, it is a challenge how to analyze the data when the degree of gene pleiotropy varies during the course of evolution. Finally, the relation between molecular adaptation and gene pleiotropy needs to be substantially investigated.

9

Modeling the Genomic Evolution of Gene Contents

Whole-genome analysis, e.g. the presence or absence of gene families over multiple genomes, is becoming an attractive approach to extract the bulk phylogenetic signals and explore the pattern of genome evolution (Snel *et al.* 1999; Lin and Gerstein 2000; Korbel *et al.* 2002; Natale *et al.* 2000; Clarke *et al.* 2002; Fitz-Gibbon and House 1999; House and Fitz-Gibbon 2002; Gu and Zhang 2004; Huson and Steel 2004; Zhang and Gu 2004; Hahn *et al.* 2005). In this chapter, we introduce several statistical models for this purpose.

9.1 The birth–death model of gene content evolution

9.1.1 Joint size distribution of gene families

Multiple-genome comparative analysis has revealed a high variation in the size of gene families among species, because a gene family can be generated, expanded, reduced, or lost during the course of genome evolution. Therefore, the joint size distribution of the gene family among genomes is useful for phylogenomic analysis.

We (Gu and Zhang 2004) developed a general stochastic model, based on the stochastic theory of birth–death process. Considering two major evolutionary processes that influence the size of gene family: gene loss (nonfunctionization or deletion) and gene proliferation (duplication). Let μ be the evolutionary rate of gene loss and λ be the evolutionary rate of gene proliferation. If each gene is subject to the same chance to be lost or duplicated, for a gene family with r member genes at $t = 0$, the number of member genes after t time units, denoted by X_t, shows the following distribution

$$P(X_t = k | X_0 = r) = \sum_{j=0}^{\min[r,k]} \binom{r}{j} \binom{r+k-j-1}{r-1} \beta^{r-j} \alpha^{k-j} (1-\alpha-\beta)^j, \quad k \geq 1$$

$$P(X_t = 0 | X_0 = r) = \beta^r \tag{9.1}$$

where the proliferation parameter α and the loss parameter β are given by

$$\alpha = \lambda \frac{1 - e^{(\lambda-\mu)t}}{\mu - \lambda e^{(\lambda-\mu)t}}$$

$$\beta = \mu \frac{1 - e^{(\lambda-\mu)t}}{\mu - \lambda e^{(\lambda-\mu)t}} \tag{9.2}$$

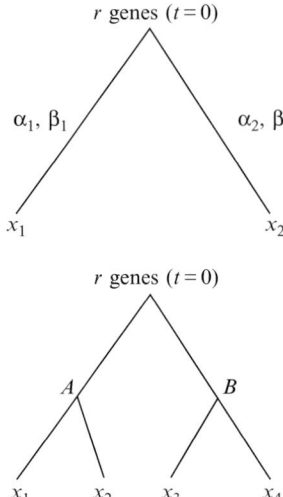

Fig. 9.1 Schematic genome evolution for two genomes and four genomes, respectively. The gene family has r member genes in the root. After t evolutionary time units, the size of the gene family is denoted by x_1 and x_2 in genomes 1 and 2, respectively. For four genomes, the size of the gene family is denoted by x_1, x_2, x_3 and x_4, respectively.

respectively. Eq. (9.2) implies $\alpha/\beta = \lambda/\mu$, as called the P/L ratio. For the size of gene family under the birth–death model is expected to be $X_0 e^{(\lambda-\mu)t}$, $\alpha > \beta$ (or $P/L > 1$) indicates, on average, the increase of gene family size during the evolution and *vice versa*.

Consider two genomes that have been diverged t time units ago (Fig. 9.1). For a given gene family, assume that there are r member genes at $t = 0$ (in the common ancestor), and X_i number genes in each genome $i = 1, 2$, respectively. Under the assumption of independent evolution between lineages, the (conditional) joint probability is given by $P(X_1, X_2|X_0 = r) = P(X_1|X_0 = r) \times P(X_2|X_0 = r)$. Since the size of a gene family in the ancestral genome is unknown, a (prior) distribution for $X_0 = r$ is assumed, denoted by $\pi(r)$. Thus, the joint probability of X_1 and X_2 is given by

$$P(X_1, X_2) = \sum_{r=1}^{\infty} \pi(r) P(X_1, X_2 | X_0 = r) = \sum_{r=1}^{\infty} \pi(r) P(X_1|r) P(X_2|r) \qquad (9.3)$$

where $P(X_i|r)$ is short for $P(X_i|X_0 = r)$ defined by Eq. (9.1).

For the general n-genomes, let X_i represent the size of a gene family in the i-th genome, $i = 1, \ldots, n$. The joint size distribution of the gene family $\mathbf{X} = (X_1, \ldots, X_n)$ can be derived according to the Markov chain model, similar to DNA sequence evolution (Felsenstein 1981). For example, for four genomes (Fig. 9.1), it is given by

$$P(\mathbf{X}) = \sum_{r_0} \sum_{r_A} \sum_{r_B} \pi(r_0) P(r_A|r_0; \alpha_5, \beta_5) P(r_B|r_0; \alpha_6, \beta_6)$$
$$\times P(X_1|r_A; \alpha_1, \beta_1) P(X_2|r_A; \alpha_2, \beta_2) P(X_3|r_B; \alpha_3, \beta_3) P(X_4|r_B; \alpha_4, \beta_4) \qquad (9.4)$$

where $P(.|.; \alpha_i, \beta_i)$ is the transition probability for branch i, defined by Eq. (9.1).

9.1.2 Genome distances and gene content information

Given the joint size distribution, say, Eq. (9.4) for four genomes, maximum-likelihood phylogeny can be implemented. Unfortunately, the complexity of transition probability (Eq. (9.1)) makes it almost intractable for the genome-level analysis. Thus, the distance method becomes highly desirable, but first one should define an additive genome distance measure. With some algebras from Eq. (9.2), two quantities, the proliferation measure d_λ and the loss measure d_μ, are given by

$$d_\lambda = \frac{\alpha}{\beta - \alpha} \ln \frac{1 - \alpha}{1 - \beta} = \lambda t$$

$$d_\mu = \frac{\beta}{\beta - \alpha} \ln \frac{1 - \alpha}{1 - \beta} = \mu t \tag{9.5}$$

respectively. For two genomes (Fig. 9.1), let λ_i, μ_i, α_i, β_i, d_{λ_i}, and d_{μ_i} be the corresponding parameters in each lineage, $i = 1, 2$; see Eqs. (9.2) and (9.5). Then, we define the proliferation genome distance between two genomes (the P-distance for short) as $G_P = d_{\lambda_1} + d_{\lambda_2} = (\lambda_1 + \lambda_2)t$; from Eq. (9.5) it is given by

$$G_P = \sum_{i=1,2} \frac{\alpha_i}{\beta_i - \alpha_i} \ln \frac{1 - \alpha_i}{1 - \beta_i} \tag{9.6}$$

In the same manner, the loss genome distance (L-distance for short) between two genomes is defined as $G_L = d_{\mu_1} + d_{\mu_2} = (\mu_1 + \mu_2)t$, given by

$$G_L = \sum_{i=1,2} \frac{\beta_i}{\beta_i - \alpha_i} \ln \frac{1 - \alpha_i}{1 - \beta_i} \tag{9.7}$$

and the general genome distance measure is defined as $G = G_P + G_L$, i.e.

$$G = \sum_{i=1,2} \frac{\alpha_i + \beta_i}{\beta_i - \alpha_i} \ln \frac{1 - \alpha_i}{1 - \beta_i} \tag{9.8}$$

Apparently, these genome distance measures are additive and $G_P/G_P = P/L$ ratio. Equations (9.6)–(9.8) provide the relationship between genome distances and parameters in the probabilistic model (Eqs. (9.1)–(9.3)). To calculate the genome distance, we shall develop a computationally efficient method for estimating the parameters (α_i and β_i).

The concept of gene content was introduced by several authors for studying the universal genome tree (e.g. Snel *et al.* 1999). For two genomes $i = 1, 2$, let Y_i be the gene content index of a gene family: $Y_i = 1$ indicates at least one member gene found in the i-th genome; otherwise $Y_i = 0$. Therefore, gene content pattern is the most degenerated size distribution of the gene family. In the following we will show that it becomes insufficient for estimating the genome distance.

From Eq. (9.3), one can show that the joint probability of Y_1 and Y_2 is given by

$$P(Y_1, Y_2) = \sum_{r=1}^{\infty} \pi(r) P(Y_1|r) P(Y_2|r) \tag{9.9}$$

Since $P(Y_i = 0|r) = \beta_i^r$, and $P(Y_i = 1|r) = 1 - \beta_i^r$, $i = 1, 2$, the analytical form of $P(Y_1, Y_2)$ can be obtained if a geometric prior is assumed, i.e. $\pi(r) = (1 - f)^{r-1} f$. For simplicity, let $P(i, j) = P(Y_1 = i, Y_2 = j)$. Then, putting $\pi(r)$ into Eq. (9.9) we have

$$P(1, 1) = 1 - Q(\beta_1) - Q(\beta_2) + Q(\beta_1 \beta_2)$$

$$P(1, 0) = Q(\beta_2) - Q(\beta_1 \beta_2)$$

$$P(0, 1) = Q(\beta_1) - Q(\beta_1 \beta_2)$$

$$P(0, 0) = Q(\beta_1 \beta_2) \tag{9.10}$$

where the function $Q(\beta)$ ($\beta = \beta_1$, β_2 or $\beta_1 \beta_2$) is defined as

$$Q(\beta) = \sum_{r=1}^{\infty} \pi(r) \beta^r = \frac{\beta f}{1 - (1 - f)\beta} \tag{9.11}$$

Since Eq. (9.10) only relies on the loss parameters β_1 and β_2, we cannot estimate the proliferation parameters (α_1 and α_2).

9.1.3   Extended gene content and genome distance estimation

The above analysis indicates that the additive genome distances defined by Eqs. (9.6)–(9.8) in general cannot be estimated by the gene content approach. Nevertheless, Gu and Zhang (2004) have found a plausible solution by further dividing the non-zero (member genes) case into two states: single-copy (one-member) or duplicates (two or more member genes). Hence, this extended gene content analysis considers three possible states: no member gene ($Z = 0$), single-copy gene ($Z = 1$), and duplicate genes ($Z = 2$). According to Eq.(9-1), their probabilities are $P(Z = 0|X_0 = r) = P(X_t = 0|X_0 = r)$, $P(Z = 1|X_0 = r) = P(X_t = 1|X_0 = r)$ and $P(Z = 2|X_0 = r) = \sum_{k \geq 2} P(X_t = k|X_0 = r)$, as given by

$$P(Z = 0|X_0 = r) = \beta^r$$

$$P(Z = 1|X_0 = r) = r\beta^{r-1}(1 - \beta)(1 - \alpha)$$

$$P(Z = 2|X_0 = r) = 1 - \beta^r - r\beta^{r-1}(1 - \beta)(1 - \alpha) \tag{9.12}$$

respectively.

The joint distribution for two genomes

Consider two genomes that have been diverged t time units ago (Fig. 9.1). Let $Z_i = 0$, 1, or 2 be the extended gene content index for a gene family in the i-th genome, $i = 1, 2$. Similar to Eq. (9.3) and Eq. (9.9), the joint distribution of Z_1 and Z_2 is given by

$$P(Z_1, Z_2) = \sum_{r=1}^{\infty} \pi(r) P(Z_1, Z_2|X_0 = r) = \sum_{r=1}^{\infty} \pi(r) P(Z_1|r) P(Z_2|r) \tag{9.13}$$

where $P(Z_i|r) = P(Z_i|X_0 = r)$. Given the geometric distribution for $\pi(r) = f(1 - f)^{r-1}$, we obtain the analytical forms of Eq. (9.13) as follows

$$P(0,0) = Q(\beta_1\beta_2)$$

$$P(0,1) = \beta_1\omega_2 R(\beta_1\beta_2)$$

$$P(0,2) = Q(\beta_1) - Q(\beta_1\beta_2) - \beta_1\omega_2 R(\beta_1\beta_2)$$

$$P(1,0) = \beta_2\omega_1 R(\beta_1\beta_2)$$

$$P(1,1) = \omega_1\omega_2 S(\beta_1\beta_2)$$

$$P(1,2) = \omega_1[R(\beta_1) - \beta_2 R(\beta_1\beta_2)] - \omega_1\omega_2 S(\beta_1\beta_2)$$

$$P(2,0) = Q(\beta_2) - Q(\beta_1\beta_2) - \beta_2\omega_1 R(\beta_1\beta_2)$$

$$P(2,1) = \omega_2[R(\beta_2) - \beta_1 R(\beta_1\beta_2)] - \omega_1\omega_2 S(\beta_1\beta_2)$$

$$P(2,2) = 1 - Q(\beta_1) - Q(\beta_2) + Q(\beta_1\beta_2) - \omega_1[R(\beta_1) - \beta_2 R(\beta_1\beta_2)]$$
$$\qquad\qquad - \omega_2[R(\beta_2) - \beta_1 R(\beta_1\beta_2)] + \omega_1\omega_2 S(\beta_1\beta_2) \tag{9.14}$$

where $\omega_1 = (1 - \beta_1)(1 - \alpha_1)$ and $\omega_2 = (1 - \beta_2)(1 - \alpha_2)$; the function $Q(\beta)$ is given by Eq. (9.11), the function $R(\beta) = \sum_{r=1}^{\infty} \pi(r)r\beta^{r-1}$ is given by

$$R(\beta) = \frac{f}{1 - (1 - f)\beta} + \frac{f(1 - f)\beta}{[1 - (1 - f)\beta]^2} \tag{9.15}$$

and the function $S(\beta) = \sum_{r=1}^{\infty} \pi(r)r^2\beta^{r-1}$ is given by

$$S(\beta) = \frac{f}{1 - (1 - f)\beta} + \frac{3f(1 - f)\beta}{[1 - (1 - f)\beta]^2} + \frac{2f(1 - f)^2\beta^2}{[1 - (1 - f)\beta]^3} \tag{9.16}$$

Here $\beta = \beta_1$, β_2 or $\beta_1\beta_2$.

Parameter estimation

When the extended gene content data matrix for any two genomes 1 and 2 is given, we develop a maximum-likelihood-based approach to estimating the genome distances. Usually the prior parameter f can be estimated from the observed size frequencies of gene families. Since the pattern of double loss (i.e. $Z_1 = 0$ and $Z_2 = 0$) is not observable, one may use the following modified joint probability

$$q(Z_1, Z_2) = \frac{P(Z_1, Z_2)}{1 - P(0,0)} = \frac{P(Z_1, Z_2)}{1 - Q(\beta_1\beta_2)} \tag{9.17}$$

for $Z_1, Z_2 = 0$, 1, or 2, except $Z_1 = Z_2 = 0$. Let n_{ij} be the number of gene families with the pattern $Z_1 = i$ and $Z_2 = j$, where $i, j = 0, 1, 2$ except $i = j = 0$. Then, the likelihood for the two genomes can be written as

$$L(\alpha_1, \alpha_2, \beta_1, \beta_2|\text{data}) = \prod_{i,j} q(i, j)^{n_{ij}} \tag{9.18}$$

We use the Newton–Raphoson numerical iteration to obtain the ML estimates of α_1, α_2, β_1, and β_2. Their sampling variance-covariance matrix is approximately computed by the inverse of Fisher's information matrix. When these parameters $(\alpha_1, \alpha_2, \beta_1, \beta_2)$ are estimated, the computation of genome distances by Eqs. (9.6) to (9.8) are straightforward, and the sampling variance of a genome distance can be obtained by the delta method.

9.1.4   Simulations and case study

Gu and Zhang (2004) have conducted extensive computer simulations to examine the performance of phylogenetic reconstruction using the extended gene content data. Here we discuss the main results briefly.

Estimation of genome distance is asymptotically unbiased

We first simulate the stochastic process according to the two-genome evolution scenario (Fig. 9.1), when the evolutionary parameters ($\lambda_i t$ and $\mu_i t$, $i = 1, 2$) are given. For each gene family, the number of genes on the root, r, is generated from a geometric distribution with the parameter $f = 0.5$. In each replicate, we implement the ML algorithm to estimate the proliferation parameters α_i and the loss parameters β_i ($i = 1, 2$), and then compute the genome distances according to Eqs. (9.6) to (9.8). The mean and squared root of variance for each estimate are used for examining the statistical properties.

We have studied four typical cases: the gene-loss model ($\lambda = 0$), the growth model ($\lambda > \mu$), the equal model ($\lambda = \mu$), and the reduction model ($\lambda < \mu$). The number of gene families (N) is set to be $N = 200, 500$, and 1000, respectively. We have examined a variety of combinations from these models in two lineages, and found that the estimates of these parameters and genome distances are asymptotically biased, which are virtually trivial when $N > 500$. The sampling variances of genome distances decrease with the increasing of the number of gene families, which are usually acceptable if $N > 500$.

Genome tree inference is efficient

Gu and Zhang (2004) have examined the tree-making performance of the extended gene content approach, using a typical four genome scenario (Fig. 9.2). After the extended gene content matrix of four genomes is simulated, we estimate the genome distance matrix and infer the tree using the neighbor-joining (NJ) algorithm. The efficiency of phylogenetic inference is then measured by the percentage of correct topology inference over 1000 replicates. It has been shown that, except for some extreme cases, the correct percentage of tree-making is satisfactory (> 70 per cent) when $N > 500$.

Table 9.1 shows the correct percentage of tree-making when the true tree has four equal external branch lengths (Fig. 9.2(A)). When the internal branch length (c) is short, the genome tree inference can be significantly improved as N becomes large. To examine the tree-making consistency, we consider two typical patterns when the external branches are highly unequal (Fig. 9.2 panels B and C). As shown in Table 9.2, the performance is poor when N is small and the internal branch length is short. Nevertheless, even in the very extreme case, the correct percentage of tree-making is close to 100 per cent for a sufficient large number of gene families.

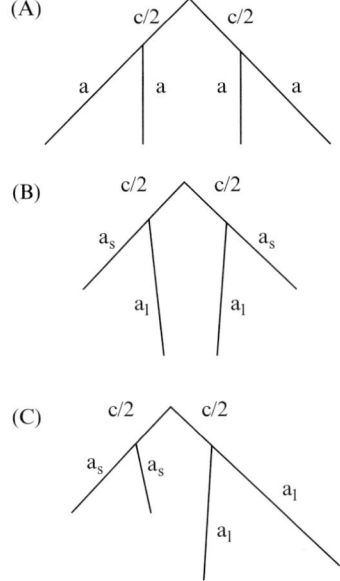

Fig. 9.2 The genome tree used for a computer simulation study. (A) Equal external branch lengths. (B). Unequal external branches (Felsenstein's zone). (C) Unequal external branches (non-Felsenstein's zone).

Software and case study: the universal genome tree of life

Gu *et al.* (2005a) developed a software to infer the genome tree from the extented gene content data. We applied this tree-making method to infer the universal genome tree of 35 complete genomes. The extended gene content data were obtained from the COG database (http://www.ncbi.nlm.nih.gov/COG/). Then, the pairwise genome distance (G) is estimated according to Eq. (9.8).

We used the NJ method (Saitou and Nei 1987) to infer the genome phylogeny. The overall genome tree based on extended gene contents (Fig. 9.3) supports the concept of universal tree, similar to previous gene content trees (Snel *et al.* 1999; Wolf *et al.* 2002) and the standard 16S RNA tree (Olsen *et al.* 1994). That is, two major lineages of cellular life, the Archaea and Bacteria, are monophyletic from the third lineage (Eukarya, representing by the yeast genome), supported by 100 per cent bootstrap values. There are a few aspects in which our tree differs from other gene content trees. We have compared our result to that of Wolf *et al.* (2002). In their study, the genome distance between species (A and B) was calculated $D_{AB} = 1 - J_{AB}$, where J_{AB} is the Jaccard coefficient, which reflects the similarity of gene contents between A and B. Consider the phylogeny of Archaea, for instance. Both studies support that Hbs (*Halobacterium sp*) appears at the root of the tree, and the Euryachaeota (Afu, Mja, Mth, and Pho; see Fig. 9.3 for species abbreviations) are clustered together. However, our genome phylogeny suggests that Crenarchaeota Ape (*Aeropyrum pernix*) may also branch-off, while Wolf *et al.* (2002) showed it was clustered with the Euryachaeota Tac (*Thermoplasma acidophilum*).

Table 9.1 Correct percentage (%) of tree making: equal external branch lengths (see Fig. 9.1(A). From Gu and Zhang (2004).

N	The c/a Ratio				
	1	1/2	1/4	1/8	1/16
(1) $a = 0.5, P/L = 0$					
100	100	95	78	55	50
500	100	100	98	85	59
2000	100	100	100	99	70
(2) $a = 0.75, P/L = 0.5$					
100	100	96	82	57	54
500	100	100	100	95	66
2000	100	100	100	100	78
(3) $a = 1.0, P/L = 1$					
100	100	100	89	63	44
500	100	100	100	88	67
2000	100	100	100	98	82
(4) $a = 0.75, P/L = 2$					
100	100	99	86	64	46
500	100	100	100	91	59
2000	100	100	100	100	73

9.2 Likelihood of four genomes under simple gene contents

Moreover, Zhang and Gu (2004) studied the likelihood function of four genomes in the case of simple gene content information, which is a binary (1 or 0) index for a gene family in a genome. Let $Y = 1$ denote the situation that this genome maintains at least one member gene of the gene family, and $Y = 0$ be the situation that all the member genes are lost. As shown in the above section, given the r member genes at $t = 0$, i.e. $X_0 = r$, the transition probabilities of $Y = 1$ and $Y = 0$ after t time units are given by

$$P(Y = 0|X_0 = r) = \beta^r$$
$$P(Y = 1|X_0 = r) = 1 - \beta^r \tag{9.19}$$

respectively, where β is given by Eq. (9.2).

Table 9.2 Correct percentage (%) of tree making: unequal branch lengths (see Fig. 9.1(B) and (C)). From Gu and Zhang (2004).

	The c/a_t ratio				
N	1	0.8	0.4	0.2	0.1
(1) Refer to Fig. 9.1(B)					
100	73	66	58	41	30
500	98	92	80	50	40
2000	100	100	95	87	78
(2) Refer to Fig. 9.1(C)					
100	79	78	75	66	60
500	100	97	92	76	78
2000	100	100	100	96	95

Note—genome branch lengths: $a_1 = 0.6$, $a_s = 0.06$, and $P/L = 0.5$.

For a gene family with four member genes, let Y_i be the status of the gene family in genome i, $i = 1, 2, 3, 4$; $Y_i = 1$ if at least one member gene is found in this genome, and otherwise $Y_i = 0$. Thus, there are 15 gene content patterns, i.e. $(Y_1, Y_2, Y_3, Y_4) = (1, 1, 1, 1)$, $(1, 1, 1, 0)$, ..., $(0, 0, 0, 0)$.

9.2.1 Likelihood function: case A

In the following we derive the likelihood function under the phylogenetic tree shown in Fig. 9.4(A). Thus, the probability of observing the status of Y_1, Y_2, Y_3, and Y_4 is given by

$$P(Y_1, Y_2, Y_3, Y_4) = \sum_{r_0=0}^{\infty} \sum_{r_A=0}^{\infty} \sum_{r_B=0}^{\infty} \pi(r_0) P(r_A|r_0; \beta_5, \alpha_5) P(r_B|r_0; \beta_6, \alpha_6)$$

$$\times P(Y_1|r_A; \beta_1) P(Y_2|r_A; \beta_2) P(Y_3|r_B; \beta_3) P(Y_4|r_B; \beta_4) \quad (9.20)$$

For instance, given the geometric distribution of $\pi(r_0)$, one can show that the probability of gene content pattern $(0, 0, 0, 0)$ is

$$P(0, 0, 0, 0) = \sum_{r_0=0}^{\infty} \pi(r_0) \sum_{r_A=0}^{\infty} P(r_A|r_0; \alpha_5, \beta_5) \beta_1^{r_A} \beta_2^{r_A} \sum_{r_B=0}^{\infty} P(r_B|r_0; \alpha_6, \beta_6) \beta_3^{r_B} \beta_4^{r_B}$$

$$(9.21)$$

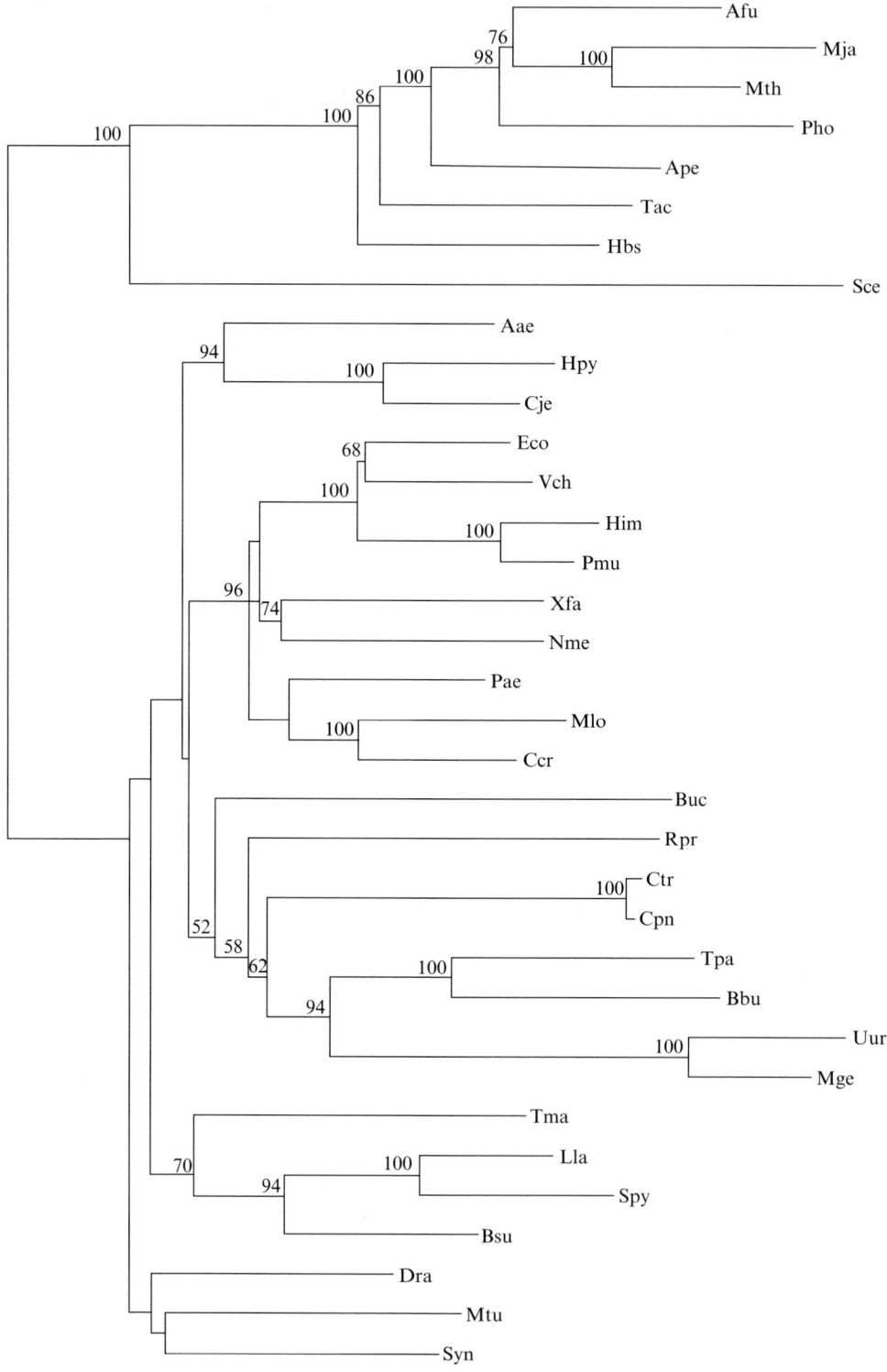

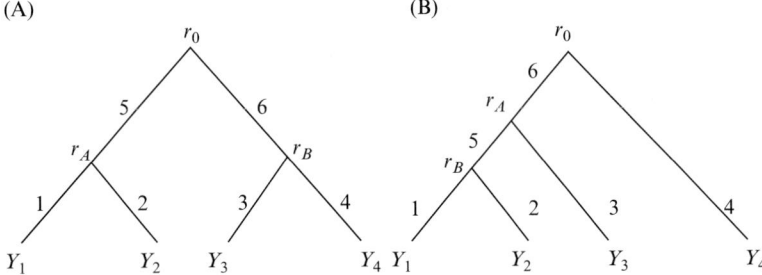

Fig. 9.4 Two types of tree topologies in the case of the rooted tree of four genomes: (A) Symmetric topology, and (B) asymmetric topology.

Noting that the probability generating function for the random variable X_t that follows a birth–death process is in the form of

$$G_X(s; \alpha, \beta, r_0) = \sum_{r=0}^{\infty} P(r|r_0; \alpha, \beta)s^r = \left[\frac{\beta + (1 - \alpha - \beta)s}{1 - \alpha s}\right]^{r_0}$$

for any variable s that is between 0 and 1, we obtain

$$P(0, 0, 0, 0) = \sum_{r_0=0}^{\infty} \pi(r_0)G_X(\beta_1\beta_2; \alpha_5, \beta_5, r_0)G_X(\beta_3\beta_4; \alpha_6, \beta_6, r_0)$$

$$= \sum_{r_0=0}^{\infty} (1 - f)^{r_0} f \times \gamma_0^{r_0} = \frac{f\gamma_0}{1 - (1 - f)\gamma_0} \tag{9.22}$$

where γ_0 is given by

$$\gamma_0 = \left[\frac{\beta_5 + (1 - \alpha_5 - \beta_5)\beta_1\beta_2}{1 - \alpha_5\beta_1\beta_2}\right]\left[\frac{\beta_6 + (1 - \alpha_6 - \beta_6)\beta_3\beta_4}{1 - \alpha_6\beta_3\beta_4}\right] \tag{9.23}$$

Fig. 9.3 The genome phylogeny of 35 microbial complete genomes, inferred by the extended gene-content dataset. Bootstrapping values greater than 50 per cent are not presented. Modified from Gu and Zhang (2004). Species abbreviations: Archaea: Afu, *Archaeoglobus fulgidus*; Hbs, *Halobacterium sp.* NRC-1; Mja, *Methanococcus jannaschii*; Mth, *Methanothermobacter thermautotrophicus*; Tac, *Thermoplasma acidophilum*; Pho, *Pyrococcus horikoshii*; Ape, *Aeropyrum pernix*. Eukaryota: Sce, *Saccharomyces cerevisiae*. Bacteria: Aae, *Aquifex aeolicus*; Tma, *Thermotoga maritime*; Dra, *Deinococcus radiodurans;* Mtu, *Mycobacterium tuberculosis H37Rv*; Lla, *Lactococcus lactis*; Spy, *Streptococcus pyogenes M1 GAS*; Bsu, *Bacillus subtilis*; Syn, *Synechocystissp.*; Eco, *Escherichia coliK12;* Buc, *Buchnera sp. APS*; Vch, *Vibrio cholerae*; Pae, *Pseudomonas aeruginosa*; Hin, *Haemophilus influenzae*; Pmu, *Pasteurella multocida*; Xfa, *Xylella fastidiosa 9a5c*; Nme, *Neisseria meningitidis MC58*; Hpy, *Helicobacter pylori 26695*; Cje, *Campylobacter jejuni*; Mlo, *Mesorhizobium*.

In the same manner, the probability function for the rest of the gene content patterns can be derived, though it could be tedious. As shown in Table 9.3, the second column displays the expressions of $\gamma_i, i = 0, 1, \cdots, 14$, which are used in the probability functions (the third column). Note that each γ_i is a function of the parameters, $\{\alpha_5, \alpha_6, \beta_1, \cdots, \beta_6\}$.

Table 9.3 Probability (mass) functions (pmf) of $Y = (Y_1, Y_2, Y_3, Y_4)$ in case A.

i	Y	γ_i	pmf
0	(0,0,0,0)	$\left[\frac{\beta_5+(1-\alpha_5-\beta_5)\beta_1\beta_2}{1-\alpha_5\beta_1\beta_2}\right]\left[\frac{\beta_6+(1-\alpha_6-\beta_6)\beta_3\beta_4}{1-\alpha_6\beta_3\beta_4}\right]$	$\frac{f\gamma_0}{1-(1-f)\gamma_0}$
1	(0,0,0,1)	$\left[\frac{\beta_5+(1-\alpha_5-\beta_5)\beta_1\beta_2}{1-\alpha_5\beta_1\beta_2}\right]\left[\frac{\beta_6+(1-\alpha_6-\beta_6)\beta_3}{1-\alpha_6\beta_3}\right]$	$\frac{f\gamma_1}{1-(1-f)\gamma_1} - \frac{f\gamma_0}{1-(1-f)\gamma_0}$
2	(0,0,1,0)	$\left[\frac{\beta_5+(1-\alpha_5-\beta_5)\beta_1\beta_2}{1-\alpha_5\beta_1\beta_2}\right]\left[\frac{\beta_6+(1-\alpha_6-\beta_6)\beta_4}{1-\alpha_6\beta_4}\right]$	$\frac{f\gamma_2}{1-(1-f)\gamma_2} - \frac{f\gamma_0}{1-(1-f)\gamma_0}$
3	(0,1,0,0)	$\left[\frac{\beta_5+(1-\alpha_5-\beta_5)\beta_1}{1-\alpha_5\beta_1}\right]\left[\frac{\beta_6+(1-\alpha_6-\beta_6)\beta_3\beta_4}{1-\alpha_6\beta_3\beta_4}\right]$	$\frac{f\gamma_3}{1-(1-f)\gamma_3} - \frac{f\gamma_0}{1-(1-f)\gamma_0}$
4	(1,0,0,0)	$\left[\frac{\beta_5+(1-\alpha_5-\beta_5)\beta_2}{1-\alpha_5\beta_2}\right]\left[\frac{\beta_6+(1-\alpha_6-\beta_6)\beta_3\beta_4}{1-\alpha_6\beta_3\beta_4}\right]$	$\frac{f\gamma_4}{1-(1-f)\gamma_4} - \frac{f\gamma_0}{1-(1-f)\gamma_0}$
5	(0,1,1,0)	$\left[\frac{\beta_5+(1-\alpha_5-\beta_5)\beta_1}{1-\alpha_5\beta_1}\right]\left[\frac{\beta_6+(1-\alpha_6-\beta_6)\beta_4}{1-\alpha_6\beta_4}\right]$	$\frac{f\gamma_5}{1-(1-f)\gamma_5} - \frac{f\gamma_3}{1-(1-f)\gamma_3} - \frac{f\gamma_2}{1-(1-f)\gamma_2} + \frac{f\gamma_0}{1-(1-f)\gamma_0}$
6	(0,0,1,1)	$\left[\frac{\beta_5+(1-\alpha_5-\beta_5)\beta_1\beta_2}{1-\alpha_5\beta_1\beta_2}\right]$	$\frac{f\gamma_6}{1-(1-f)\gamma_6} - \frac{f\gamma_2}{1-(1-f)\gamma_2} - \frac{f\gamma_1}{1-(1-f)\gamma_1} + \frac{f\gamma_0}{1-(1-f)\gamma_0}$
7	(0,1,0,1)	$\left[\frac{\beta_5+(1-\alpha_5-\beta_5)\beta_1}{1-\alpha_5\beta_1}\right]\left[\frac{\beta_6+(1-\alpha_6-\beta_6)\beta_3}{1-\alpha_6\beta_3}\right]$	$\frac{f\gamma_7}{1-(1-f)\gamma_7} - \frac{f\gamma_3}{1-(1-f)\gamma_3} - \frac{f\gamma_1}{1-(1-f)\gamma_1} + \frac{f\gamma_0}{1-(1-f)\gamma_0}$
8	(1,0,0,1)	$\left[\frac{\beta_5+(1-\alpha_5-\beta_5)\beta_2}{1-\alpha_5\beta_2}\right]\left[\frac{\beta_6+(1-\alpha_6-\beta_6)\beta_3}{1-\alpha_6\beta_3}\right]$	$\frac{f\gamma_8}{1-(1-f)\gamma_8} - \frac{f\gamma_4}{1-(1-f)\gamma_4} - \frac{f\gamma_1}{1-(1-f)\gamma_1} + \frac{f\gamma_0}{1-(1-f)\gamma_0}$
9	(1,1,0,0)	$\left[\frac{\beta_6+(1-\alpha_6-\beta_6)\beta_3\beta_4}{1-\alpha_6\beta_3\beta_4}\right]$	$\frac{f\gamma_9}{1-(1-f)\gamma_9} - \frac{f\gamma_3}{1-(1-f)\gamma_3} - \frac{f\gamma_4}{1-(1-f)\gamma_4} + \frac{f\gamma_0}{1-(1-f)\gamma_0}$
10	(1,0,1,0)	$\left[\frac{\beta_5+(1-\alpha_5-\beta_5)\beta_2}{1-\alpha_5\beta_2}\right]\left[\frac{\beta_6+(1-\alpha_6-\beta_6)\beta_4}{1-\alpha_6\beta_4}\right]$	$\frac{f\gamma_{10}}{1-(1-f)\gamma_{10}} - \frac{f\gamma_4}{1-(1-f)\gamma_4} - \frac{f\gamma_2}{1-(1-f)\gamma_2} + \frac{f\gamma_0}{1-(1-f)\gamma_0}$
11	(0,1,1,1)	$\left[\frac{\beta_5+(1-\alpha_5-\beta_5)\beta_1}{1-\alpha_5\beta_1}\right]$	$\frac{f\gamma_{11}}{1-(1-f)\gamma_{11}} - \frac{f\gamma_7}{1-(1-f)\gamma_7} - \frac{f\gamma_5}{1-(1-f)\gamma_5} + \frac{f\gamma_3}{1-(1-f)\gamma_3} - $ $P(0,0,1,1)$

12	(1,1,1,0)	$\left[\dfrac{\beta_6+(1-\alpha_6-\beta_6)\beta_4}{1-\alpha_6\beta_4}\right]$	$\dfrac{f\gamma_{12}}{1-(1-f)\gamma_{12}} - \dfrac{f\gamma_5}{1-(1-f)\gamma_5} - $ $\dfrac{f\gamma_{10}}{1-(1-f)\gamma_{10}} + \dfrac{f\gamma_2}{1-(1-f)\gamma_2} - $ $P(1,1,0,0)$
13	(1,1,0,1)	$\left[\dfrac{\beta_6+(1-\alpha_6-\beta_6)\beta_3}{1-\alpha_6\beta_3}\right]$	$\dfrac{f\gamma_{13}}{1-(1-f)\gamma_{13}} - \dfrac{f\gamma_8}{1-(1-f)\gamma_8} - $ $\dfrac{f\gamma_7}{1-(1-f)\gamma_7} + \dfrac{f\gamma_1}{1-(1-f)\gamma_1} - $ $P(1,1,0,0)$
14	(1,0,1,1)	$\left[\dfrac{\beta_5+(1-\alpha_5-\beta_5)\beta_2}{1-\alpha_5\beta_2}\right]$	$\dfrac{f\gamma_{14}}{1-(1-f)\gamma_{14}} - \dfrac{f\gamma_{10}}{1-(1-f)\gamma_{10}} - $ $\dfrac{f\gamma_8}{1-(1-f)\gamma_8} + \dfrac{f\gamma_4}{1-(1-f)\gamma_4} - $ $P(0,0,1,1)$
15	(1,1,1,1)	—	$1 - \left(\dfrac{f\gamma_{13}}{1-(1-f)\gamma_{13}} + \dfrac{f\gamma_{12}}{1-(1-f)\gamma_{12}}\right.$ $\left. - \dfrac{f\gamma_9}{1-(1-f)\gamma_9}\right) - P(0,1,1,1) - $ $P(1,0,1,1) - P(0,0,1,1)$

9.2.2 Likelihood function: case B

Under the topology indicated in Fig. 9.4(B), the joint probability of simple gene content pattern $\{Y_1, Y_2, Y_3, Y_4\}$ is given by

$$P(Y_1,Y_2,Y_3,Y_4) = \sum_{r_0=0}^{\infty}\sum_{r_A=0}^{\infty}\sum_{r_B=0}^{\infty}\pi(r_0)P(r_A|r_0;\beta_6,\alpha_6)P(r_B|r_A;\beta_5,\alpha_5)$$

$$\times P(Y_1|r_B;\beta_1)P(Y_2|r_B;\beta_2)P(Y_3|r_A;\beta_3)P(Y_4|r_0;\beta_4).$$

$$(9.24)$$

For $\{Y_1, Y_2, Y_3, Y_4\} = \{0,0,0,0\}$, we then have

$$P(0,0,0,0) = \sum_{r_0=0}^{\infty}\pi(r_0)\beta_4^{r_0}\sum_{r_A=0}^{\infty}P(r_A|r_0;\alpha_6,\beta_6)\beta_3^{r_A}\sum_{r_B=0}^{\infty}P(r_B|r_A;\alpha_5,\beta_5)\beta_1^{r_B}\beta_2^{r_B}$$

$$= \sum_{r_0=0}^{\infty}(1-f)^{r_0}f\beta_4^{r_0}\sum_{r_A=0}^{\infty}P(r_A|r_0;\alpha_6,\beta_6)\beta_3^{r_A}\left[\frac{\beta_5+(1-\alpha_5-\beta_5)\beta_1\beta_2}{1-\alpha_5\beta_1\beta_2}\right]^{r_A}$$

$$= \sum_{r_0=0}^{\infty}(1-f)^{r_0}f\beta_4^{r_0}\left[\frac{\beta_6+(1-\alpha_6-\beta_6)\beta_3\delta_0}{1-\alpha_6\beta_3\delta_0}\right]^{r_0}$$

$$= \frac{f\gamma_0}{1-(1-f)\gamma_0},$$

where δ_0 is given by

$$\delta_0 = \frac{\beta_5+(1-\alpha_5-\beta_5)\beta_1\beta_2}{1-\alpha_5\beta_1\beta_2}$$

and γ_0 is a function of the parameters $\{\alpha_5, \alpha_6, \beta_1, \cdots, \beta_6\}$ such that

$$\gamma_0 = \beta_4 \times \left[\frac{\beta_6 + (1 - \alpha_6 - \beta_6)\beta_3\delta_0}{1 - \alpha_6\beta_3\delta_0} \right]$$

Hence, the probability functions for the rest of gene content patterns can be derived in the same way. These probabilities have the same formula as those under the topology in Fig. 9.4A (Table 9.3) but with different $\gamma_i, i = 0, \cdots, 14$, as listed in Table 9.4.

9.2.3 Likelihood function

The gene content pattern of a gene family having been lost in all genomes cannot be observed, i.e. (0,0,0,0). To avoid this problem, we have to standardize these probability

Table 9.4 Probability (mass) functions of $Y = (Y_1, Y_2, Y_3, Y_4)$ in case B.

$\mathbf{Y}$	i	γ_i
(0,0,0,0)	0	$\beta_4 \times \left[\dfrac{\beta_6 + (1 - \alpha_6 - \beta_6)\beta_3 \frac{\beta_5+(1-\alpha_5-\beta_5)\beta_1\beta_2}{1-\alpha_5\beta_1\beta_2}}{1 - \alpha_6\beta_3 \frac{\beta_5+(1-\alpha_5-\beta_5)\beta_1\beta_2}{1-\alpha_5\beta_1\beta_2}} \right]$
(0,0,0,1)	1	$\dfrac{\beta_6 + (1 - \alpha_6 - \beta_6)\beta_3 \frac{\beta_5+(1-\alpha_5-\beta_5)\beta_1\beta_2}{1-\alpha_5\beta_1\beta_2}}{1 - \alpha_6\beta_3 \frac{\beta_5+(1-\alpha_5-\beta_5)\beta_1\beta_2}{1-\alpha_5\beta_1\beta_2}}$
(0,0,1,0)	2	$\beta_4 \times \left[\dfrac{\beta_6 + (1 - \alpha_6 - \beta_6) \frac{\beta_5+(1-\alpha_5-\beta_5)\beta_1\beta_2}{1-\alpha_5\beta_1\beta_2}}{1 - \alpha_6 \frac{\beta_5+(1-\alpha_5-\beta_5)\beta_1\beta_2}{1-\alpha_5\beta_1\beta_2}} \right]$
(0,1,0,0)	3	$\beta_4 \times \left[\dfrac{\beta_6 + (1 - \alpha_6 - \beta_6)\beta_3 \frac{\beta_5+(1-\alpha_5-\beta_5)\beta_1}{1-\alpha_5\beta_1}}{1 - \alpha_6\beta_3 \frac{\beta_5+(1-\alpha_5-\beta_5)\beta_1}{1-\alpha_5\beta_1}} \right]$
(1,0,0,0)	4	$\beta_4 \times \left[\dfrac{\beta_6 + (1 - \alpha_6 - \beta_6)\beta_3 \frac{\beta_5+(1-\alpha_5-\beta_5)\beta_2}{1-\alpha_5\beta_2}}{1 - \alpha_6\beta_3 \frac{\beta_5+(1-\alpha_5-\beta_5)\beta_2}{1-\alpha_5\beta_2}} \right]$
(0,1,1,0)	5	$\beta_4 \times \left[\dfrac{\beta_6 + (1 - \alpha_6 - \beta_6) \frac{\beta_5+(1-\alpha_5-\beta_5)\beta_1}{1-\alpha_5\beta_1}}{1 - \alpha_6 \frac{\beta_5+(1-\alpha_5-\beta_5)\beta_1}{1-\alpha_5\beta_1}} \right]$
(0,0,1,1)	6	$\dfrac{\beta_6 + (1 - \alpha_6 - \beta_6) \frac{\beta_5+(1-\alpha_5-\beta_5)\beta_1\beta_2}{1-\alpha_5\beta_1\beta_2}}{1 - \alpha_6 \frac{\beta_5+(1-\alpha_5-\beta_5)\beta_1\beta_2}{1-\alpha_5\beta_1\beta_2}}$
(0,1,0,1)	7	$\dfrac{\beta_6 + (1 - \alpha_6 - \beta_6)\beta_3 \frac{\beta_5+(1-\alpha_5-\beta_5)\beta_1}{1-\alpha_5\beta_1}}{1 - \alpha_6\beta_3 \frac{\beta_5+(1-\alpha_5-\beta_5)\beta_1}{1-\alpha_5\beta_1}}$

(1,0,0,1)	8	$\dfrac{\beta_6 + (1 - \alpha_6 - \beta_6)\beta_3 \frac{\beta_5 + (1 - \alpha_5 - \beta_5)\beta_2}{1 - \alpha_5\beta_2}}{1 - \alpha_6\beta_3 \frac{\beta_5 + (1 - \alpha_5 - \beta_5)\beta_2}{1 - \alpha_5\beta_2}}$
(1,1,0,0)	9	$\beta_4 \times \left[\dfrac{\beta_6 + (1 - \alpha_6 - \beta_6)\beta_3}{1 - \alpha_6\beta_3} \right]$
(1,0,1,0)	10	$\beta_4 \times \left[\dfrac{\beta_6 + (1 - \alpha_6 - \beta_6)\frac{\beta_5 + (1 - \alpha_5 - \beta_5)\beta_2}{1 - \alpha_5\beta_2}}{1 - \alpha_6 \frac{\beta_5 + (1 - \alpha_5 - \beta_5)\beta_2}{1 - \alpha_5\beta_2}} \right]$
(0,1,1,1)	11	$\dfrac{\beta_6 + (1 - \alpha_6 - \beta_6)\frac{\beta_5 + (1 - \alpha_5 - \beta_5)\beta_1}{1 - \alpha_5\beta_1}}{1 - \alpha_6 \frac{\beta_5 + (1 - \alpha_5 - \beta_5)\beta_1}{1 - \alpha_5\beta_1}}$
(1,1,1,0)	12	β_4
(1,1,0,1)	13	$\dfrac{\beta_6 + (1 - \alpha_6 - \beta_6)\beta_3}{1 - \alpha_6\beta_3}$
(1,0,1,1)	14	$\dfrac{\beta_6 + (1 - \alpha_6 - \beta_6)\frac{\beta_5 + (1 - \alpha_5 - \beta_5)\beta_2}{1 - \alpha_5\beta_2}}{1 - \alpha_6 \frac{\beta_5 + (1 - \alpha_5 - \beta_5)\beta_2}{1 - \alpha_5\beta_2}}$

functions as follows

$$q(Y_1, Y_2, Y_3, Y_4) = \frac{P(Y_1, Y_2, Y_3, Y_4)}{1 - P(0,0,0,0)}, \quad (Y_1, Y_2, Y_3, Y_4) \neq (0,0,0,0) \tag{9.25}$$

If there are M gene families in the case of four genomes, the likelihood for the parameters, $\alpha_5, \alpha_6, \beta_1, \cdots, \beta_6$, can be written as

$$L(\beta_1, \ldots, \beta_6 | \text{data}) = \prod_{m=1}^{M} q(Y_{m1}, Y_{m2}, Y_{m3}, Y_{m4}) = \prod_{i=1}^{15} q_i^{n_i}, \tag{9.26}$$

where q_i denotes the standardized probability for the i-th gene content pattern as indicated in the second column of Table 9.3, and n_i denotes the number of gene families that has pattern i; $\sum_i^{15} n_i = M$.

These unknown parameters can be estimated based on the maximum-likelihood approach. We have demonstrated that in the case of four genomes, there are only two different types of likelihood functions. We can find the value of the maximum likelihood (ML) in each case, e.g. using the Newton–Raphson algorithm, and then we use the topology which provides the maximum ML among all the MLs as the selected phylogenetic tree for the four genomes (Zhang and Gu 2004).

9.3 Birth–death model with lateral gene transfer

The stochastic models for gene content evolution we have discussed above do not explicitly involve the effect of lateral gene transfer (LGT), mainly due to the complexity of modeling. As LGT becomes an important mechanism in the evolution of microorganisms (Lawrence, 1999; Logsdon and Faguy, 1999), we attempt to address this theoretical issue (Zhang, Hemasinha, and Gu, unpublished results).

9.3.1 General birth–death process considering LGT

We use a linear birth and death process with the consideration of lateral gene transfer, namely, the average numbers of births and deaths per gene family in the genome are proportional to the number of existed member genes. Formally, it can be written as

$$\lambda_i = \lambda i + a$$

$$\mu_i = \mu i,$$

where λ and μ are the birth and death rates, respectively, and a is the rate of increase of the member genes due to the lateral gene transfer. If $a = 0$, the model is reduced to previous models based on the pure birth and death process.

The transition probability can be derived using the probability generating function. For an arbitrary but fixed initial state i, define the probability generating function $G_k(s,t)$ by the power series

$$G_k(s,t) = \sum_{r=0}^{\infty} s^r P_{k,r}(t),$$

where $0 < |s| < 1$. Since $P_{k,r}(t) \le 1$, this power series has a radius of convergence at least 1 and hence defines an analytic function of s. Differentiating $G_k(s,t)$ with respect to t leads to

$$\frac{\partial G_k(s,t)}{\partial t} = \sum_{r=0}^{\infty} s^r P_{k,r}^r(t)$$

With the help of the Kolmogorov forward equation and solving partial differential equations, we obtain analytical forms for $G_k(s,t)$ under different situations. When $\lambda \ne \mu$, the probability generating function $G_k(s,t)$ is derived as $G_k(s,t)$

$$G_k(s,t) = \frac{(\lambda - \mu)^{a/\lambda}}{(\lambda e^{(\lambda-\mu)t} - \mu)^{a/\lambda}} \frac{(\beta + (1 - \beta - \alpha)s)^k}{(1 - \alpha s)^{k+a/\lambda}},$$

where α and β are given by Eq. (9.2).

The transition probability is then obtained after applying general binomial expansion to the terms in $G_k(s,t)$. We have

$$P_{k,r}(t) = (1-\alpha)^{\delta} \sum_{i=0}^{min(k,r)} \binom{r}{i} \frac{(r+\delta)_{k-i}}{(k-i)!} \beta^{r-i}(1-\alpha-\beta)^i \alpha^{k-i}, \qquad (9.27)$$

where $\delta = a/\lambda$ which is the ratio of LGT rate and birth rate and $(r+\delta)_{k-i} = (r+\delta) \times (r+\delta+1) \cdots \times (r+\delta+k-i+1)$.

When the gene birth and death have the same rates, i.e. $\lambda = \mu$, we have

$$G_k(s,t) = \frac{(\lambda t + (1 - \lambda t)s)^k}{(1 + \lambda t - \lambda ts)^{k+a/\lambda}}$$

and

$$P_{k,r}(t) = \frac{(\eta)^{r+k}}{(1+\eta)^{r+k+\delta}} \sum_{i=0}^{min(k,r)} \binom{r}{i} \frac{(r+\delta)_{k-i}}{(k-i)!} \left(\frac{1}{(\eta)^2} - 1\right)^i, \qquad (9.28)$$

where $\eta = \lambda t = \mu t$. It should be noted that the assumption of equal gene birth and loss rates is over-simplified, because in practice it is usually difficult to know whether the gene birth and loss rates are the same.

9.3.2 Extended gene content under LGT

We use the extended gene content to estimate the parameters α, β, and δ, which considers three possible states of a gene family: no member gene present, single-copy gene present, and duplicates in the current genome. Let $Z = 0, 1, 2$ stand for the situations of complete loss of a member gene ($X_t = 0$), single-copy of a member gene ($X_t = 1$), and duplicated member gene ($X_t \geq 2$), respectively. Using Eq. (9.27), we have the transition probabilities

$$P(Z = 0|X(0) = r) = (1 - \alpha)^\delta \beta^r$$

$$P(Z = 1|X(0) = r) = (1 - \alpha)^\delta (r + \delta)\beta^r \alpha + (1 - \alpha)^\delta r \beta^{r-1}(1 - \alpha - \beta)$$

$$P(Z = 2|X(0) = r) = 1 - (1 - \alpha)^\delta \beta^r (1 + (r + \delta)\alpha + r\beta^{-1}(1 - \alpha - \beta)),$$

$$(9.29)$$

Consider two genomes that have been diverged t time units ago, each of extented gene content pattern is denoted by Z_1 and Z_2, respectively. Apparently, there are 9 combinations of patterns. Similar to the above derivation in the case of no LGT (Gu and Zhang 2004), we obtain

$$P(0,0) = (1 - \alpha_1)^{\delta_1}(1 - \alpha_2)^{\delta_2} Q(\beta_1 \beta_2)$$

$$P(0,1) = (1 - \alpha_1)^{\delta_1}(1 - \alpha_2)^{\delta_2}(1 - \alpha_2)(1 - \beta_2)\beta_1 R(\beta_1 \beta_2)$$
$$+ (1 - \alpha_1)^{\delta_1}(1 - \alpha_2)^{\delta_2} \alpha_2 \delta_2 Q(\beta_1 \beta_2)$$

$$P(0,2) = (1 - \alpha_1)^{\delta_1} Q(\beta_1) - P(0,0) - P(0,1)$$

$$P(1,0) = (1 - \alpha_1)^{\delta_1}(1 - \alpha_2)^{\delta_2}(1 - \alpha_1)(1 - \beta_1)\beta_2 R(\beta_1 \beta_2)$$
$$+ (1 - \alpha_1)^{\delta_1}(1 - \alpha_2)^{\delta_2} \alpha_1 \delta_1 Q(\beta_1 \beta_2)$$

$$P(1,1) = (1 - \alpha_1)^{\delta_1}(1 - \alpha_2)^{\delta_2}(1 - \alpha_1)(1 - \beta_1)(1 - \alpha_2)(1 - \beta_2)S(\beta_1 \beta_2)$$
$$+ (1 - \alpha_1)^{\delta_1}(1 - \alpha_2)^{\delta_2}(\alpha_1 \beta_1 \delta_1 (1 - \alpha_2)(1 - \beta_2)$$
$$+ \alpha_2 \beta_2 \delta_2 (1 - \alpha_1)(1 - \beta_1))R(\beta_1 \beta_2)$$
$$+ (1 - \alpha_1)^{\delta_1}(1 - \alpha_2)^{\delta_2} \alpha_1 \alpha_2 \delta_1 \delta_2 Q(\beta_1 \beta_2)$$

$$P(1,2) = (1-\alpha_1)^{\delta_1}(1-\alpha_1)(1-\beta_1)R(\beta_1) + (1-\alpha_1)^{\delta_1}\alpha_1\delta_1 Q(\beta_1) - P(1,0) - P(1,1)$$

$$P(2,0) = (1-\alpha_2)^{\delta_2} Q(\beta_2) - P(0,0) - P(1,0)$$

$$P(2,1) = (1-\alpha_2)^{\delta_2}(1-\alpha_2)(1-\beta_2)R(\beta_2)$$
$$+(1-\alpha_2)^{\delta_2}\alpha_2\delta_2 Q(\beta_2) - P(0,1) - P(1,1)$$

$$P(2,2) = 1 - (1-\alpha_2)^{\delta_2}(1+\alpha_2\delta_2)Q(\beta_2)$$
$$-(1-\alpha_2)^{\delta_2}(1-\alpha_2)(1-\beta_2)R(\beta_2) - P(0,2) - P(1,2), \qquad (9.30)$$

where

$$Q(\beta) = \frac{f}{1-\beta(1-f)}, \quad R(\beta) = \frac{f(1-f)}{(1-(1-f)\beta)^2}$$

and

$$S(\beta_1\beta_2) = \frac{f(1-f)}{(1-\beta_1\beta_2(1-f))^2} + \frac{2\beta_1\beta_2(1-f)}{(1-\beta_1\beta_2(1-f))^3}.$$

Since double losses are not observable, we modify Eq.(9-30) as

$$q(Z_1 = i, Z_2 = j|\phi) = \frac{P(Z_1 = i, Z_2 = j)}{P(Z_1 = 0, Z_2 = 0)}, \quad i,j = 0,1,2 \text{ and } (i,j) \neq (0,0), \qquad (9.31)$$

and the likelihood function for the parameters is then given as

$$L(\phi|\mathbf{Z_1}, \mathbf{Z_2}) = \prod_{l=1}^{N} q(Z_{1_l} = i, Z_{2_l} = j|\phi) = \prod_{\mathbf{i,j}=\{0,1,2\},\{ij\}\neq\{00\}} \mathbf{q(i,j)^{n_{ij}}}, \qquad (9.32)$$

if there are N gene families in the two genomes. $\mathbf{Z_1}, \mathbf{Z_2}$ is used to denote all the extended gene content information on N gene families. n_{ij} is the correspondent observed number of gene families, and we have $\sum_{i,j=\{0,1,2\},\{ij\}\neq\{00\}} n_{ij} = N$. The parameters α_i, β_i, and δ_i can be estimated using the maximum-likelihood approach.

9.3.3 Simple gene content and LGT

Gene content is a binary (0 or 1) index of a gene family in a genome. Let $Y = 1$ denote the case that at time t this genome maintains at least one member gene of the gene family, i.e. $X_t \geq 1$, and $Y = 0$ be the case that all the member genes are lost, i.e. $X_t = 0$. Though only using simple gene content information is not enough to infer possible lateral gene transfer and gene birth–death, it is theoreticaly interesting to study the effect of the LGT on the evolution of gene content in the genome. Under this simplest case, the transition probability Eq. (9.29) becomes

$$P(Y = 0|X_0 = r) = P(X_t = 0|X_0 = r)$$
$$= (1-\alpha)^{\delta}\beta^r$$
$$P(Y = 1|X_0 = r) = 1 - (1-\alpha)^{\delta}\beta^r. \qquad (9.33)$$

Hence, the joint probability of two genomes with $Y_1 = i$ and $Y_2 = j$ $(i, j = 0, 1)$ is given by

$$q(Y_1 = i, Y_2 = j|\phi) = \frac{P(Y_1 = i, Y_2 = j)}{P(Y_1 = 0, Y_2 = 0)}, \ i, j = 0, 1, \ \text{and} \ (i, j) \neq (0, 0),$$

where $\phi = \{\beta_1, \beta_2, \delta_1, \delta_2\}$, since $\{0, 0\}$ is not observable. Specifically, for every possible combination of Y_1 and Y_1, we have the modified probabilities

$$q(0, 1) = \frac{(1 - \alpha_1)^{\delta_1} Q(\beta_1) - (1 - \alpha_1)^{\delta_1} (1 - \alpha_2)^{\delta_2} Q(\beta_1 \beta_2)}{1 - (1 - \alpha_1)^{\delta_1} (1 - \alpha_2)^{\delta_2} Q(\beta_1 \beta_2)}$$

$$q(1, 0) = \frac{(1 - \alpha_2)^{\delta_2} Q(\beta_2) - (1 - \alpha_2)^{\delta_2} (1 - \alpha_1)^{\delta_1} Q(\beta_1 \beta_2)}{1 - (1 - \alpha_1)^{\delta_1} (1 - \alpha_2)^{\delta_2} Q(\beta_1 \beta_2)}$$

$$q(1, 1) = \frac{1 - (1 - \alpha_2)^{\delta_2} Q(\beta_2) - (1 - \alpha_1)^{\delta_1} Q(\beta_1) + (1 - \alpha_1)^{\delta_1} (1 - \alpha_2)^{\delta_2} Q(\beta_1 \beta_2)}{1 - (1 - \alpha_1)^{\delta_1} (1 - \alpha_2)^{\delta_2} Q(\beta_1 \beta_2)},$$

$$(9.34)$$

where $Q(z) = f/[1 - z(1 - f)]$. Note that parameters α_i are not in the above expressions, which implies the inestimability of α_i.

9.3.4 Comments

Simulation study has shown that overall, based on the extended gene content, the LGT parameters are difficult to estimate, while the estimates for the gene birth and loss parameters seem to be fine. These observations are not surprising, as we can only evaluate the combined effect of gene birth and LGT rather than independent effects because we lack sufficient information. It remains a challenging issue in phylogenomic inference under the lateral gene transfer.

9.4 Other Models

9.4.1 Blocks model

Spencer *et al.* (2006) studied the blocks model for the evolution of gene content, assuming that duplications, deletions, and transfers can affect multiple genes within a family. In the birth–death model, we know that duplication and deletion events operate independently and at a constant rate on each member gene. In the blocks models, the units may be larger than one gene. Spencer *et al.* (2006) modeled the innovation, deletions, and duplications of single genes exactly as in the birth–death model, with the exception of the transition from state 1 to state 0 (deletion process). They considered the loss process from single-gene losses to deletions that result in the loss of an entire gene family. Spencer *et al.* (2006) analyzed two pairs of genomes: two *E. coli* strains, and the distantly-related *Archaeoglobus fulgidus* (archaea) and *Bacillus subtilis* (gram positive bacteria) and concluded that blocks models described the data better than birth–death models.

9.4.2   Equal birth–death rate model

Hahn *et al.* (2005) developed an analytical pipeline for estimating the tempo and mode of gene family evolution from the size distribution of gene families. They implemented a special birth–death model that assumed equal birth and death rates in gene family evolution and showed that it can be efficiently applied to multi-species genome comparisons. This model took into account the lengths of branches on phylogenetic trees, as well as duplication and deletion rates, and hence provides expectations for divergence in gene family size among lineages. The model is useful for identifying large-scale patterns in genome evolution and its ability to make stronger inferences regarding the role of natural selection in gene family expansion or contraction. However, the remaining question is whether the key assumption, equal birth and death rates, is biologically realistic. In fact, as shown above, in principle one can easily develop an analytical pipeline without this assumption, but the computational complexity is the major problem.

9.4.3   Constant-birth, proportional-death model

Huson and Steel (2004) considered the following model. Genomes evolve according to a constant-birth, proportional-death Markov process. That is, at each instant each gene in the genome can be independently deleted with the death rate, or the genome can acquire a new gene (gene genesis) independently, at a constant rate. This model actually is a special case of the general birth–death model with lateral gene transfer when the birth rate from gene duplications is negligible. Under this model, Huson and Steel (2004) derived a maximum-likelihood estimation of evolutionary distance between species under a simple model of gene genesis and gene loss. Using simulated data, they compared the accuracy of the tree reconstruction using this ML distance measure to an earlier *ad hoc* distance, as well as the character-based Dollo parsimony method. Their results showed a consistent trend, with the character-based method and ML distance measure outperforming the earlier *ad hoc* distance method.

10

Advanced Topics in Systems Biology and Network Evolution

As evolutionary biologists have always been concerned with the genetic basis for the emergence of complex phenotypes, advances in genomics and systems biology are facilitating a paradigm shift of molecular evolutionary biology toward a better understanding of the relationship of genotypes and phenotypes (Gerhart and Kirschner 1997; Elena and Lenski 2003; Lynch 2007c; Wagner *et al.* 2007). In spite of high controversies, this subject, featured by some buzzwords in the literature such as modularity, evolvability, robustness, and complexity, has recently received a broad array of novel insights from nearly every branch of biological science (Medina 2005; Lynch 2007c; Pal *et al.* 2006a, 2006b; Koonin and Wolf 2006; McGuigan and Sgro 2009). For instance, one of the research hotspots in evolutionary genomics was the pervasive, weak and sophisticated genomic correlations between the sequence conservation, expression level, protein connectivity, and gene essentiality (Hirsh and Fraser 2001; Fraser *et al.* 2002; Wall *et al.* 2005). These genomic correlations have generated a number of interesting yet controversial issues about the pattern of genomic evolution (Duret and Mouchiroud 2000; Krylov *et al.* 2003; Pal *et al.* 2003; Jordan *et al.* 2003; Yang *et al.* 2003; Rocha and Danchin 2004; Drummond *et al.* 2005; Drummond *et al.* 2006; Salathe *et al.* 2006; Batada *et al.* 2006; Wolf *et al.* 2006; Wolf 2006).

From the evolutionary perspective, the central question is whether natural selection is a necessary and/or sufficient force to explain the emergence of genomic and cellular features that underlie the building of complex organisms. Lynch (2007c) has criticized the adaptive hypothesis for the origins of organismal complexity (True and Haag 2001; Alon 2003; Carroll 2005; Aharoni *et al.* 2005; Adami 2006; Tsong *et al.* 2006), claiming that nothing in evolution makes sense in light of population genetics that takes the effects of mutation, genetic drift, and natural selection into account. The importance of mutation types and genetic drifts on the phenotype evolution has also been emphasized by Nei and his associates (e.g. Nei *et al.* 1997, 2008; Nei 2005, 2007; Nozawa *et al.* 2007) as well as Sole and Valverde (2006). One plausible approach to resolving these fundamental issues is to model the features of biological complexity as parameters instead of emerged properties, under the principle of population genetics and molecular evolution. In this chapter, we discuss some mostly recent results in this trend.

10.1   GC mutational bias rather than adaptation driving tyrosine loss in metazoan genome evolution

The origin and evolution of cellular signaling has recently been a hotspot in evolutionary genomics (Manning *et al.* 2008; Collins 2009; Tan *et al.* 2009; Holt *et al.* 2009; Landry *et al.* 2009). Tan *et al.* (2009) observed a negative correlation between the genomic frequency of the amino acid tyrosine (Y) and the number of cell types, as well as the number of tyrosine protein kinases in yeast and 15 metazoan model organisms. They further claimed that spurious and deleterious tyrosine phosphorylations may have been effectively removed by loss-of-tyrosine mutations that are selectively beneficial. Consequently, natural selection may shape the signaling network complexity for adaptive increase of cell types of metazoans. We (Su, Huang, and Gu, unpublished results) challenge the validity of Tan *et al.*'s adaptive hypothesis, and present strong evidence supporting the biased mutational hypothesis (Gu *et al.* 1998). That is, directional mutational pressure toward high genomic GC content (guanine plus cytosine) during the metazoan evolution may be the main driving force for the loss of genome-wide tyrosine, simply because tyrosine is encoded by two AU(T)-rich codons (UAU and UAC). Hence, a more plausible evolutionary scenario would be that metazoans may utilize the GC mutational bias to remove spurious tyrosine-related phosphorylations, and to facilitate functional specificity of signaling pathways.

We first notice that genomic analysis of choanoflagellate *Monosiga brevicollis*, a unicellular species close to the metazoan lineage that contains canonical tyrosine kinases (Manning *et al.* 2008), did not support Tan *et al.*'s (2009) adaptive hypothesis. This is because the genomic tyrosine frequency of choanoflagellate is extraordinarily low, which would nullify the claim that organisms with few cell types are expected to have a high tyrosine frequency (Fig. 10.1(A)). This dilemma can be easily resolved from the hypothesis of GC mutational bias, by the fact that the genomic GC content of choanoflagellate is as high as 54 per cent (Manning *et al.* 2008), which is much higher than that (38 per cent) of the yeast, another simple organism, and those multi-cellular metazoans (35 per cent–46 per cent). To test whether the variation of genomic GC content determines the variation of tyrosine frequency in yeast, choanoflagellate, and metazoan genomes, we have reanalyzed the genome data used in Tan *et al.* (2009) and the choanoflagellate genome. As expected, Fig. 10.1(B) shows a highly significant negative correlation between the genomic GC content and the tyrosine frequency (Spearman's $R = 0.92$, $P < 1 \times 10^{-6}$). However, such simple regression analysis, as well as all analyses conducted in Tan *et al.* (2009), neglects the effect of phylogenetic tree. The consequence is to inflate the significance level, especially in the case of small sample size. Nevertheless, after correcting this effect as Gu *et al.* (1998), the negative correlation between GC content and tyrosine frequency remains valid ($P < 1 \times 10^{-4}$).

Further, the GC mutational bias hypothesis predicts a similar trend in other amino acids encoded by AU(T)-rich codons, such as phenylalanine (F), asparagine (N), lysine (K), isoleucine (I), and methionine (M), and an opposite trend in amino acids encoded by GC-rich codons, such as proline (P), alanine (A), glycine (G), and tryptophan (W). These two patterns as exactly predicted by the mutational bias hypothesis are consistent with the previous result (Gu *et al.* 1998) in bacteria. Since there is no need

(A)

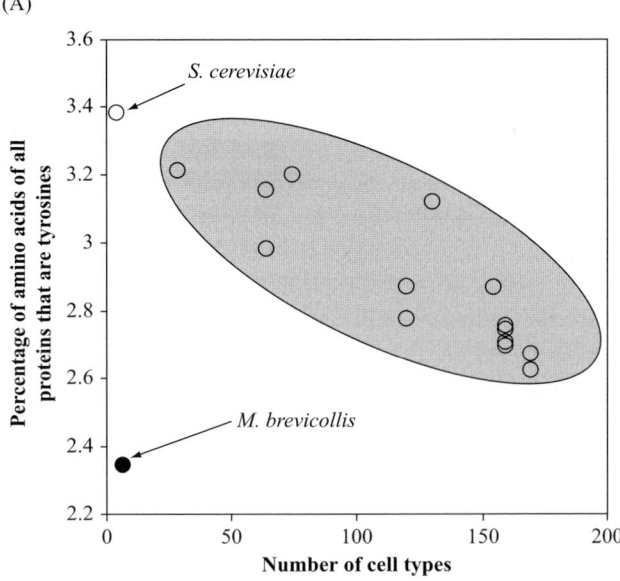

(B)

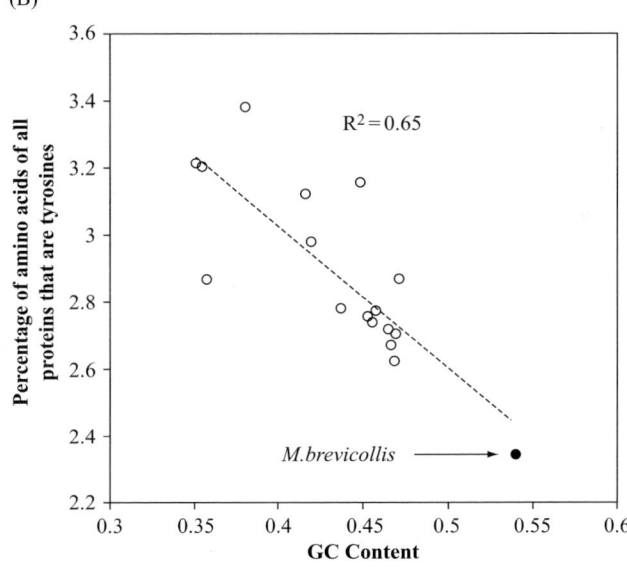

Fig. 10.1 Correlations of organism complexity or genomic GC content with the genomic frequency of tyrosine. (A) The choanoflagellate *M. brevicollis* (solid dot) is an apparent outlier in the correlation between the frequency of genome-encoded tyrosine residues and the number of cell types of organism. (B) The genomically tyrosine content decreases as the genomic GC content increases. The empty dots represent the yeast and other 15 metazoans analyzed in the Tan *et al.*'s study. Here we use the mean GC content of upstream 2kb and downstrean 2kb sequences for all protein-coding genes of the genome to represent the GC content bias. Using the GC content of the intron region, GC content at the four-fold degenerate sites of genes (GC4), or limiting the analysis to sets of one-to-one orthologous sequences do not change our results virtually. Su *et al.* (personal communication).

to assume an additional, specific mechanism to explain the loss of tyrosine in these organisms, the basis of Tan *et al.*'s adaptive hypothesis (2009) is unlikely to be sound.

Tan *et al.* (2009) observed that the tyrosine loss rate in tyrosine phosphorylated proteins (pTyr) is slower than that in non-pTyr proteins, and attributed this difference to the positive selection in these non-pTyr proteins to remove potentially deleterious phosphorylation sites. Under the biased mutational hypothesis, a much simpler explanation invokes to some weak selection constraints imposed on functional phosphorylated tyrosine sites in pTyr proteins, consistent with the view (Landry *et al.* 2009) that phosphoproteomes are weakly constrained. On the other hand, in these non-pTyr proteins, gain and loss of tyrosine are mainly determined by the genetic drifts under the theme of neutral evolution (Kimura 1983).

Our analysis may provide some new insights on the evolutionary interplay between biased mutational pressure, signaling network, and multi-cellularity (Lynch 2007c) in metazoans and unicellular choanoflagellate. After their split over one billion years ago, genomic GC contents have increased independently in both lineages, likely driven by the directional mutational bias (Sueoka 1988; Gu *et al.* 1998). In addition, GC isochores in warm-blood animals may further increase the GC content in the coding region (Bernardi *et al.* 1985). The GC mutational bias may initiate the parallel, non-adaptive processes of tyrosine loss in both metazoan and choanoflagellate lineages. Interestingly, both lineages have independently acquired almost the same number of tyrosine kinases by continuous gene duplications. Tan *et al.* (2009) speculated that removing nonfunctional kinase-tyrosine interactions may help to facilitate the evolution of modular-like signaling network through gene duplications. Our recent theoretical network analysis (Gu 2009) indicates that it might have occurred. Unlike Tan *et al.* (2009), we emphasize that emergence of a tyrosine kinase-related signaling network would be mainly driven by the GC-biased mutational pressure and gene duplications, rather than adaptation. Natural selection, as an efficient opportunist, may have successfully taken advantage of this non-adaptive, mutation-driven genomic dynamics to enhance the organismal complexity (multi-cellularity) in the metazoan lineage. In choanoflagellate, the non-adaptive origin of tyrosine kinases did not result in any dramatic change in phenotypic complexity, though the exact role of these species-specific tyrosine kinases remains unknown. In a broad sense, how mutational bias may affect the origin and evolution of protein interactions by controlling amino acid compositions and abundance of specific interaction motifs should become an interesting issue.

10.2   Contribution of duplicate genes to genetic robustness

Genetic robustness has been a fundamental issue in the study of biological complexity (Wilson *et al.* 1977; de Visser *et al.* 2003; Yang *et al.* 2003; Hirsh and Fraser 2001; Wagner 1999, 2000a, 2005a, 2005b, 2008). Functional compensation between duplicate genes has been thought to be an important factor in genetic robustness (Conant and Wagner 2004; Dean *et al.* 2008; Gu 2003; Gu *et al.* 2003; Harrison *et al.* 2007; Ihmels *et al.* 2007; Kamath *et al.* 2003; Winzeler *et al.* 1999). This is because the existence of a close duplicate in the same genome could decrease the chance of being essential (indispensable), that is, a null mutation of the gene that has a lethal effect on the organismal fitness. Indeed, duplicate genes are less essential than single-copy genes in several model organisms (Gu *et al.* 2003; Conant and Wagner 2004; Kamath *et al.* 2003). However, the relative contribution of duplicate genes to genetic robustness remains a subject open to debate (Gu *et al.* 2003; Liang and Li 2007; Liao and Zhang 2007; Su and Gu 2008). In the following, we model the evolution of functional compensation between duplicate genes, which may provide some new insights about this problem.

10.2.1   Functional compensation between duplicate genes

According to the subfunctionalization framework (Prince and Pickett 2002), the function of a protein and the expression pattern can be conceptually viewed as a sum of several (m) independent subcomponents. After the gene duplication, two duplicate copies are initially identical in each subcomponent. The follow-up functional divergence between them can be described as the complementary loss of subcomponents (subfunctionalization) (Force *et al.* 1999, 2005), or neo-functionalization (Ohno 1970). For simplicity, we assume that the fitness effect by the single gene deletion can only be any of two states: (*i*) dispensable or nonessential, denoted by d^+, which means that the fitness of the mutant is the same as that of the wild-type; and (*ii*) indispensable or essential, denoted by d^-, which means that the fitness effect of the mutant is lethal. Wagner (2000b, 2001) has conducted a population genetics analysis to show how the effective population size and pleiotropic mutations may affect the evolution of functional overlap between duplicate genes.

In the case of two-member gene family, let f_A or f_B be the fitness of the organism when duplicate gene A or B is deleted, respectively. Our purpose is to derive $Q(f_A, f_B)$, the probability of being the pattern (f_A, f_B), where $f_A, f_B = d^+, d^-$. That is, $Q(d^+, d^+)$ is the probability that both duplicates are dispensable. Similarly, $Q(d^-, d^+)$ is the probability that gene A is essential but gene B is dispensable, and so is the $Q(d^+, d^-)$. Finally, $Q(d^-, d^-)$ is the probability that both duplicates are indispensable or essential.

For a duplicate pair with m subfunctional components, the property of multi-functionality ($m > 1$) is a prerequisite for subfunctionalization (Prince and Pickett 2002). In each subfunctional component, we assume a two-state model: One state is 'active' ('1'), which means fully functional, and the other one is 'inactive' ('0'). Let U_{ij} ($i, j = 0, 1$) be the probability that a given subfunctional component is in state-i in gene A and state-j in gene B. For instance, U_{11} is the probability that a given component is in the state of '1' (active) in both genes; U_{01} is that of '0' (inactive) in gene A but '1' (active) in gene B; and so forth. Obviously, $U_{11} + U_{10} + U_{01} + U_{00} = 1$.

10.2.2   When an essential gene is duplicated

Suppose that the ancestral gene (O) of duplicates A and B was essential, a status denoted by O^-. In this case, duplicate A is dispensable if all its m functional components can be compensated by duplicate B, and *vice versa*. Obviously, a nonfunctional 'pseudogene' (ψ) is dispensable. To help derivation of $Q(f_A, f_B|O^-)$, the probability of duplicates A and B being the status (f_A, f_B) when their ancestral gene was essential, we first consider the combined status $d_\psi^+ = (d^+ \text{ or } \psi)$ called the (broad sense) dispensability that includes being a functional gene or a nonfunctional pseudogene. We therefore claim the following relationships

$$Q(d_\psi^+, d_\psi^+|O^-) = (U_{11} + U_{00})^m$$

$$Q(d_\psi^+, d^-|O^-) = (1 - U_{10})^m - (U_{11} + U_{00})^m$$

$$Q(d^-, d_\psi^+|O^-) = (1 - U_{01})^m - (U_{11} + U_{00})^m$$

$$Q(d^-, d^-|O^-) = 1 - Q(d_\psi^+, d_\psi^+|O^-) - Q(d_\psi^+, d^-|O^-) - Q(d^-, d_\psi^+|O^-) \qquad (10.1)$$

The rationale for the first equation in Eq. (10.1) is that both duplicates are (broad sense) dispensable if each subfunctional component is active (functional compensation) or inactive (functional loss) in both duplicate genes. The second equation means that the probability of gene A to be (broad sense) dispensable but B to be essential equals to the probability of (broad sense) dispensable gene A (the first term of the right hand) minus the probability of both being (broad sense) dispensable (the second term). Similarly we have the third equation. The fourth is from the sum over all probabilities to be one.

Next we distinguish between the dispensable functional duplicate and the pseudogene. Let ϕ be the probability that one of duplicate genes becomes nonfunctional by null mutations such as stop-codon mutations. On the other hand, null subfunctionalization may cause one duplicate gene to be nonfunctional. It includes three types, i.e. one of the copies, or both, have lost all functional components, with a probability of U_{01}^m, U_{10}^m or $1 - (1 - U_{00})^m$, respectively. Under some moderate assumptions, the probability that at least one of two duplicates being a pseudogene, denoted by ψ^*, is given by

$$Q(\psi^*|O^-) = \phi + (1 - \phi)\left(U_{01}^m + U_{10}^m + 1 - (1 - U_{00})^m\right)$$

Moreover, when an essential gene is duplicated, each subcomponent must be functionally active at least in one duplicate gene, the condition of *not-all-loss* that implies $U_{00} = 0$. Put together, Eq. (10.1) can be rewritten as follows

$$Q(d^+, d^+|O^-) = (1 - \phi)U_{11}^m$$

$$Q(d^+, d^-|O^-) = (1 - \phi)\left[(1 - U_{10})^m - U_{11}^m - U_{01}^m\right]$$

$$Q(d^-, d^+|O^-) = (1 - \phi)\left[(1 - U_{01})^m - U_{11}^m + U_{10}^m\right]$$

$$Q(d^-, d^-|O^-) = (1 - \phi)\left[1 - (1 - U_{10})^m - (1 - U_{01})^m + U_{11}^m\right]$$

$$Q(\psi^*|O^-) = \phi + (1 - \phi)\left(U_{01}^m + U_{10}^m\right) \qquad (10.2)$$

In practice, the probability of a duplicate gene being dispensable has been used to measure the duplicate effect on genetic robustness, after excluding pseudogenes. Note that the probability of being dispensable given the essential ancestral gene, is $P(d^+|O^-) = [Q(d^+, d^+|O^-) + Q(d^+, d^-|O^-)]/[1 - Q(\psi^*|O^-)]$. From Eq. (10.2) we have

$$P(d^+|O^-) = \frac{(1 - U_{10})^m - U_{01}^m}{1 - U_{01}^m - U_{10}^m} \tag{10.3}$$

For simplicity, one may assume that the gain and loss of subfunctional components are symmetric between duplicates, i.e. $U_{10} = U_{01} = U$ and $U_{11} = 1 - 2U$. Hence, Eq. (10.3) can be simplified as follows

$$P(d^+|O^-) = \frac{(1 - U)^m - U^m}{1 - 2U^m} \tag{10.4}$$

where $0 \leq U \leq 1/2$. At the initial stage, $U_{11} = 1$ so that $U = 0$ and $P(d^+|O^-) = 1$. When all subfunctional components are complements between duplicates, $U = 1/2$ and $P(d^+|O^-) = 0$. which means that both duplicates becomes essential. To determine U we make the following assumptions:

(1) With a probability of f_1, a single component in the current duplicate pair has existed in the ancestral gene prior to the gene duplication, called the ancestral component, while with a probability of $f_0 = 1 - f_1$, it has a recent origin after the gene duplication, called the derived component.

(2) With a probability of ϵ, a functional component cannot be subfunctionalized; one may call the basic component. With a probability of $1 - \epsilon$, a functional component can be subfunctionalized between duplicate genes. Therefore, an ancestral component remains active in both duplicates with a probability of $\epsilon + (1 - \epsilon)e^{-2\lambda t}$, while it remains active in only one of two duplicates with a probability of $(1 - \epsilon)(1 - e^{-2\lambda t})$, where λ is the loss rate of an ancestral component. Under this model, double-loss of an ancient functional component is negligible.

(3) The probability of being a derived component is given by $f_0 = a_0(1 - e^{-2\rho t})$, where ρ is the rate of acquiring a new functional component and a_0 is the up-bound of f_0. Simultaneous acquisition of the same functional component in both duplicates is negligible.

Based on these assumptions, the probability of a subfunctional component being active in both duplicates is given by

$$U_{11} = f_1 \left[\epsilon + (1 - \epsilon)e^{-2\lambda t} \right]$$

Noting that $U = U_{10} = U_{01} = (1 - U_{11})/2$, we obtain

$$U = 1/2 - \left(1 - a_0 + a_0 e^{-2\rho t}\right) \left[\epsilon + (1 - \epsilon)e^{-2\lambda t} \right]/2 \tag{10.5}$$

Eqs. (10.3) to (10.5) show that $P(d^+|O^-)$ decays with the divergence time t as a result of subfunctionalization and neofunctionalization. At $t \to \infty$, one can easily show $U \to U_\infty = 1/2 - r_0$, where $r_0 = \epsilon(1 - a_0)/2$. Hence, when the ancestral gene was essential, the probability of a duplicate remaining dispensable reaches an equilibrium, denoted by $P_\infty(d^+|O^-)$, which is positive when $r_0 > 0$.

10.2.3 When a dispensable gene is duplicated

The subfunctionalization framework (Force *et al.* 1999; Prince and Pickett 2002) for duplicate gene preservation does not explictly take the effect of ancient genetic robustness, i.e. ancient genetic buffering or ancient paralogs prior to the duplication, into account. If a dispensable gene is duplicated, whether these two copies can be functionally preserved remains open to question. We argue that subfunctionalization would be much less efficient when a dispensable gene is to be duplicated. To avoid one of these duplicates being a pseudogene, neofunctionalization (Ohno 1970) must be invoked as an escaping mechanism from the ancestral genetic buffering or the ancestral duplicate compensation.

Let W be the probability of a duplicate gene escaping from the ancestral genetic robustness. Hence, the ancient genetic buffering or compensation may allow the divergence between duplicate genes to be nearly neutral until one of them has escaped from the ancient buffering or compensation. Since the fate of a genetically-buffered duplicate is almost independent of the other copy, it is sufficient to only consider the case of one duplicate thereafter. Given the dispensable ancestral gene (O^+), let $Q(d^+|O^+)$ or $Q(d^-|O^+)$ be the probability of a duplicate copy being dispensable or essential, respectively, and $Q(\psi|O^+)$ be that of being nonfunctional (pseudogene). Let μ be the rate of being a pseudogene by null mutations. Together, we thus obtain

$$Q(d^+|O^+) = e^{-\mu t}(1 - W)(1 - \phi)$$

$$Q(d^-|O^+) = W(1 - \phi)$$

$$Q(\psi|O^+) = (1 - e^{-\mu t})(1 - W)(1 - \phi) + \phi \qquad (10.6)$$

One may choose a simple form $W = \beta(1 - e^{-bt})$, indicating that the escaping probability increases with time t toward β as $t \to \infty$; b is the escaping rate. Apparently, $Q(d^+|O^+) = 1$ when $t = 0$. After a sufficiently long time, a duplicate gene without escaping the ancestral buffering would become a pseudogene and lost from the genome. Hence, when $t \to \infty$, we have $Q_\infty(d^+|O^+) = 0$, $Q_\infty(d^-|O^+) = \beta$ or $Q_\infty(\psi|O^+) = 1 - \beta$.

10.2.4 Hypothesis: duplication of dispensable genes to maintain genetic buffering

An important feature of biological complexity is the genetic robustness against null mutations (Gerhart and Kirschner 1997; Lynch 2007c; Wagner *et al* 2007). Current wisdom suggests two principle mechanisms. The first is derived from overlapping gene functions (Gu *et al.* 2003; Ihmels *et al.* 2007). Because most eukaryotes have a significant portion of duplicate genes and because many of these duplicate genes have at least partial redundant functions, functional compensation apparently is one plausible mechanism for the genetic resilience. The second mechanism against null mutations, known as genetic buffering, is based on specified alternative pathways or backup circuits that minimize or even remove the deleterious effects of mutations. Though both mechanisms are well documented in the literature, which mechanism is more prevalent has been controversial (Liang and Li 2007; Liao and Zhang 2007; Su and Gu 2008). On the other hand, the preservation of duplicate genes is coupled with the loss of overlapping gene functions between duplicates, via either subfunctionalization or neofunctionalization (Force *et al.* 1999). As a result, the contribution of duplicate genes to genetic robustness appears to be evolutionarily transient.

Here we argue that genetic buffering is also evolutionarily transient. Consider a single copy gene (A) that has been genetically buffered. Since any null mutation of gene A can result in virtually no or very weak phenotypic effect, it is actually nearly-neutral to the organism, which can be fixed by the genetic drifts particularly when the effective population size (N_e) is small. In this case, the related genetic buffering would be diminished because the backup circuit now has to play a primary role in performing the function that was previously carried out by gene A. In this sense, the lifetime of a genetic buffering mechanism may be largely determined by the (nearly-neutral) age of the buffered gene in the population. The transient nature of genetic robustness may provide an alternative view to the theory that robustness as a phenotype may be adaptive and maintained by strong natural selection.

We propose that duplication of dispensable genes may provide a mechanism to maintain the genetic buffering. This is simply because multiple copies of a dispensable gene can prolong the lifetime of genetic buffering that is subject to the loss of buffered gene during the long-term evolution. From Eq. (10.6), the probability for protecting the underlying genetic buffering through the preservation of at least one copy of the buffered gene (A) is roughly given by

$$q = 1 - \left[1 - Q(d^+|Q^+)\right]^n \tag{10.7}$$

where n is the number of duplicate genes. Apparently, the chance to maintain the genetic buffering mechanism can be significantly increased by the number of duplicates of the buffered gene. In spite of potential dosage and stoichiometric effects, duplicate genes under the ancestral genetic buffering may have largely undergone independent neutral evolution in protein sequence and expression levels. This may help to understand why predictions from the theory of subfunctionalization are not usually elaborated by functional genomics data, because subfunctionalization is unlikely to occur after a dispensable gene is duplicated. In some cases, neutral divergence between duplicates may provide opportunities for acquiring some new functions. Moreover, for any genetic buffering mechanism that may buffer multiple genes, the degree of 'pleiotropic-buffering' can be maintained by simultaneous gene duplications, such as genome-wide duplications, even though most of the redundant gene finally become nonfunctional.

In short, our hypothesis demonstrates that consecutive gene/genome duplication events may provide a flux of raw genetic materials for the maintenance of genetic buffering. Though most duplications of dispensable genes have finally become pseudogenes, this nonadaptive, mutation-driven process may protect the underlying genetic buffering mechanisms by keeping the genetically buffered genes in the genome during the long-term evolution. Clearly, this view is consistent with the mutational hypothesis for the origins of organismal complexity, without invoking natural selection as an adaptive force to maintain the genetic robustness.

10.3   Evolution of gene–gene interactions

Initiated by an influential, but controversial study (Fraser *et al.* 2002) about the effect of protein–protein interactions on protein sequence evolution, interactivity at the DNA, protein, and genetic level has been a major topic in systems biology and evolution (von Mering *et al.* 2002; Wagner 2001; Jordan *et al.* 2003, 2004; Agrafioti *et al.* 2005; Coulomb *et al.* 2005; Wall *et al.* 2005; Hahn and Kern 2005; Mintseris and Weng 2005; Wuchty *et al.* 2006; Yu *et al.* 2007; Zou *et al.* 2009). In this section, we tackle a specific issue, i.e. how to measure the rate of interaction gains and losses during evolution. For simplicity, we use two duplicate genes as illustration (Wagner 2001).

Consider the functional interaction of gene A to any other gene X. In the simplest case, the status of this A-X interaction is $r = 1$ if these two genes are connected, or $r = 0$ otherwise. For two duplicate genes A and B, an important measure for their functional overlapping is the number (n_{11}) of other (X) genes that interact with both duplicate genes A and B $(A - X$ and $B - X)$. One may utilize a simple Poisson model to estimate the interaction distance (D_I) between duplicates, i.e. the average number of interaction losses and gains. Under this model, double-loss of an ancient interaction in both duplicate lineages is unlikely. For a set (n) of genes, each of which interacts with A or B or both, one can calculate the proportion of interaction divergence $\hat{q} = 1 - n_{11}/n$, and estimate the interaction distance by

$$\hat{D} = -n \ln(1 - \hat{q}) \tag{10.8}$$

However, because the high throughput functional genomic data involve high level noises, statistical evaluation of inferred interactions becomes a big challenge in the genomic analysis. We shall address this issue, i.e. how to take the statistical uncertainty in the evolutionary analysis.

10.3.1   *p*-Value representation of gene–gene interaction

Many studies have adopted the p-value approach to measure the statistical significance of a gene–gene interaction. That is, instead of a binary $(r = 1$ or $0)$ status of an $A - X$ interaction, a p-value is assigned for the interaction between any two genes; a small p-value, e.g. $p = 0.001$, means that the interaction is highly statistically significant, and *vice versa*. Inference of a gene–gene interaction depends on the cutoff (α): the status of an interaction is positive if $p < \alpha$, or negative otherwise. As a result, the interaction distance between duplicate genes can be sensitive to the cutoff selected. One feasible solution to overcome this cutoff problem is to treat these p-values as observations.

For the p-value presentation of gene–gene interactions, we have to develop an explicit model for the interaction evolution. In the case of two duplicates A and B, there are four combined patterns (r_A, r_B): (1, 1), (1, 0), (0, 1), and (0, 0), respectively. For instance, (1, 1) means that both duplicate genes have the interaction with the same gene X; (1, 0) means duplicate gene A has the interaction but gene B does not, and so forth. Next we make the following assumptions

(1) Let θ_0 be, at the genome level, the proportion of genes that have no interaction with A nor with B.

(2) With a probability of f_1, the ancestral gene prior to the duplication has interacted with gene X, called the ancestral interaction. With a probability of $f_0 = 1 - f_1$, the ancestral gene did not interact with gene X.

(3) Let λ be the loss rate of an interaction during the evolution. Tentatively, we assume that an ancestral interaction remains in both duplicates with a probability of $e^{-2\lambda t}$; double-loss of an ancestral interaction in both duplicates is unlikely.

(4) The probability of acquiring a new interaction, called the derived interaction, is $f_0 = 1 - e^{-2\rho t}$, where ρ is the rate of acquiring a new interaction. Simultaneous acquisition of an interaction to the same gene X in both duplicates is unlikely.

Therefore, one can show that the probability $P(r_A, r_B)$ is given by

$$P(1, 1) = (1 - \theta_0)e^{-2(\lambda+\rho)t}$$

$$P(1, 0) + P(0, 1) = (1 - \theta_0)\left[1 - e^{-2(\lambda+\rho)t}\right]$$

$$P(0, 0) = \theta_0 \tag{10.9}$$

respectively. In practice, one may assume $P(1, 0) = P(0, 1)$.

10.3.2   General framework

For two duplicate genes A and B with any other gene X, let p_A and p_B be the p-values for gene interactions $A - X$ and $B - X$, respectively. Let y_A and y_B be any given transformations of p_A and p_B, respectively. Two simple forms that are practically useful are $y = -\ln p$ and $y = p$, respectively. Then, we define the expectation of squared y-score differences between genes A and B as follows

$$\delta_{AB}^2 = E[(y_A - y_B)^2] \tag{10.10}$$

where E is for expectation.

To calculate δ_{AB}^2, we use the conditional expectations with respect to the interaction patterns (r_A, r_B). To simplify notations, let $\gamma_{11} = E[(y_A - y_B)^2 | r_A = 1, r_B = 1]$, $\gamma_{00} = E[(y_A - y_B)^2 | r_A = 0, r_B = 0]$, $\gamma_{10} = E[(y_A - y_B)^2 | r_A = 1, r_B = 0]$, and $\gamma_{01} = E[(y_A - y_B)^2 | r_A = 0, r_B = 1]$. Therefore, we have

$$E[(y_A - y_B)^2] = \sum_{r_A, r_B = 0, 1} E[(y_A - y_B)^2 | r_A, r_B]P(r_A, r_B)$$

$$= \gamma_{11}P(1, 1) + \gamma_{10}P(1, 0) + \gamma_{01}P(0, 1) + \gamma_{00}P(0, 0)$$

Together with the probability of each (r_A, r_B) given by Eq. (10.10), and assuming $P(1, 0) = P(0, 1)$, we obtain

$$\delta_{AB}^2 = \delta_\infty^2 - (\delta_\infty^2 - \delta_0^2)e^{-2(\lambda+\rho)t} \tag{10.11}$$

where δ_∞^2 and δ_0^2 are given by

$$\delta_\infty^2 = (1 - \theta_0)(\gamma_{10} + \gamma_{01})/2 + \theta_0 \gamma_{00}$$

$$\delta_0^2 = (1 - \theta_0)\gamma_{11} + \theta_0 \gamma_{00}$$

respectively. When $t = 0$, $\delta_{AB}^2 = \delta_0^2$, and δ_{AB}^2 increases with t and ultimately reaches δ_∞^2 as $t \to \infty$. Moreover, one may define the effective proportion of different interactions between duplicate genes A and B

$$q_e = \frac{\delta_{AB}^2 - \delta_0^2}{\delta_\infty^2 - \delta_0^2} \tag{10.12}$$

such that q_e satisfies

$$q_e = 1 - e^{-2(\lambda + \rho)t}$$

Then, given the sample size (N) of functional interactions, the functional interaction distance defined by $D_I = 2n(\lambda + \rho)t$, is given by

$$D_I = -n \ln (1 - q_e) \tag{10.13}$$

If we further assume that the variance of q_e is approximated by a binomial distribution, i.e. $Var(q_e) \approx q_e(1 - q_e)/N$, the sampling variance of $\hat{D}_I$ can be calculated as follows

$$Var(\hat{D}_I) \approx \frac{N\hat{q}_e}{1 - \hat{q}_e}$$

In short, estimation of D_I turns out to estimate the effective proportion of different interactions between duplicate genes, which can be achieved when the transformed p-value (the y-score) is specified.

10.3.3   Estimation of γ_{11}, γ_{10}, γ_{01}, and γ_{00}

Let $\gamma_{r_A, r_B} = E[(y_A - y_B)^2 | r_A, r_B]$. Under the independent assumption of y_A and y_B, we have

$$\gamma_{r_A, r_B} = E[y_A^2 | r_A] + E[y_B^2 | r_B] - 2E[y_A | r_A]E[y_B | r_B]$$

It is well-known that the p-value follows a uniform distribution under the null hypothesis of $r = 0$ (no interaction). In this case, the mean and variance of $y = f(p)$ are denoted by $\bar{y}_0$ and σ_0^2, respectively. Thus, $E[y | r = 0] = \bar{y}_0$ and $E[y^2 | r = 0] = \sigma_0^2 + (\bar{y}_0)^2$. On the other hand, the mean and variance of y under the alternative hypothesis of $r = 1$ are usually unknown but can be estimated from the observed genome-wide p-value distribution. Concisely, we denote the mean by $\bar{y} = E[y | r = 1]$ and the variance by $\sigma^2 = Var(y | r = 1)$ so that $E[y^2 | r = 1] = \sigma^2 + (\bar{y})^2$. Put together, we have derived

$$\gamma_{11} = 2\sigma^2$$

$$\gamma_{10} = \gamma_{01} = \sigma^2 + \sigma_0^2 + (\bar{y} - \bar{y}_0)^2$$

$$\gamma_{00} = 2\sigma_0^2$$

In particular, we discuss two cases that may be useful in practice.

The p-based method The simplest case is to use the p-value directly, i.e. $y = p$. Noting that under the null hypothesis of no interaction $(r = 0)$, p follows a uniform distribution in $[0, 1]$ with the mean $1/2$ and the variance $1/12$, we obtain

$$\gamma_{11} = 2\sigma_p^2$$

$$\gamma_{10} = \sigma_p^2 + 1/12 + (\bar{p} - 1/2)^2$$

$$\gamma_{00} = 1/6$$

where $\bar{p}$ and σ_p^2 is the mean and variation of p-values under the alternative hypothesis of $r = 1$, respectively.

The $-\ln p$-based method We suggest a (negative) log-transformation score (y) for the p-value of an interaction, i.e. $y = -\ln p$, because of some good sampling properties. For instance, an interaction associated with a p-value of 0.001 is statistically sound, while that with more than 0.05 is usually considered as non-significance. Consider a hypothetical example for two duplicate genes A and B. Assume $p_A = 0.001$ and $p_B = 0.5$ for the functional interactions $A - X$ and $B - X$, respectively, so that $(p_A - p_B)^2 = 0.4999^2$. Secondly, for another interaction pair $A - X'$ and $B - X'$, we assume that the p-values are $p'_A = 0.30$ and $p'_B = 0.8$ so that $(p'_A - p'_B)^2 = 0.5^2$. The virtually the same score between the two cases is apparently counter-intuitive, because one may statistically infer that gene A interacts with X but not for gene B, whereas both A and B are unlikely to interact with gene X'.

The log-transformed score may avoid this problem. In the above case, $(y_A - y_B)^2 = 6.22^2$ is much higher than $(y'_A - y'_B)^2 = 0.98^2$. Since the p-value follows a uniform distribution under the null hypothesis of $r = 0$ (no interaction), one can show that y follows an exponential distribution with the mean $\bar{y}_0 = 1$ and variance $\sigma_0^2 = 1$. Therefore, we have

$$\gamma_{11} = 2\sigma^2$$

$$\gamma_{10} = \sigma^2 + 1 + (\bar{y} - 1)^2$$

$$\gamma_{11} = 2$$

10.3.4 Comments

The p-value-based approach may provide a general framework in evolutionary genomics because of two advantages: First, it takes into account the statistical uncertainty of gene–gene interactions at the genome level. And second, it avoids the detail of statistical procedure that is data-specific and/or technology-specific. When it is implemented, however, the main problem is the accuracy of p-value estimation. Since the statistical method for calculating these p-values is approximate, the estimated p-value is always biased. From the view of conventional statistics, people are only concerned about the p-value accuracy in the range of $0.001 - 0.10$. So the genome-wide $p - value$ distribution could be considerably deviated from a uniform distribution under the null hypothesis. We shall address these issues (Gu, unpublished).

10.4 Origin of modularity and complexity

10.4.1 Some backgrounds

One of the central issues in systems biology is to understand the origin of gene network complexity (Hartwell *et al.* 1999; Barabasi and Oltvai 2004; Wagner *et al.* 2007). Many biological systems, from metabolic pathway to protein–protein interactions, can be represented as (gene) networks (Barabasi *et al.* 2003; Barabasi and Oltvai 2004; Barabasi 2009). In a typical network, many nodes are organized into a complex topology by massive interactions (links) between nodes. For example, in the yeast protein–protein interaction network (Jeong *et al.* 2001), a node represents a protein, and a link between two proteins indicates their interaction. The high complexity of a gene network can be characterized by a *power-law* degree, i.e. the number of links for a node (called degree) is right-skewedly distributed; many nodes with low degrees and a small number of nodes with high degrees (Albert and Barabasi 2002). Substantial genome-wide analyses (Barabasi and Albert 1999; Jeong *et al.* 2000, 2001; Hahn *et al.* 2004. Han *et al.* 2004) have shown that cellular networks can be featured by (*i*) the power-law of degree distribution; (*ii*) the small-world property with occasional long-range links, and (*iii*) network centrality with hubs (nodes with many links). They are called *scale-free* because these network features are independent of the size of the network.

Another important feature of a cellular network is *modularity*, which refers to a network of interactions if it can be subdivided into relatively autonomous, internally highly connected components (Hartwell *et al.* 1999; Fraser 2005; Chen and Dokholyan 2006). Though there is a general agreement that organisms have a modular organization as a fundamental rule, current study of modularity seeks to capture the various levels and kinds of functional heterogeneity found in organisms, which has emerged as a hot research subject in developmental and evolutionary biology, as well as systems biology (Wagner *et al.* 2007; Gu and Su 2007; Su *et al.* 2007). One main open problem is about the origin and evolution of modularity, e.g. it has been highly controversial whether modules arise through the action of natural selection or because of non-adaptive processes such as biased mutational mechanisms (Lynch 2007c; Wagner *et al.* 2007).

The pioneering work of Barabasi and Albert (1999), referred to as the BA model below, provides an elegant yet simple evolutionary model for the origin of scale-free network (power-law). It was based on two mechanisms: (*i*) Network growth is a continuous process during the course of evolution, and (*ii*) hubs that already have many interactions have high chances to increase connectivity, called the *attachment preference* (Jeong *et al.* 2003). Though the BA model successfully predicts the scale-free property, it was unable to predict the property of modularity (Barabasi and Oltvai 2004). On the other hand, Wagner and Mezey (2004) suggested that the evolution of modularity could be driven by the selection for robustness, such as differential elimination of pleiotropic effects. This idea can be generalized to a 'differential erosion model' for the evolution of modularity (Wagner *et al.* 2007). Recently, Gu (2009)has recognized that from the gene network point of view, elimination of pleiotropic effects can be viewed as the process of link losses. Since the origin of new links has the attachment preference to hubs that leads to the scale-free property, one may speculate that a random link-loss (no hub preference) may provide a mechanism for the origin of modularity. The article of Gu (2009) proposed an evolutionary model of gene network, which may lead to the emergence of scale-free property and modularity simultaneously.

10.4.2   Scale-free network and modularity

Network biology offers a quantifiable description of the networks that characterize various biological systems. These measures are useful to compare and characterize different complex networks, which help to explain the origin of observed network characteristics (Fig. 10.2).

Scale-free property and power-law   In a scale-free network characterized by highly connected genes (nodes), or hubs, the number of genes with a given degree (number of links) follows a power law. That is, the probability that a chosen gene node has exactly k links follows

$$P(k) \sim k^{-\gamma} \tag{10.14}$$

where γ is the degree component, which is usually in the range of $1 < \gamma < 3$ (Albert and Barabasi 2002). The power-law indicates there is a high heterogeneity of node degrees and no typical value in the network such as the mean degree that could be used to characterize the level (or scale) of node connectivity. Thus, a network with the absence of a typical degree is usually called 'scale-free', in which a few hubs dominate network features. For instance, a common feature of scale-free networks is the small-world property that any two nodes can be connected with a much shorter chain of links (through hubs) than that expected by the random network. Analysis of the yeast protein–protein interaction network indicated that about 60 per cent of the hub proteins (defined as > 15 interactions) are essential, whereas only 10 per cent of few connected proteins with less than 5 links (Jeong et al 2001) are essential. Though the exact biological meaning of these network features remains for further investigation (Keller 2005), we found the power-law helpful to understand the origin of modularity, as shown below.

Modularity and clustering coefficient   Meanwhile, many studies (e.g. Ravasz *et al.* 2002; Giot, *et al.* 2003; Barabasi and Oltvai 2004; Vazque *et al.* 2004; Wagner *et al.* 2007) have unveiled the fact that almost all cellular networks from protein–protein interactions, metabolic to regulatory networks, show the feature of modularity. Moreover, several authors (Ravasz *et al.* 2002; Barabasi and Oltvai 2004; Vazque *et al.* 2004) used the node-specific clustering coefficient, the cohesiveness of the neighborhood of a node (A) that has k links, to examine the modularity in a scale-free network. From this view, gene A's neighbor is the set (k) of genes that are directly linked to gene A. Mathematically, the clustering coefficient $C(k)$ is defined by

$$C(k) = 2T(k)/[k(k-1)] \tag{10.15}$$

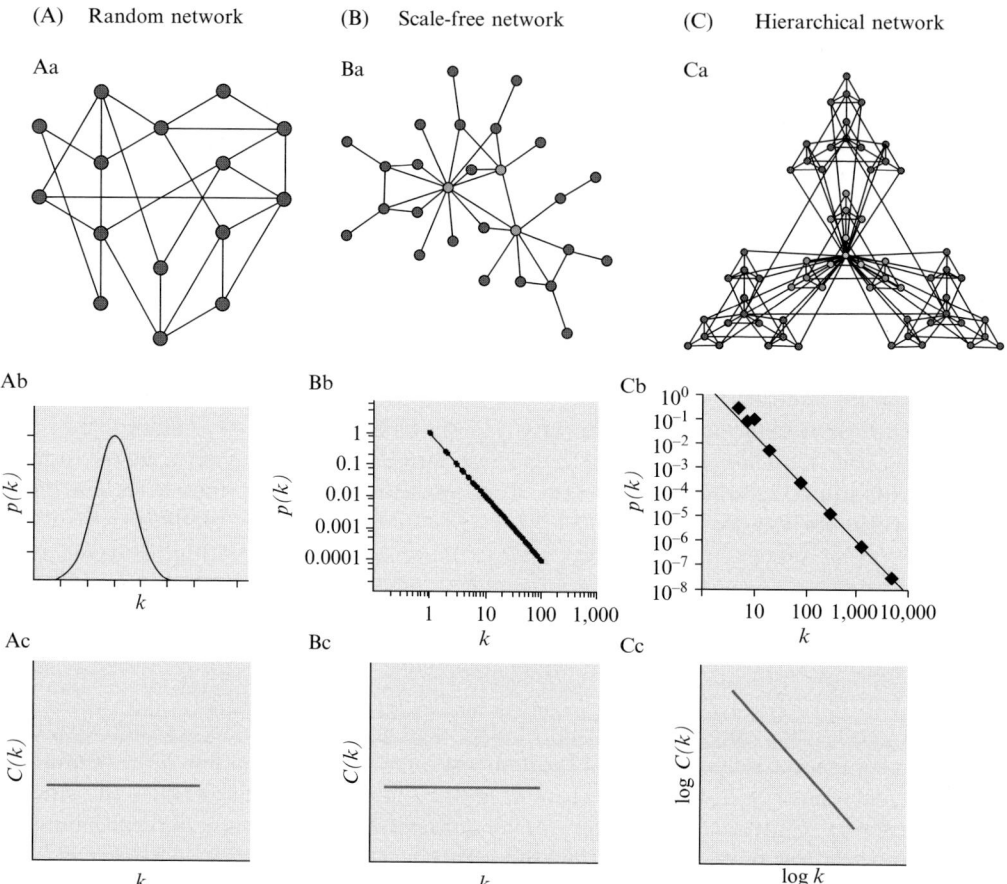

(A)   Random network   (B)   Scale-free network   (C)   Hierarchical network

Fig. 10.2 Three types of network models crucial for explaining the origin of observed network characteristics. (A) Random networks, the Erdös–Rènyi (ER) model starts with N nodes and connects each pair of nodes with probability p, which creates a graph with randomly placed links (panel Aa). The node degrees follow a Poisson distribution (see panel Ab). The clustering coefficient is independent of a node's degree, so $C(k)$ appears as a horizontal line if plotted as a function of k (panel Ac). (B) scale-free networks are characterized by a power-law degree distribution. The probability that a node is highly connected is statistically more significant than in a random graph, so that the network's properties often being determined by a relatively small number of highly connected nodes that are known as hubs (panel Ba). The scale-free network that is created by the Barabàsi–Albert model does not have an inherent modularity, so $C(k)$ is independent of k (panel Bc). (C) Hierarchical network seamlessly integrates a scale-free topology with an inherent modular structure by generating a network that has a power-law degree distribution (panel Cb). The most important signature of hierarchical modularity is the scaling of the clustering coefficient, which follows $C(k) \sim k^{-1}$ (panel Cc). From Barabasi and Oltvai (2004).

where $T(k)$ is the number of direct links between any two gene A's neighbors, and $k(k-1)/2$ is the total possible number for links among neighbors. In fact, $C(k)$ measures how close the local neighborhood of a node is to be part of a clique (module), a region of the graph where every node is connected to every other node. In practice, we used the averaged clustering coefficients of nodes that have the same degree k to characterize the network modularity.

It has been claimed (Barabasi and Oltvai 2004) that in a typical scale-free network, the mean clustering coefficient $C(k)$ is independent of the node degree k, i.e. $C(k) \sim C_0$, a feature of no modularity that is similar to a random network. This indicates that, on average the size of a local neighborhood, determined by the node degree (k), does not influence the cohesiveness. Using the power-law as a measure of scale-free property and the clustering coefficient as modularity, Ravasz *et al.* (2002) found that the prediction of constant clustering coefficient was rejected by several cellular networks. This finding implies that a pure scale-free network generated by the BA model (Barabasi and Albert 1999) may not be sufficient to explain the complexity of cellular networks. Further, Ravasz *et al.* (2002) observed a log-log inverse relationship between $C(k)$ and the node degree (k), indicating that nodes with small links (k) and a high $C(k)$ belong to highly interconnected small modules, while highly connected hubs with a low $C(k)$ link different modules. Then, they proposed the concept of hierarchical networks that integrate a scale-free topology with an inherent modular structure so that it not only has a power-law degree distribution $P(k)$, but also a power-law scaling of the clustering coefficient with node degree k, that is

$$C(k) \sim k^{-\theta}$$

In short, in a hierarchical network, sparsely connected nodes are part of highly clustered subnets, with communication between the different highly clustered neighborhoods around a few hubs. However, it seems a puzzle how evolution can lead to various biological networks that are characterized by the 'contrasted' scale-free and modular features.

10.4.3 Origin of modularity in a scale-free network

How a scale-free network with modularity to be evolved?

Cellular networks have long been thought to be modular, composed of functionally separable subnetworks corresponding to specific biological functions (Hartwell *et al.* 1999; Wagner *et al.* 2007). In addition, there is a high degree of overlap and crosstalk between modules (Han *et al.* 2004). Hence, the hierarchical network, or the scale-free network with modularity, is biologically sound. A fundamental issue then is how a hierarchical network evolves, which remains unsolved.

Barabasi and Albert (1999) developed an elegant and concise framework (the BA model for short) to describe the origin of a scale-free network. It incorporates two mechanisms: the growth (i.e. an increase in the numbers of nodes and links over the time) and the preferential attachment (i.e. a new gene favorably linking to the existed 'hub' gene). Barabasi and Albert (1999) have shown that the generated network has a power-law $P(k) \sim k^{-3}$, while the clustering coefficient is a constant (Ravasz *et al.* 2002). A number of modified BA models that included various network constraints and system-specific preferential attachments have been proposed, see Albert and Barabasi (2002) for a review, to explain gene networks with degree component $\gamma = 2 \sim 3$ but cannot predict the modularity.

Alternatively, Ravasz *et al.* (2002) proposed a growing network model by implementing iterative subnetwork duplications and integrations to its original seed core. This growth algorithm leads to a power-law of node degree with $\gamma \approx 2$, and a scaling power function of clustering coefficient $C(k) \sim k^{-1}$ when the size of the seed graph of duplication is four. However, Ravasz *et al.*'s (2002) model has no intrinsic relationship with the BA model, and the biological meaning of the proposed mechanism was unclear. In short, it is desirable to develop a unified evolutionary model that can explain the following network features: (*i*) the scale-free property characterized by a power-law of Eq. (10.14), agreeing with the range of observed degree exponent $\gamma = 1.5 \sim 2.5$; and (*ii*) the decreasing of clustering coefficient $C(k)$ with node degree k, agreeing with the range of observed clustering-degree exponent $\theta = 0.5 \sim 1.5$. As will be discussed below, Gu (2009) provided a surprisingly simple solution for this problem. That is, by adding the random link-loss (no preference) mechanism to the classical BA model, the network can evolve to be scale-free with the feature of modularity.

Biological BA model (BBA): evolution by preference attachment and random loss of links

Gu (2009) proposed a revised BA model, and coined the term biological BA (BBA). Under the new BBA model, network evolution is driven by three parameters: λ and μ are the gain rate and loss rate of a link, respectively, and g is the origin rate of a new gene (node). Starting from the original BA model (Barabasi and Albert 1999), the evolutionary mechanism for a gene network can be described as follows: (1) *Gene Network Growth*: At the dawn of the origin of life, there was a small number (n_0) of nodes (RNA/DNA/proteins) to form the most preliminary genetic network. These initial connections, in spite of small numbers (I_0), should be very fundamental for the meaning of life. Since then, as new genes continuously added into the genome, new connections were created with a rate of λ between new and existed genes. (2) *Gene Connectivity Preference:* The rule of 'rich-gets-richer' means that a new node may have a higher probability to be linked to a node that already has a large number of connections.

The original BA model results in a fixed power-law degree, i.e. $\gamma = 3$. We have noticed that the BA model assumes that a connection in the gene network cannot be lost once it is generated. In fact, loss of connections between nodes in a gene network (or more generally, loss-of-function) has played an important role during the course of evolution. Therefore, we make the following additional assumption: (3) *Random Loss of Connections.*

Let x_i be the number of interactions of node i at time t. Under the assumptions (1)–(3), the original BA model can be extended as the following differential equation

$$\dot{x}_i = \lambda(1 - \omega)\pi(x_i, t) + \frac{\omega\lambda}{n(t)} - \frac{\mu}{n(t)} \qquad (10.16)$$

where $\dot{x} = dx_i/dt$, λ and μ are the growth rate and loss rate of interactions, respectively, the probability $\pi(x_i, t)$ corresponds to the preference of node i to be linked at time t, and $n(t)$ is the total number of genes (nodes) in the gene network at time t. The constant ω $(0 \leq \omega \leq 1)$ represents the chance for the random selection of node i to be the interaction-generator; here we set $\omega = 1/2$.

BA (1999) suggested a linear function for $\pi(x_i, t)$ to measure the preferential attachment, i.e.

$$\pi(x_i) = \frac{x_i}{\sum_{j=1}^{n(t)} x_j} = \frac{x_i}{2I(t)}$$

where $I(t)$ is the total number of interactions in the gene network at time t. Since an interaction between two genes (nodes) will be counted twice, we have $2I(t) = \sum_{j=1}^{n(t)} x_j$. From Eq. (10.16), one can easily verify that $2\dot{I} = \sum_{i=1}^{n(t)} \dot{x}_i$, leading to $\dot{I} = (\lambda - \mu)/2$. Moreover, the simplest evolutionary model for the network growth is given by $\dot{n}(t) = g$, where g is the origin rate of a new node. Together, the growth of gene network and the growth of network interactions are both linear with respect to the evolutionary time, that is

$$n(t) = gt + n_0$$

$$I(t) = (\lambda - \mu)t/2 + I_0$$

where n_0 and I_0 are the initial numbers of genes and interactions, respectively. Since the initial numbers n_0 and I_0 are usually small, they can be neglected for the long-term evolution (large t). Together, Eq. (10.16) can be approximately expressed as follows

$$\dot{x}_i = \frac{x_i}{bt} + \frac{h}{t} \tag{10.17}$$

where $b = 2(1 - \mu/\lambda)$ and $h = (\lambda/2 - \mu)/g$.

The solution of Eq. (10.17) relies on the corresponding initial condition: when gene (node) i was generated at the evolutionary time τ_i, this newly-born gene i had x_0 interactions, i.e. $x_i(\tau_i) = x_{0,i}$, leading to the result $x_i(t) = (x_{0,i} + a)(t/\tau_i)^{1/b} - a$, where $a = bh$. It indicates that the expected number of interactions of a gene increases with the evolutionary time t. On the other hand, given the current time-point by setting $t = T$, the expected interaction difference between genes is determined by their difference in age. Since the gene age is measured by the initial time-point τ_i, the frequency of gene connectivity, $P(k)$, can be derived from the age distribution of τ_i. To this end, we rewrite the solution of Eq. (10.17), $x_i(t)$, as follows

$$\tau_i = \left[\frac{x_{0,i} + a}{x_i + a} \right]^b T$$

where x_i is short for $x(T)$. In fact, the linear growth of the gene network means that genes are added constantly to the network during the course of evolution from the time period $[0, T]$, implying a uniform distribution of gene age (τ_i), that is, for any given r $(0 < r < 1)$, we have $P(\tau_i < rT) = r$. Consider the probability of a node i whose interactions are smaller than a given number k, $P(x_i < k)$. From the expression of τ_i, one can verify that $x_i < k$ means $\tau_i > [(x_{0,i} + a)/(k + a)]^b T$. Under the assumption of uniform distribution for τ_i in the time period $[0, T]$, we obtain

$$P(x_i < k) = P(\tau_i > \left[\frac{x_{0,i} + a}{k + a}\right]^b T) = 1 - P(\tau_i < \left[\frac{x_{0,i} + a}{k + a}\right]^b T) = 1 - \left[\frac{x_{0,i} + a}{k + a}\right]^b$$

It follows that the probability $P(k) = P(x_i < k + 1) - P(x_i < k)$ can be approximately obtained by $P(k) \sim \partial P(x_i < k)/\partial k$, resulting in a general form of the power-law,

$$P(k) \sim (k + a)^{-\gamma} \tag{10.18}$$

with the power-law degree $\gamma = b + 1$, that is

$$\gamma = 3 - 2\mu/\lambda \tag{10.19}$$

The constant $a = bh$ is called the shift factor; when $a = 0$, the power-law is reduced to the canonical form.

Therefore, the classical BA model is a special case of $\mu = 0$, i.e. no link loss, resulting in $\gamma = 3$. Note that the assumption of network growth requires the constraint that the loss-rate of links must be less than the gain-rate, i.e. $\mu < \lambda$, consistent to the fact of $\gamma > 1$ for most biological networks. Interestingly, the loss-gain ratio of connectivity evolution (μ/λ) determines the shape of any scale-free network with $1 < \gamma \leq 3$. For cellular networks with $\gamma \approx 2$, Eq. (10.19) indicates that $\mu \approx 0.5\lambda$, i.e. the rate of link gain is approximately two-fold larger than the rate of link loss.

Random link-loss mechanism generating modularity of gene network

We envisage a scenario of complex network evolution to generate the modularity through the random process of link losses. The key is that link loss has no preference between high-degree nodes and low-degree nodes. Consequently, while the number of links (degree) in hubs is less affected by the random link loss, the degree in some non-hub nodes can be reduced to a very few number of links, creating a hierarchical network randomly.

In the following we derive the power-scaling of clustering coefficient $C(k)$, with the assumption that gene duplication is the main mechanism for the origin of new genes. First we consider $T(k)$, the number of direct links between any two of k-neighbors of gene A. Suppose that one of gene A's neighbors, say, the i-th neighbor, was duplicated; the duplicated copy was numbered as the $k + 1$-th neighbor of gene A. Since the clustering coefficient $C(k)$ is the probability that two neighbors are linked, the expected number of links between the i-th neighbor and other neighbors is therefore given by $(k - 1)C(k)$. After the gene duplication, the increased number of direct links between neighbors is thus expected to be $(k - 1)C(k)$. On the other hand, random link loss may reduce the number of direct links between neighbors. During the time period (Δt) for gene duplication and preservation, the number of link losses is expected to be $(k - 1)C(k) \times \mu\Delta t$. Typically, Δt can be characterized by $1/g$; here g is the rate of gene duplication, i.e. $\Delta t \approx 1/g$. Therefore, the expected loss number of direct links is given by $(k-1)C(k) \times \mu/g$. Together, the net change in the number of direct links, denoted by $\Delta T(k) = T(k+1) - T(k)$, is given by

$$\Delta T(k) = (k-1)C(k) - (k-1)[\mu/g]C(k) = 2(1 - \mu/g)\frac{T(k)}{k}$$

Using the continuous approximation, we claim that $T(k)$ satisfies the following differential equation

$$\frac{dT(k)}{dk} = 2(1 - \mu/g)\frac{T(k)}{k} \tag{10.20}$$

Solving with the initial condition $T(2) = 1/2$ results in

$$T(k) = 0.5 \times k^{2(1-\mu/g)}$$

Hence, we obtain a power-law relation between $C(k)$ and k as follows

$$C(k) = \frac{2T(k)}{k(k-1)} \sim k^{-2\mu/g}$$

In other words, we have derived that the clustering coefficient has a power-law, i.e. $C(k) \sim k^{-\theta}$, where θ is given by

$$\theta = 2\mu/g \tag{10.21}$$

In the case of the classical BA model with no link loss ($\mu = 0$), $\theta = 0$ that indicates no inherent modularity of the scale-free gene network. When highly clustered neighborhoods around several hubs can be generated by preferential linking, random losses of links may provide an evolutionary mechanism to create sparsely connected genes that connect between the different highly clustered subnets of hubs (see Fig. 10.2).

10.4.4 Protein–protein interaction data analysis

Gu (2009) has analyzed protein–protein interaction networks from ten organisms (Rain *et al.* 2001; Giot *et al.* 2003; Han *et al.* 2004) (see Table 10.1 and Fig. 10.3). In each dataset, we counted the distribution of node degrees, $P(k)$, and calculated the average clustering coefficients, $C(k)$, for each node degree $k \geq 2$. Then we estimated both degree components (γ, θ) by the $\ln P(k) \sim \ln k$ and $\ln C(k) \sim \ln k$ regressions, respectively. Other methods including nonlinear approaches gave similar results. For instance, Fig. 10.3 shows the $P(k) \sim k$ and $C(k) \sim k$ relationships in the human protein–protein interaction network. In addition, we included estimates of two other cellular networks (metabolic and regulatory) in two organisms (E.coli and yeast) (Vazque *et al.* 2004).

It is impressive that all these gene or cellular networks show scale-free property and modularity. The range of degree component (γ) is from 1.40 to 2.53, while the range of clustering component (θ) is from 0.63 to 1.50. Roughly speaking, the power-law degree of $P(k)$ is around 2 ($\gamma = 1.86 \pm 0.07$), while that of $C(k)$ is around 1 ($\theta = 0.96 \pm 0.08$). We found no evidence for any difference due to the taxon level, network type, or network size. Furthermore, according to Eqs. (10.19) and Eq. (10.21),

Table 10.1 Estimated evolutionary parameters of gene networks under the BBA (Biological BA) model.

Network types	Organisms	γ	θ	λ/g	μ/g	λ/μ
Protein interaction	Yeast	1.80	1.02	0.85	0.51	1.67
	Human	1.73	1.07	0.84	0.54	1.57
	Worm	1.59	0.63	0.45	0.32	1.42
	house mouse	2.53	0.91	1.94	0.46	4.26
	Helicobactor	1.64	0.45	0.33	0.23	1.47
	E.coli	1.84	1.38	1.19	0.69	1.72
	Norway rat	2.03	1.19	1.23	0.60	2.06
	Arabidopsis	1.63	1.50	1.09	0.75	1.46
	Rice	1.77	0.63	0.51	0.32	1.63
	Cow	1.40	1.12	0.70	0.56	1.25
Metabolic	Yeast	2.00	0.70	0.70	0.35	2.00
	E.coli	2.00	0.80	0.80	0.40	2.00
Regulatory	Yeast	2.00	1.00	1.00	0.50	2.00
	E.coli	2.10	1.00	1.11	0.50	2.22
mean $\pm$ *s.e.*		1.86 ± 0.07	0.96 ± 0.08	0.91 ± 0.11	0.48 ± 0.04	1.91 ± 0.20

Note: γ: power-law degree component; θ: clustering coefficient component; λ/g: the rate of new link relative to the rate of new gene; μ/g: the rate of link loss relative to the rate of new gene; and λ/μ: the rate ratio of new links and link losses.

we estimated the relative gain rate of links (λ/g) and the relative loss rate of links (μ/g). Taking an average over 14 network datasets, we obtained $\lambda/g = 0.91 \pm 0.11$ and $\mu/g = 0.48 \pm 0.04$, respectively. We interpret these data as follows: When the network increases by two nodes (genes), on average two new links are created and one existed link is lost.

10.4.5   Hypothesis: random loss of interactions may shape modularity in a complex gene network

We have developed the BBA model for gene network evolution, providing a unifying explanation for the origin of modularity and scale-free property. That is, the scale-free property is the result of preferential new-link origins, while the modularity is the result of random-loss of existed links. Given the complexity of gene networks, our BBA model reveals the connection between network features and basic evolutionary mechanisms that can be potentially tested by the high throughput genomics data.

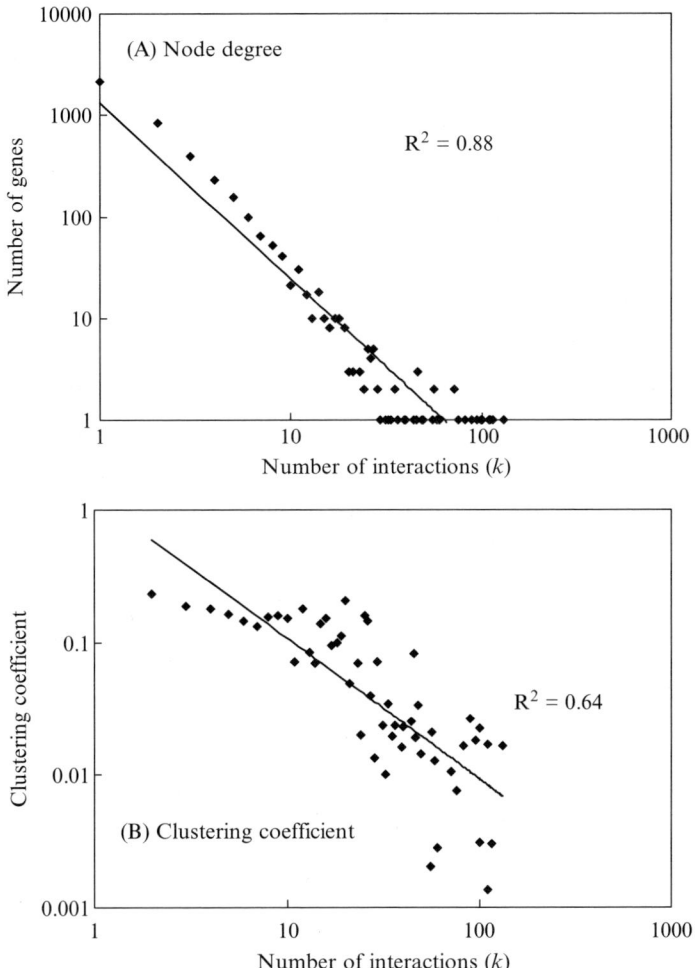

Fig. 10.3 The $P(k)$-$C(k)$ analysis for human protein-protein interactions. (A) Number of genes plotting against the number of interactions (links), showing a power-law of $P(k)$. (B) (Mean) clustering coefficient plotting against the number of interactions (links), showing a power-law of $C(k)$. From Gu (2009)

Gene network growth The BBA model assumes that the growth of the gene network (through gene duplications) is a constant. Therefore, the number of genes in a genome increases linearly with evolutionary time. This assumption is only the first-order approximation for the trend of long-term gene network evolution. For an evolutionary lineage during a particular geological time period, however, the size of the gene network can be either static, reduced, or expanded (Gu and Zhang 2004). In this sense, like the original BA model, our BBA model only represents a theoretical theme of gene network evolution from the simple life forms to advanced organisms.

Attachment preference and random loss of links   The central premise of the BA model is the attachment preference for new links. Under the BBA model, we tentatively demonstrate its biological interpretations. If the new link is through the process of developmental rewiring, the attachment preference means that this rewiring is to enhance the existing major developmental pathways. Though new genes were mainly from gene duplications, follow-up functional divergence, either subfunctionalization (complementary loss of links between two duplicate copies) or neo-functionalization (creating some new links) cannot directly result in the scale-free property (Pastor-Satorras *et al.* 2003; Wu and Gu 2005). We suggest that neo-functionalization after gene duplication may tend to have a new link to an existing molecular pathway characterized by some hubs. On the other hand, subfunctionalization resulting in the loss of existed links may be random without any preference. In fact, random link losses in duplicate genes can effectively mimic the DDC process (duplication, degeneration, and complementation).

The random-loss links and the erosion model of modularity   We have realized the intrinsic relationship between our BBA model (the random-loss mechanism of existed links) and the erosion model of differential elimination of pleiotropic effects (Wagner and Mezey 2004). Under BBA, emergence of modularity is the result of random link-loss (reducing the gene pleiotropy), while hubs are formed by the attachment preference. The difference between these two models is that the erosion model implies a transition from relatively uniform interactions to modularity by differential link-losses, while the BBA model suggests a gradual process for the origin of modularity in evolution. Further research will be focused on how to test these theoretical models. For instance, we recently developed a statistical method for estimating gene pleiotropy from protein sequence analysis (Gu 2007a), which may provide some insights into the relationship between modularity, pleiotropy, and network complexity.

The 2-2-1 hypothesis   The pattern for gene network evolution we found can be concisely called the 2-2-1 hypothesis, short for two new genes, two new links, one lost link, and may articulate the evolutionary pace of gene networks. One may argue that this hypothesis may lead to many isolated nodes in the network. In practice, however, it will not happen if new genes have been mostly generated through gene duplications with existed link. Moreover, we found that two degree components that characterize the power-law (γ) and modularity (θ) are influenced by the same underlying evolutionary mechanisms. Hence, when the relative gain rate of links $\lambda/g \approx 1$, one can easily show from Eqs. (10.19) and (10.21) the following evolutionary constraint of the scale-free property and modularity:

$$\gamma + \theta \approx 3 \tag{10.22}$$

Hence, it is desirable to have a coherent understanding of the emergence of gene network and functional organization from the insight of evolutionary biology.

Transition between scale-free and random networks   In the above analysis, we assume $\omega = 1/2$. In general, one can show that the power-law $P(k) \sim (k + a)^{-(b+1)}$ holds, whereas $b = (1 - \mu/\lambda)/(1 - \omega)$, $h = (\lambda\omega - \mu)/g$, and $a = bh$. In the special case when $\omega \to 1$, we demonstrate that $P(k) \sim e^{-k/h}$. Since $\omega = 1$ means no effect of attachment preference, the number of hubs decays exponentially. Thus, the BBA model integrates the scale-free and random networks, leading to the following formula

$$P(k) \sim (k + k_c)^{-\gamma} \tag{10.23}$$

where the threshold k_c is given by $k_c = (\gamma - 1)h$. Under this representation, the power-law is valid for $k > k_c$. A high k_c means that only a small number of highly connected nodes are scale-free. Increasing k_c by the degree component γ could dramatically decrease the hub frequency; in this case, the scale-free property is vanishing and the network quickly becomes random, that is, $P(k) \sim e^{-k/h}$ when $\gamma \to \infty$ and so $k_c \to \infty$.

10.5 Network motif analysis and yeast genome duplication

In the evolution of complex biological networks, gene duplication is thought to be a key mechanism by which networks evolve and new components are added (Aury *et al.* 2006; Barabasi and Albert 1999; Prince and Pickett 2002; Ispolatov *et al.* 2005; Presser *et al.* 2008; Evlampiev and Isambert 2008; Gu 2009). However, little is known about the evolutionary pattern of interactions after the gene duplication, or the effects of gene interaction on the fate of duplicate genes. Presser *et al.* (2008) have proposed a mathematical framework for describing the protein interaction network after the duplication, which decomposes a protein interaction network into a vector of network motifs. Here we introduce this theory briefly.

Network motifs are small subgraphs, or interaction patterns, which have been a valuable tool in identifying functional structure in many biological networks including transcriptional, neural, and developmental networks (Milo *et al.* 2002; Conant and Wagner 2003a; Wuchty *et al.* 2003; Shen-Orr *et al.* 2002; Sole and Valverde 2006; Cordero and Hogeweg 2006). Presser *et al.* (2008) applied the concept of network motifs to WGD (Whole Genome Duplication) genes in *S. cerevisiae* (yeast) and analyzed network motifs composed of two duplicate pairs, each of which was generated by WGD (namely, motifs of interactions within four proteins). Figure 10.4 presents two hypothetical cases to illustrate how current network motifs can be formed through a process of duplication and divergence. Self-interacting proteins lead to a post-duplication interaction between duplicate genes. If two ancestral genes interacted, there are four interactions formed between their pairs of descendants. The duplication step thus yields an initial motif (called a zero-order motif). During the divergence step, interactions might be gained, lost, or retained.

There are $4 \times (4 - 1)/2 = 6$ possible interactions between any four proteins (Fig. 10.4). As each interaction has two states (yes or no), it hence results in $2^6 = 64$ possible motifs. This number can be further reduced to 19 different motif classes after accounting for symmetry (Table 10.2). Presser *et al.* (2008) used 450 WGD duplicate pairs (Kellis *et al.* 2004), whereas interactions between these proteins were from the Database of Interacting Proteins (DIP). Then, the modern distribution (m_{modern}), or frequencies, of these 19 motif classes, is given by Table 10.2. Presser *et al.* (2008) developed an evolutionary model describing the protein connectivity, which consists of two steps: duplication and divergence.

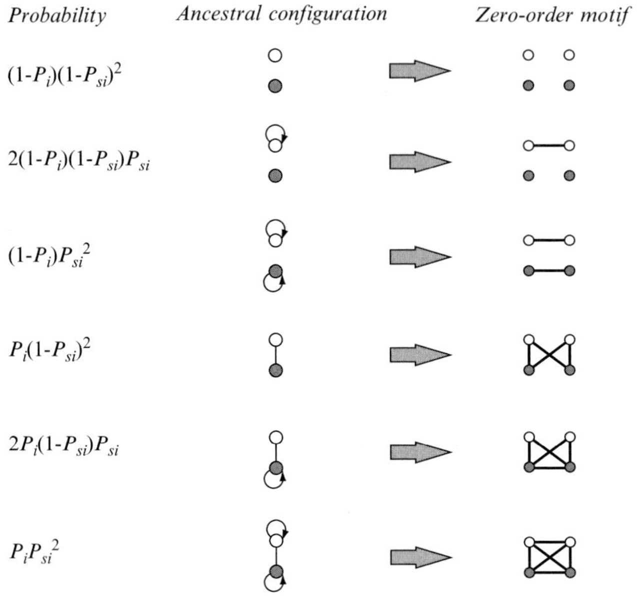

Probability	Ancestral configuration		Zero-order motif
$(1\text{-}P_i)(1\text{-}P_{si})^2$			
$2(1\text{-}P_i)(1\text{-}P_{si})P_{si}$			
$(1\text{-}P_i)P_{si}^2$			
$P_i(1\text{-}P_{si})^2$			
$2P_i(1\text{-}P_{si})P_{si}$			
$P_iP_{si}^2$			

Fig. 10.4 Immediately after the genome duplication, motifs of can be one of six zero-order motifs with probability vector $\boldsymbol{m}_0$ (row vector shown as its transpose). The probabilities of observing each ancestral configuration, and hence each zero-order motif, are listed as functions of the ancestral interaction (Pi) and self-interaction (Psi) probabilities. From Presser *et al.* (2008).

(1) The duplication step assumes that each protein is duplicated along with all its interactions. Because the two duplicated proteins are initially identical, the resulting interaction sets are identical. If a protein was self-interacting, each of its duplicates will be self-interacting, and an interaction will exist between the duplicates. This duplication process can generate only 6 different motifs of the possible 19 (Fig. 10.4). Frequencies of these initial patterns, termed as 'zero-order motifs', are represented by $\mathbf{m}_0$, which are determined by P_{si} and P_i, the probabilities of protein self-interaction and of interaction between two different proteins, respectively. Presser *et al.* (2008) have shown that

$$m_{0,1} = (1 - P_i)(1 - P_{si})^2$$
$$m_{0,2} = 2(1 - P_i)(1 - P_{si})P_{si}$$
$$m_{0,3} = (1 - P_i)P_{si}^2$$
$$m_{0,4} = P_i(1 - P_{si})^2$$
$$m_{0,5} = 2P_i(1 - P_{si})P_{si}$$
$$m_{0,6} = P_iP_{si}^2 \tag{10.24}$$

Table 10.2 Motif distribution in the modern protein interaction network.

Motif class no.	Motif class	No. of motifs present in today's yeast proteome	Modern motif frequency (m_{modem})
1		81,983	8.15×10^{-1}
2		17,748	1.76×10^{-1}
3		215	2.13×10^{-3}
4		925	9.16×10^{-2}
5		14	1.39×10^{-4}
6		2	1.98×10^{-5}
7		93	9.21×10^{-4}
8		15	1.48×10^{-4}
9		6	5.94×10^{-5}
10		0	0
11		16	1.58×10^{-4}
12		0	0
13		1	9.90×10^{-6}
14		1	9.90×10^{-6}
15		0	0
16		4	3.96×10^{-5}
17		0	0
18		1	9.90×10^{-6}
19		1	9.90×10^{-6}

Note that at the initial conditions, 13 of the 19 motifs can only be generated by the follow-up process of divergence.

(2) The second step in the model encompasses the evolutionary process after the genome duplication. Mutations leading to the addition or deletion of an interaction are assumed to occur with probabilities P_+ and P_-, respectively. It follows that, mathematically, the frequencies of the current 19 motifs are determined by a transition matrix $\mathbf{T}$, whose elements are the probabilities of evolution from the initial, 6-element condition vector $\mathbf{m}_0$, to an observed, 19-element vector, that is (see Eq. (10.24)),

$$\mathbf{m}_0\mathbf{T} = \mathbf{m}_{modern} \tag{10.25}$$

where the transition matrix elements are functions of P_+ and P_- as derived by Presser *et al.* (2008); also see below for a brief comment, and the initial condition zero-order motif vector $\mathbf{m}_0$ is a function of the preduplication parameters P_i and P_{si}.

Presser *et al.* (2008) solved the problem of Eq. (10.25) for the best-fit values of P_i, P_{si}, P_+, and P_- (Table 10.3). Figure 10.5(A) shows that the observed number of motifs is in good agreement with the predictions of the model given the best-fit parameters obtained. As shown in Table 10.3, post-duplication rewiring of the network involved a high probability of interaction loss, whereas the likelihood of gaining an interaction was small. They found that the predicted frequency of self-interactions in the preduplication network is significantly higher than that observed in today's network. This could suggest a structural difference between the modern and ancestral networks, preferential addition or retention of interactions, or selective pressure to preserve duplicates of self-interacting proteins.

Comments on the transition probability matrix T First we consider the transition probability for any of the 64 original motifs. Let n_G, n_L, n_R, and n_A represent the number of edges that are gained, lost, retained, or remain absent for

Table 10.3 Best-fit values of preduplication network connectivity and postduplication dynamics inferred from the proteomic network motif distribution of *Saccharomyces cerevisiae*.

Parameter	Parameter value ± SD
P_i	0.0023 ± 0.0003
P_{si}	0.25 ± 0.04
P_+	0.0007 ± 0.0001
P_-	0.61 ± 0.03

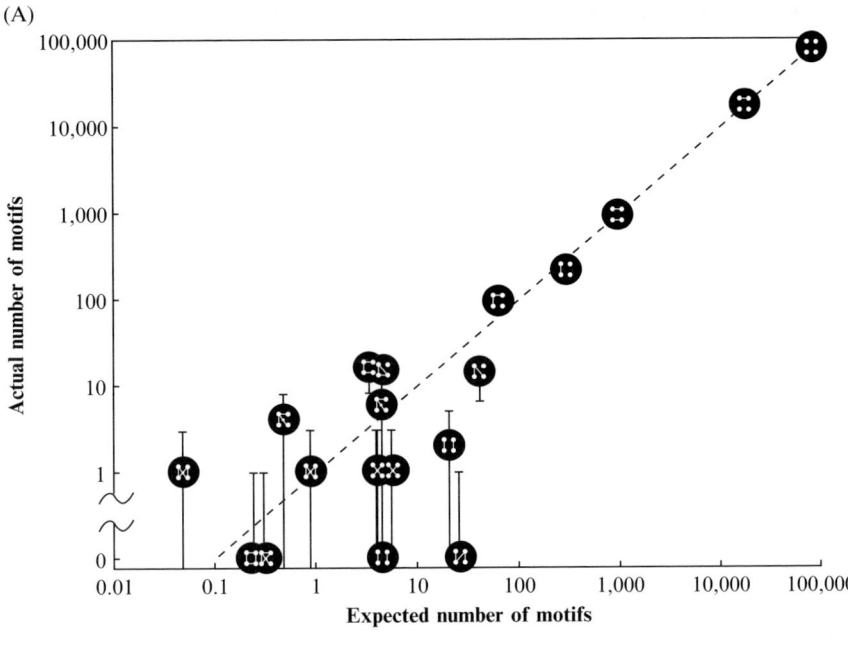

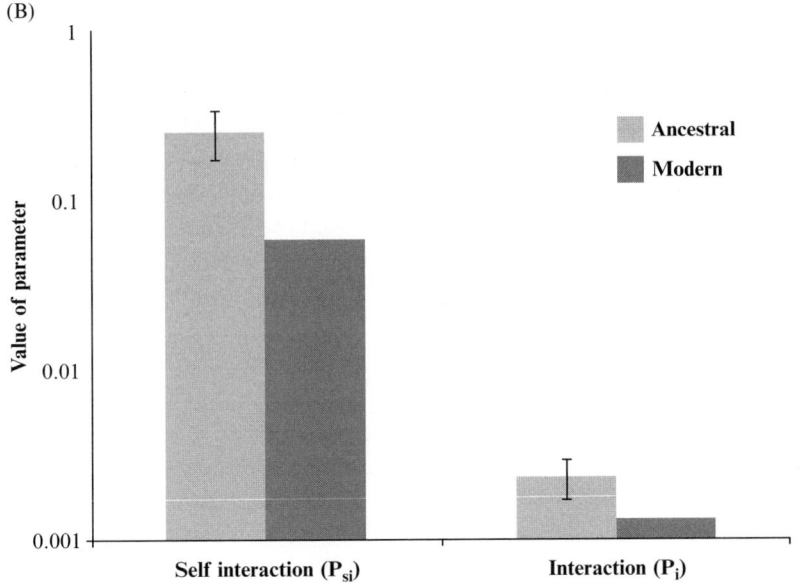

Fig. 10.5 The modern motif distribution closely resembles the expected distribution. (A) We solved the system of 19 equations in 4 unknowns to compute the best-fit network. The expected number of motifs given the best-fit parameters Pi, Psi, $P+$, and $P-$ (x axis) is plotted against actual motif data (Table 10.2). (B) Observed values for Pi and Psi in the modern network are compared with the inferred Pi and Psi parameters for the ancestral preduplication network. Although the intergene connectivity (Pi) is very similar, the inferred self-interaction frequency (Psi) of that network differs by a factor of five from the equivalent modern value. From Presser *et al.* (2008).

a given motif. Then, under the assumption of independence in interaction gain and loss, we have

$$U = P_+^{n_G} P_-^{n_L} (1 - P_-)^{n_R} (1 - P_+)^{n_A}$$

Next consider each of the 19 motif classes that may have a number of equivalent motifs. The transition probabilities are given by a matrix $\mathbf{T}$, whose entries T_{ij} represent the probability of a member of the motif class in row i becoming a member of the motif class in column j. Then we have

$$T_{ij} = \sum_{k \in j} U_{ik}$$

where the subscript $k \in j$ is for all original motifs that belong to the same j-th motif class.

10.6 Evolutionary kinetic (EK) analysis of duplicate genes

Functional redundancies, generated by gene duplications, are widespread in all known genomes, resulting in a tremendous increase to the robustness of organisms (Ohno 1970; Kirschner and Gerhart 1998; Wagner 2000; Conant and Wagner 2004; Gu *et al.* 2003; Su and Gu 2008). Yet, this type of genetic robustness also renders redundancy evolutionarily unstable (Nowark *et al.* 1997; Gu 2003), which has only a transient lifetime because functional overlaps between duplicates would be rapidly lost because of divergence (Force *et al.* 1999; Lynch and Conery 2000; Hughes 2004). In contrast, numerous reports describe instances of functional overlaps that have been conserved throughout extended evolutionary periods. Hence, at least for some duplicate pairs, redundancies are conserved throughout evolution. Kafri *et al.* (2005; 2006) argued that although retention of duplicate redundancy is much less frequent than its loss, its widespread existence is nontrivial and cannot be dismissed as the leftovers of recent duplication events. One such example is two duplicate O-acyl-transferases isozymes, redundantly catalyzing the conjugation of sterols to fatty acids, for which functional overlap has been conserved from yeast (Are1 and Are2) to mammals (ACAT1 and ACAT2). Instead, they suggested that redundancies may be selected for their contributions to genetic robustness and evolvability, as discussed below.

10.6.1   Reprogramming in duplicate backup circuits

Kafri *et al.* (2005) proposed that backup (functional compensation) among differentially expressed duplicate genes (A and B) may suggest that, upon null mutation in gene A, expression of gene B is reprogrammed to acquire an expression profile similar to the wild-type expression profile of gene A. This theory is based on the premise that the promoter architecture may partially overlap between backup-providing paralogs. For instance, such reprogramming has been experimentally values for Acs1 and Acs2 isoenzymes in the yeast. Wild-type Acs1 is subject to glucose repression, but upon deletion of Acs2, the repression of Acs1 is relieved, and Acs1 acquires an Acs2-like responsiveness to glucose. Despite dissimilar expression, the two duplicate genes share a promoter motif (CSRE) and also have unique motifs.

A crucial question is what controls the reprogramming process. Kafri *et al.* (2005) proposed a kinetic model, or reprogramming switch, consisting of two duplicate genes, A and B, that encode enzymes E_A and E_B, both of which interconvert metabolite M_1 into metabolite M_2, i.e. $M_1 \rightarrow M_2$. They proposed that backup between duplicates may use alternative architectures of transcription:

(1) Suppose that only duplicate gene A is active in the wild-type, though two duplicate genes contain the binding sites for a shared transcription factor (TF). It is possible that gene A can be induced by the TF under a low level of metabolite M_1, while gene B can be induced only by a high level of M_1.

(2) Upon knockout of gene A, at the initial stage, metabolite M_1 has been accumulated quickly because it cannot be efficiently interconverted to M_2. On the other hand, M_1 accumulation and the increase of TF concentration eventually result in an efficient activation of duplicate gene B. Consequently, the level of enzyme E_B increases to interconvert $M_1 \rightarrow M_2$. This model provides an appropriate control of backup as it couples response of gene B to an environmental condition (i.e. the accumulation of metabolite M_1) with response to an internal perturbation (i.e. silencing of gene A).

10.6.2 Responsive backup circuits (RBC) and regulatory designs

Through literature surveys, Kafri *et al.* (2006) compiled a list of examples of functional overlaps between duplicate genes that have been conserved in long-term evolution. Many such backed-up genes were shown to be transcriptionally responsive to their redundant duplicate partner and are up-regulated if the latter is mutationally silenced, this is called 'responsive backup circuits' (RBC). The concept of RBC can be illustrated by the yeast Hxt gene family that encodes a redundant set of membrane hexose transporters with varying affinities toward glucose and, consequently, different transport efficiencies (Fig. 10.6(A)).

In yeast, glucose serves as a regulatory input for alternating between aerobic and anaerobic growth. There are two independent signaling pathways, one probing intracellular glucose concentrations and the other probing extracellular concentrations. This differential sensing shows effect in the responsive backup circuit composed of Hxt1 and Htx2. In this case, feedback is made possible by having Hxt2 controlled by two opposing signals. One is its induction by extracellular glucose and the second is its repression by intracellular glucose (Fig. 10.6A). The consequence of this distinction is that although high glucose concentrations result in repression of Hxt2 expression, its induction could be triggered either by low environmental sugar, or alternatively, by mutations in genes responsible for glucose influx. Hence, one of the two duplicates is called the responsive gene because it is under repression in wild-type and that repression is relieved upon its partner's (called the controller) mutation.

Further, Kafri *et al.* (2006) proposed three possible regulatory schemes that could answer the question of what regulatory mode could account for a gene sensing and

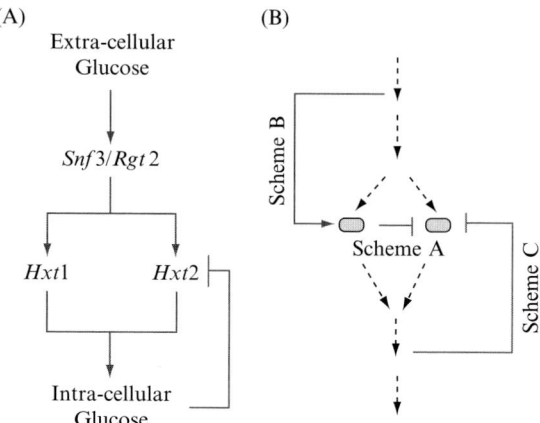

Fig. 10.6 Specific (*A*) and general (*B*) responsive back up circuitries. (*A*) The *Hxt1_Hxt2* responsive backup circuit. Extracellular glucose concentration is sensed by two membrane receptors on the outer yeast membrane, *Rgt2* and *Snf3*. These receptors, once activated by glucose, initiate a signal cascade that induces the transcription of the *Hxt* gene family of hexose transporters encoding membrane channels for glucose intake. The flux of incoming glucose generates an increase in intracellular glucose concentration that represses the transcription of *Hxt2*. (*B*) Three possibilities for feedback in RBCs. For one duplicate gene to sense and respond to its partners' intactness, feedback mechanisms must be at play. In this diagram, duplicates are represented as ovals that lie embedded within a reaction pathway illustrated by the consecutive arrows. Lines A, B, and C represent the three feedback possibilities, namely, simple negative regulation (A), substrate induction (B), and endproduct regulation (C). From Kafri *et al.* (2006).

responding to its redundant partner's intactness. Scheme A (Fig. 10.6B) entails a direct negative regulation of a gene by its functionally redundant partner. Scheme B uses the substrate abundance as a proxy for its partner's activity. In other words, overaccumulation of a substrate, potentially caused by reduced or abolished efficiency of one of the RBC pair members, signals for overproduction of the second member. In particular, scheme C employs the end product-inhibition. Assuming that an endproduct may inhibit both redundant partners, the lack of function of one of the partners would result in the absence of the product and, hence, relief of repression from the second partner. In this model, Kafri *et al.* (2006) argued that the sum (independent functions) of the concentrations of the two redundant proteins ($G_1 + G_2$) should be the biological functions that exploit redundancies between RBC pair members. Examples include reactions that are catalyzed by two independently functioning isozymes. In such cases, the total rate of product production catalyzed by the pair of isozymes would be equal to the production rate contributed by the first isozyme plus that of the second (Fig. 10.7).

After conducting extensive steady-state analyses, they (Kafri *et al.* 2006) concluded two fundamental advantages of this RBC: First, the strength of the restoration

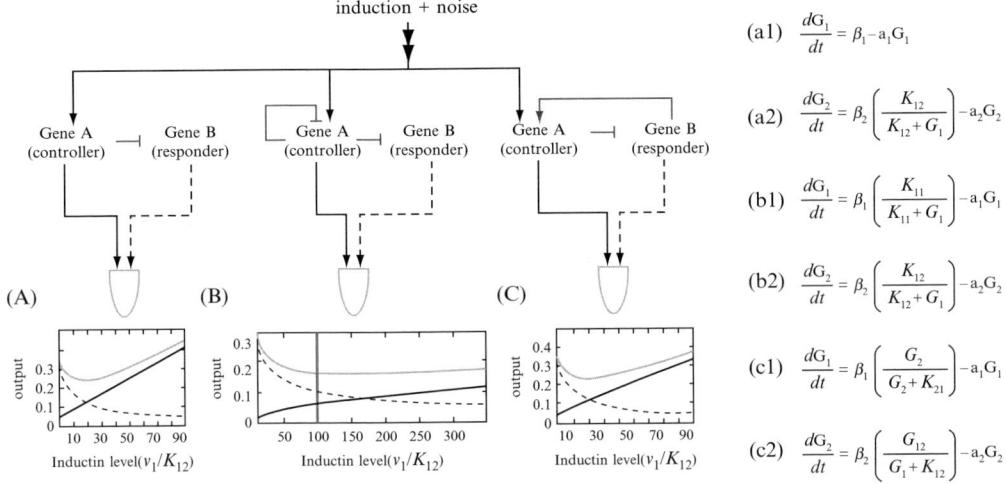

Fig. 10.7 Signal robustness provided by RBCs. Three general responsive backup circuitries are examined as follows: simple repression, modeled by equations a1 and a2 (*A*); dampened controller, modeled by equations b1 and b2 (*B*); and cycled feedback modeled by equations c1 and c2 (*C*). β and α represent the rates of protein synthesis and protein degradation, respectively; Kij is a constant quantifying the regulatory control *i* has over *j*. The RBCs are examined for their efficiency in filtering variations of the regulatory input, v_1, of the controller gene, G_1. For each RBC, a diagram is shown describing the regulatory interactions between the responsive and controlling gene. The plots show the dependency of the controller (solid), the responder (broken), and their sum, $G_1 + G_2$, (gray) on G_1's induction level, v_1. See Kafri *et al.* (2006).

response, but not the inductive response, can be fine tuned by the level of induction of the responsive gene. Second, it has the additional advantage of the negative autoregulation of the controller. In short, the responsive backup circuit (RBC) theory has demonstrated the existence of a variety of kinetic mechanisms for functional compensation between duplicate genes. It attempted to challenge the view that compensation between duplicates is evolutionarily instable. Instead, the RBC theory implies that compensation for the duplicate loss is a natural, derivative effect of the sophisticated design principles using functional redundancy.

10.6.3 Expression-triggered backup circuit hypothesis

However, the RBC theory does not address how two duplicate genes with genetic redundancy can be preserved against null mutations. In the case of a naturally-occurred null mutation in Hxt1, the backup of Hxt2 would result in virtually no fitness loss such that it behaves like a nearly-neutral mutation that can be fixed by the genetic drift. From the view of population genetics, purifying selection in Hxt1 plays a crucial role for gene preservation by eliminating deleterious mutations from the population. In this section we provide some insights about this issue.

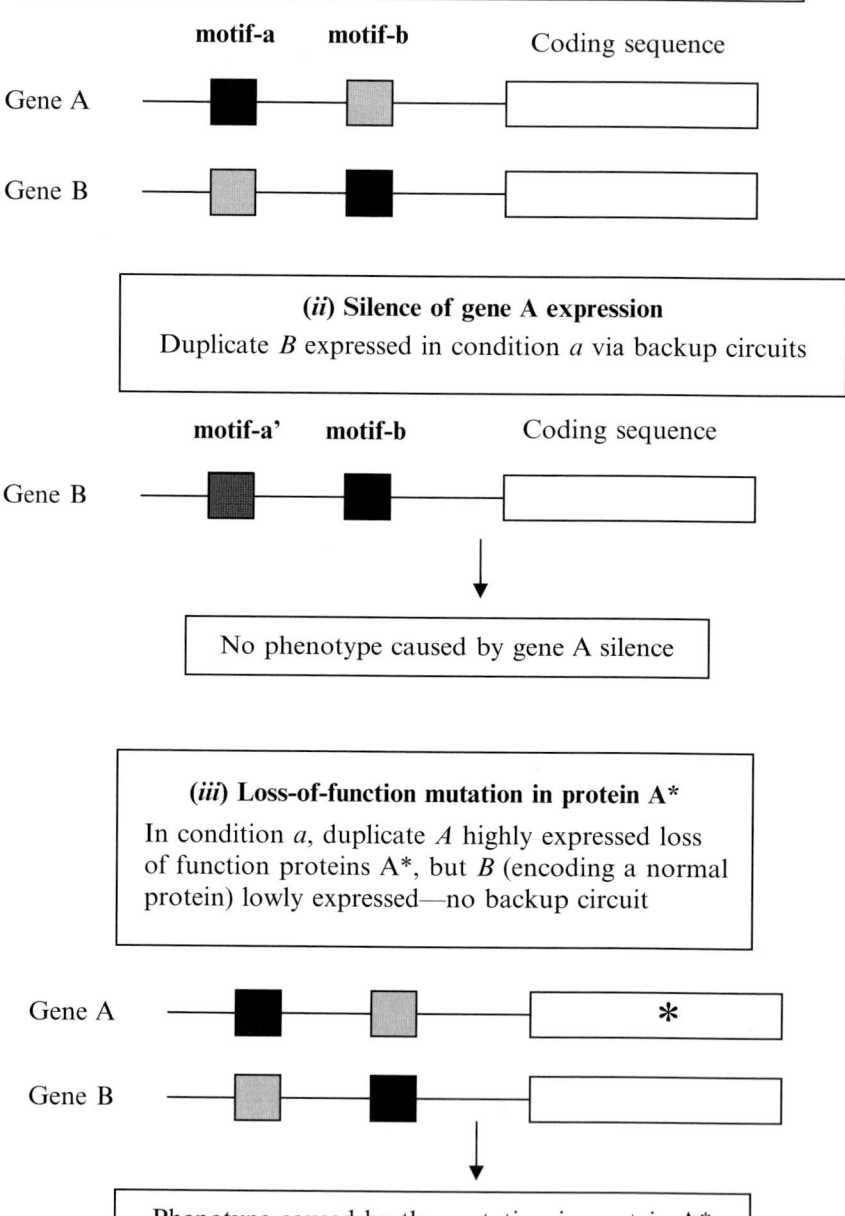

Fig. 10.8 A schematic illustration for the EEE kinetic model to explain functional compensation, expression divergence, and duplication preservation. See the context for the detail.

The first question we may ask is what may trigger the backup circuits between duplicate genes? Kafri *et al.* (2005; 2006) attributed this to the loss of protein function such as enzyme activity. The RBC modeling implies that the expressed proteins are completely functional. We call this function (F)-triggered backup circuits. Alternatively, the expression (E)-triggered backup circuits indicate that the abundance of protein molecules or mRNAs plays a key role, regardless of whether it is functional or not.

The difference between F and E-triggered backup circuits for the duplicate compensation is the outcome when the highly-expressed gene, say, gene A, encodes a nonfunctional protein caused by a nonsynonymous null mutation in the protein sequence. From the theory of RBC (Kafri *et al.* 2006), the consequence should be similar to the silence of gene A, in which the F-triggered backup circuits should be activated by the signal of loss-of-function of the encoded protein A. By contrast, the E-triggered backup circuits would not be activated in this case. Therefore, this deleterious mutation in a protein sequence cannot be functionally compensated and must be eliminated by the purifying selection.

As schematically shown in Fig. (10.8), we tentatively propose an evolutionary scenario for the retention of duplicate genes. After gene duplication, divergence in the expression may occur rapidly, driven by genetic and/or epigenetic factors. Yet, such expression divergence is kinetically reversible, that is, expression compensation between duplicates (backup circuits such as RBC) can be activated by the silence (no expression) of one duplicate gene, i.e. the E-triggered backup circuits. The key for duplicate preservation is that this E-triggered backup circuit cannot be activated by deleterious nonsynonymous mutations in the protein sequence. In such a case accumulation of a large amount of loss-of-function proteins can effectively prevent the activation of its backup duplicate gene. Since this resulting loss-of-function phenotype is selectively harmful, it can be eliminated from the population by strong purifying selection. Though further exploration is needed, the E-trigger hypothesis for duplicate backup circuits may be useful for understanding some unsolved issues in gene duplication. We further speculate that the E-trigger mechanism can be also applied to explain how genetic buffering can be initiated.

References

Abhiman, S., Daub, C.O., and Sonnhammer, E.L. (2006). Prediction of function divergence in protein families using the substitution rate variation parameter alpha. *Mol Biol Evol* **23**, 1406–1413.

Abhiman, S. and Sonnhammer, E.L. (2005a). FunShift: a database of function shift analysis on protein subfamilies. *Nucleic Acids Res* **33**, D197–200.

Abhiman, S. and Sonnhammer, E.L. (2005b). Large-scale prediction of function shift in protein families with a focus on enzymatic function. *Proteins* **60**, 758–768.

Adachi, J. and Hasegawa, M. (1996). Model of amino acid substitution in proteins encoded by mitochondrial DNA. *J Mol Evol* **42**, 459–468.

Adami, C. (2006). Digital genetics: unravelling the genetic basis of evolution. *Nat Rev Genet* **7**, 109–118.

Agrafioti, I., Swire, J., Abbott, J., Huntley, D., Butcher, S., and Stumpf, M.P. (2005). Comparative analysis of the Saccharomyces cerevisiae and Caenorhabditis elegans protein interaction networks. *Bmc Evol Biol* **5**, 23.

Aharoni, A., Gaidukov, L., Khersonsky, O., Mc, Q.G.S., Roodveldt, C., and Tawfik, D.S. (2005). The 'evolvability' of promiscuous protein functions. *Nat Genet* **37**, 73–76.

Akashi, H. and Gojobori, T. (2002). Metabolic efficiency and amino acid composition in the proteomes of Escherichia coli and Bacillus subtilis. *Proc Natl Acad Sci USA* **99**, 3695–3700.

Albert, R., and Barabasi, A.L. (2002). Statistical mechanics of complex networks. *Rev Mod Phys* **74**, 47–97.

Albert, R., Jeong, H., and Barabasi, A.L. (2000). Error and attack tolerance of complex networks. *Nature* **406,** 378–382.

Alon, U. (2003). Biological networks: the tinkerer as an engineer. *Science* **301**, 1866–1867.

Altschul, S.F. and Gish, W. (1996). Local alignment statistics. Computer Methods for Macromolecular *Sequence Analysis* **266,** 460–480.

Altschul, S.F., Gish, W., Miller, W., Myers, E.W., and Lipman, D.J. (1990). Basic local alignment search tool. *J Mol Biol* **215**, 403–410.

Altschul, S.F., Madden, T.L., Schaffer, A.A., Zhang, J., Zhang, Z., Miller, W., and Lipman, D.J. (1997). Gapped BLAST and PSI-BLAST: a new generation of protein database search programs. *Nucleic Acids Res* **25**, 3389–3402.

Atchley, W.R., Fitch, W.M., and Bronner-Fraser, M. (1994). Molecular evolution of the MyoD family of transcription factors. *Proc Natl Acad Sci USA* **91**, 11522–11526.

Audic, S. and Claverie, J.M. (1997). The significance of digital gene expression profiles. *Genome Res* **7**, 986–995.

Aury, J.M., Jaillon, O., Duret, L., Noel, B., Jubin, C., Porcel, B.M., Segurens, B., Daubin, V., Anthouard, V., Aiach, N., *et al.* (2006). Global trends of whole-genome duplications revealed by the ciliate Paramecium tetraurelia. *Nature* **444**, 171–178.

Bailey, T., and Elkan, C. (1995). Unsupervised learning of multiple motifs in biopolymers using expectation maximization. *Mach Learn* **21**, 51–80.

Balaji, S., Iyer, L.M., Aravind, L., and Babu, M.M. (2006). Uncovering a hidden distributed architecture behind scale-free transcriptional regulatory networks. *J Mol Biol* **360**, 204–212.

Balwierz, P.J., Carninci, P., Daub, C.O., Kawai, J., Hayashizaki, Y., Van Belle, W., Beisel, C., and van Nimwegen, E. (2009). Methods for analyzing deep sequencing expression data: constructing the human and mouse promoterome with deepCAGE data. *Genome Biol* **10**, R79.

Barabasi, A.L. (2009). Scale-free networks: a decade and beyond. *Science* **325**, 412–413.

Barabasi, A.L. and Albert, R. (1999). Emergence of scaling in random networks. *Science* **286**, 509–512.

Barabasi, A.L., and Oltvai, Z.N. (2004). Network biology: understanding the cell's functional organization. *Nature Reviews Genetics* **5**, 101–U115.

Barabasi, A.L., Ravasz, E., and Oltvai, Z. (2003). Hierarchical organization of modularity in complex networks. *Lect Notes Phys* **625**, 46–65206.

Barry, D. and Hartigan, J.A. (1987). Asynchronous distance between homologous DNA sequences. *Biometrics* **43**, 261–276.

Barton, N.H. (1990). Pleiotropic models of quantitative variation. *Genetics* **124**, 773–782.

Batada, N. N., T. Reguly, A., Breitkreutz, L., Boucher, B-J Breitkreutz, L.D., Hurst, and M., Tyers *et al.* (2006). Stratus not altocumulus: a new view of the yeast protein interaction network. *Plos Biology* 4(10): e317.

Batzoglou, S., Pachter, L., Mesirov, J.P., Berger, B., and Lander, E.S. (2000). Human and mouse gene structure: comparative analysis and application to exon prediction. *Genome Res* **10**, 950–958.

Benjamini, Y. and Hochberg, Y. (1995). Controlling the false discovery rate - a practical and powerful approach to multiple testing. *J Roy Stat Soc B Met* **57**, 289–300.

Bernardi, G., B. Olofsson, J., Filipski, M., Zerial, J., Salinas, G., Cuny, M., Meunier-Rotival, and F., Rodier *et al.* (1985). The mosaic genome of warm-blooded vertebrates. *Science* 228(4702): 953–958.

Bielawski, J.P. and Yang, Z. (2004). A maximum likelihood method for detecting functional divergence at individual codon sites, with application to gene family evolution. *J Mol Evol* **59**, 121–132.

Blanc, G. and Wolfe, K.H. (2004). Widespread paleopolyploidy in model plant species inferred from age distributions of duplicate genes. *Plant Cell* **16**, 1667–1678.

Blanchette, M., and Tompa, M. (2003). FootPrinter: A program designed for phylogenetic footprinting. *Nucleic Acids Res* **31**, 3840–3842.

Bloom, J.D., Silberg, J.J., Wilke, C.O., Drummond, D.A., Adami, C., and Arnold, F.H. (2005). Thermodynamic prediction of protein neutrality. *Proc Natl Acad Sci USA* **102**, 606–611.

Bourque, G. and Pevzner, P.A. (2002). Genome-scale evolution: reconstructing gene orders in the ancestral species. *Genome Res* **12**, 26–36.

Bouxsein, M.L., Rosen, C.J., Turner, C.H., Ackert, C.L., Shultz, K.L., Donahue, L.R., Churchill, G., Adamo, M.L., Powell, D.R., Turner, R.T., *et al.* (2002). Generation of a new congenic mouse strain to test the relationships among serum insulin-like growth factor I, bone mineral density, and skeletal morphology in vivo. *Journal of Bone and Mineral Research* **17**, 570–579.

Bradley, K.L., Damschen, E.I., Young, L.M., Kuefler, D., Went, S., Wray, G., Haddad, N.M., Knops, J.M.H., and Louda, S.M. (2003). Spatial heterogeneity, not visitation bias, dominates variation in herbivory. *Ecology* **84**, 2214–2221.

Bray, N. and Pachter, L. (2003). MAVID multiple alignment server. *Nucleic Acids Res* **31**, 3525–3526.

Brown, P.O. and Botstein, D. (1999). Exploring the new world of the genome with DNA microarrays. *Nature Genetics* **21**, 33–37.

Brudno, M., Do, C.B., Cooper, G.M., Kim, M.F., Davydov, E., Green, E.D., Sidow, A., and Batzoglou, S. (2003). LAGAN and Multi-LAGAN: efficient tools for large-scale multiple alignment of genomic DNA. *Genome Res* **13**, 721–731.

Burge, C., and Karlin, S. (1997). Prediction of complete gene structures in human genomic DNA. *J Mol Biol* **268**, 78–94.

Bustamante, C.D., Fledel-Alon, A., Williamson, S., Nielsen, R., Hubisz, M.T., Glanowski, S., Tanenbaum, D.M., White, T.J., Sninsky, J.J., Hernandez, R.D., *et al.* (2005). Natural selection on protein-coding genes in the human genome. *Nature* **437**, 1153–1157.

Bustamante, C.D., Townsend, J.P., and Hartl, D.L. (2000). Solvent accessibility and purifying selection within proteins of Escherichia coli and Salmonella enterica. *Mol Biol Evol* **17**, 301–308.

Caceres, M., Lachuer, J., Zapala, M.A., Redmond, J.C., Kudo, L., Geschwind, D.H., Lockhart, D.J., Preuss, T.M., and Barlow, C. (2003). Elevated gene expression levels distinguish human from non-human primate brains. *Proc Natl Acad Sci USA* **100**, 13030–13035.

Caprara, A. (1999). Formulations and hardness of multiple sorting by reversals. In *Proceedings Of The Third Annual International Conference On Computational Molecular Biology* , pp. 84–93 (ACM, Lyon, France, ACM).

Carroll, S.B. (2005). Evolution at two levels: on genes and form. *PLoS Biol* **3**, e245.

Casari, G., Sander, C., and Valencia, A. (1995). A method to predict functional residues in proteins. *Nat Struct Biol* **2**, 171–178.

Cavalli-Sforza, L.L., and Edwards, A.W. (1967). Phylogenetic analysis. Models and estimation procedures. *Am J Hum Genet* **19**, 233–257.

Cavender, J.A., and Felsenstein, J. (1987). Invariants of phylogenies in a simple case with discrete states. *J Classif* **4**, 57–71.

Chamary, J.V., Parmley, J.L., and Hurst, L.D. (2006). Hearing silence: non-neutral evolution at synonymous sites in mammals. *Nat Rev Genet* **7**, 98–108.

Chan, E.Y. (2009). Next-generation sequencing methods: impact of sequencing accuracy on SNP discovery. *Methods Mol Biol* **578**, 95–111.

Chen, Y.W. and Dokholyan, N.V. (2006). The coordinated evolution of yeast proteins is constrained by functional modularity. *Trends Genet* **22**, 416–419.

Cheng, Q., Su, Z., Zhong, Y., and Gu, X. (2009). Effect of site-specific heterogeneous evolution on phylogenetic reconstruction: a simple evaluation. *Gene* **441**, 156–162.

Clarke, G.D., Beiko, R.G., Ragan, M.A., and Charlebois, R.L. (2002). Inferring genome trees by using a filter to eliminate phylogenetically discordant sequences and a distance matrix based on mean normalized BLASTP scores. *J Bacteriol* **184**, 2072–2080.

Collins, M.O. (2009). Cell biology. Evolving cell signals. *Science* **325**, 1635–1636.

Conant, G.C. and Wagner, A. (2003a). Asymmetric sequence divergence of duplicate genes. *Genome Res* **13**, 2052–2058.

Conant, G.C., and Wagner, A. (2003b). Convergent evolution of gene circuits. *Nat Genet* **34**, 264–266.

Conant, G.C., and Wagner, A. (2004). Duplicate genes and robustness to transient gene knock-downs in Caenorhabditis elegans. *Proc Biol Sci* **271**, 89–96.

Cordero, O.X., and Hogeweg, P. (2006). Feed-forward loop circuits as a side effect of genome evolution. *Mol Biol Evol* **23**, 1931–1936.

Coulomb, S., Bauer, M., Bernard, D., and Marsolier-Kergoat, M.C. (2005). Gene essentiality and the topology of protein interaction networks. *Proc Biol Sci* **272**, 1721–1725.

Couronne, O., Poliakov, A., Bray, N., Ishkhanov, T., Ryaboy, D., Rubin, E., Pachter, L., and Dubchak, I. (2003). Strategies and tools for whole-genome alignments. *Genome Res* **13**, 73–80.

Coventry, A., Kleitman, D.J., and Berger, B. (2004). MSARI: multiple sequence alignments for statistical detection of RNA secondary structure. *Proc Natl Acad Sci USA* **101**, 12102–12107.

Davis, J.C. and Petrov, D.A. (2004). Preferential duplication of conserved proteins in eukaryotic genomes. *PLoS Biol* **2**, E55.

Dayhoff, M.O. (1972). Atlas of Protein Sequence and Structure. **5**, *Natl. Biomed. Res. Found.*, Washington, DC.

Dayhoff, M.O. (1978). Atlas of Protein Sequence and Structure. **5**, Suppl.3. *Natl. Biomed. Res. Found.*, Washington, DC.

de Visser, J.A., Hermisson, J., Wagner, G.P., Ancel Meyers, L., Bagheri-Chaichian, H., Blanchard, J.L., Chao, L., Cheverud, J.M., Elena, S.F., Fontana, W., *et al.* (2003). Perspective: Evolution and detection of genetic robustness. *Evolution* **57**, 1959–1972.

Dean, A.M. and Golding, G.B. (1997). Protein engineering reveals ancient adaptive replacements in isocitrate dehydrogenase. *Proc Natl Acad Sci USA* **94**, 3104–3109.

Dean, A.M., Neuhauser, C., Grenier, E., and Golding, G.B. (2002). The pattern of amino acid replacements in alpha/beta-barrels. *Mol Biol Evol* **19**, 1846–1864.

Dean, E.J., Davis, J.C., Davis, R.W., and Petrov, D.A. (2008). Pervasive and persistent redundancy among duplicated genes in yeast. *Plos Genetics* **4** (7), e1000113.

Denver, D.R., Morris, K., Streelman, J.T., Kim, S.K., Lynch, M., and Thomas, W.K. (2005). The transcriptional consequences of mutation and natural selection in Caenorhabditis elegans. *Nat Genet* **37**, 544–548.

DePristo, M.A., Weinreich, D.M., and Hartl, D.L. (2005). Missense meanderings in sequence space: a biophysical view of protein evolution. *Nat Rev Genet* **6**, 678–687.

Dermitzakis, E.T., and Clark, A.G. (2001). Differential selection after duplication in mammalian developmental genes. *Mol Biol Evol* **18**, 557–562.

Dermitzakis, E.T., Reymond, A., Lyle, R., Scamuffa, N., Ucla, C., Deutsch, S., Stevenson, B.J., Flegel, V., Bucher, P., Jongeneel, C.V., *et al.* (2002). Numerous potentially functional but non-genic conserved sequences on human chromosome 21. *Nature* **420,** 578–582.

di Bernardo, D., Down, T., and Hubbard, T. (2003). ddbRNA: detection of conserved secondary structures in multiple alignments. *Bioinformatics* **19**, 1606–1611.

Dickerson, R.E. (1971). The structures of cytochrome c and the rates of molecular evolution. *J Mol Evol* **1**, 26–45.

Dokholyan, N.V., and Shakhnovich, E.I. (2001). Understanding hierarchical protein evolution from first principles. *J Mol Biol* **312,** 289–307.

Doolittle, R.F., Feng, D.F., Tsang, S., Cho, G., and Little, E. (1996). Determining divergence times of the major kingdoms of living organisms with a protein clock. *Science* **271**, 470–477.

Drummond, D.A., Bloom, J.D., Adami, C., Wilke, C.O., and Arnold, F.H. (2005). Why highly expressed proteins evolve slowly. *Proc Natl Acad Sci USA* **102,** 14338–14343.

Drummond, D. A., Raval, A., and Wilke, C.O. (2006). A single determinant dominates the rate of yeast protein evolution. *Mol Biol Evol* 23(2): 327–337

Dudley, A.M., Janse, D.M., Tanay, A., Shamir, R., and Church, G.M. (2005). A global view of pleiotropy and phenotypically derived gene function in yeast. Mol Syst Biol *1*, 2005.0001.

Duret, L., and Mouchiroud, D. (2000). Determinants of substitution rates in mammalian genes: expression pattern affects selection intensity but not mutation rate. *Mol Biol Evol* 17, 68–74.

Edward, A.W.F., and L.L., Cavalli-Sforza (1964). Reconstruction of evolutionary trees. pp. 67–76. In: Heyhood, V.H. and J. McNeill (eds.) Phenetic and Phylogenetic Classification. Systematics Association Publ. No.6.

Edwards, R.J., and Shields, D.C. (2005). BADASP: predicting functional specificity in protein families using ancestral sequences. *Bioinformatics* **21**, 4190–4191.

Eisen, M. B., Spellman, P. T., Brown, P O., and Botstein, D. (1998). Cluster analysis and display of genome-wide expression patterns. *Proc Natl Acad Sci U S A* 95(25): 14863–14868.

Eisen, J.A. (1998). Phylogenomics: improving functional predictions for uncharacterized genes by evolutionary analysis. *Genome Res* 8, 163–167.

Eisen, J.A. and Fraser, C.M. (2003). Phylogenomics: intersection of evolution and genomics. *Science* **300,** 1706–1707.

Elena, S.F. and Lenski, R.E. (2003). Evolution experiments with microorganisms: the dynamics and genetic bases of adaptation. *Nat Rev Genet* **4**, 457–469.

Enard, W., Khaitovich, P., Klose, J., Zollner, S., Heissig, F., Giavalisco, P., Nieselt-Struwe, K., Muchmore, E., Varki, A., Ravid, R., *et al.* (2002). Intra- and interspecific variation in primate gene expression patterns. *Science* **296**, 340–343.

Evangelisti, A.M., and Wagner, A. (2004). Molecular evolution in the yeast transcriptional regulation network. *J Exp Zool B Mol Dev Evol* **302**, 392–411.

Evens, W.J., and Grant, G.R. (2005). Statistical Methods in Bioinformatics: An Introduction (second edition). Springer, New York, USA.

Evlampiev, K. and Isambert, H. (2007). Modeling protein network evolution under genome duplication and domain shuffling. *Bmc Syst Biol* **1**, 49.

Evlampiev, K., and Isambert, H. (2008). Conservation and topology of protein interaction networks under duplication-divergence evolution. *P Natl Acad Sci USA* **105**, 9863–9868.

Ewing, R.M. and Claverie, J.M. (2000). EST databases as multi-conditional gene expression datasets. *Pac Symp Biocomput*, 430–442.

Eyre-Walker, A., Woolfit, M., and Phelps, T. (2006). The distribution of fitness effects of new deleterious amino acid mutations in humans. *Genetics* **173**, 891–900.

Felsenstein, J. (1978). Cases in which parsimony or compatibility methods will be positively misleading. *Syst Zool* **27**, 401–410.

Felsenstein, J. (1981). Evolutionary trees from DNA sequences: a maximum likelihood approach. *J Mol Evol* **17**, 368–376.

Felsenstein, J. (1985). Confidence limits on phylogenies: an approach using the bootstrap. *Evolution* **39,** 783–791.

Felsenstein, J. (1988). *Phylogenies and quantitative characters. Annu Rev Ecol Syst* **19**, 445–471.

Fisher, R.A. (1930). *The Genetical Theory Of Natural Selection* (Oxford, The Clarendon Press).

Fitch, W.M. (1971). Toward defining the course of evolution: minimum change for a specific tree topology. *Syst Zool* **20**, 406–416.

Fitch, W.M. (1981). A non-sequential method for constructing trees and hierarchical classifications. *J Mol Evol* **18,** 30–37.

Fitch, W.M., and Margoliash, E. (1967). Construction of phylogenetic trees. *Science* **155**, 279–284.

Fitz-Gibbon, S.T. and House, C.H. (1999). Whole genome-based phylogenetic analysis of free-living microorganisms. *Nucleic Acids Research* **27**, 4218–4222.

Force, A., Cresko, W.A., Pickett, F.B., Proulx, S.R., Amemiya, C., and Lynch, M. (2005). The origin of subfunctions and modular gene regulation. *Genetics* **170**, 433–446.

Force, A., Lynch, M., Pickett, F.B., Amores, A., Yan, Y.L., and Postlethwait, J. (1999). Preservation of duplicate genes by complementary, degenerative mutations. *Genetics* **151**, 1531–1545.

Forsberg, R. and Christiansen, F.B. (2003). A codon-based model of host-specific selection in parasites, with an application to the influenza A virus. *Molecular Biology and Evolution* **20**, 1252–1259.

Fraser, H.B. (2005). Modularity and evolutionary constraint on proteins. *Nat Genet* **37,** 351–352.

Fraser, H.B., Hirsh, A.E., Steinmetz, L.M., Scharfe, C., and Feldman, M.W. (2002). Evolutionary rate in the protein interaction network. *Science* **296**, 750–752.

Galtier, N. and Gouy, M. (1995). Inferring phylogenies from DNA sequences of unequal base compositions. *Proc Natl Acad Sci USA* **92**, 11317–11321.

Galtier, N. and Gouy, M. (1998). Inferring pattern and process: maximum-likelihood implementation of a nonhomogeneous model of DNA sequence evolution for phylogenetic analysis. *Mol Biol Evol* **15**, 871–879.

Gascuel, O. (1997). BIONJ: an improved version of the NJ algorithm based on a simple model of sequence data. *Mol Biol Evol* **14**, 685–695.

Gaucher, E.A., Miyamoto, M.M., and Benner, S.A. (2001). Function-structure analysis of proteins using covarion-based evolutionary approaches: Elongation factors. *Proc Natl Acad Sci USA* **98**, 548–552.

Gauzzi, M.C., Velazquez, L., McKendry, R., Mogensen, K.E., Fellous, M., and Pellegrini, S. (1996). Interferon-alpha-dependent activation of Tyk2 requires phosphorylation of positive regulatory tyrosines by another kinase. *Journal of Biological Chemistry* **271**, 20494–20500.

Ge, N., and Epstein, C.B. (2004). An empirical Bayesian significance test of cDNA library data. *J Comput Biol* **11**, 1175–1188.

Gerhart, J., and Kirschner, M. (1997). *Cells, Embryos, And Evolution : Toward a Cellular and Developmental Understanding of Phenotypic Variation and Evolutionary Adaptability* (Malden, Mass., Blackwell Science).

Giardine, B., Elnitski, L., Riemer, C., Makalowska, I., Schwartz, S., Miller, W., and Hardison, R.C. (2003). GALA, a database for genomic sequence alignments and annotations. *Genome Res* **13**, 732–741.

Gibbs, W.W. (2003). The unseen genome: gems among the junk. *Sci Am* **289**, 26–33.

Gilad, Y., Oshlack, A., and Rifkin, S.A. (2006). Natural selection on gene expression. *Trends Genet* **22**, 456–461.

Gillespie, J.H. (1991). The causes of molecular evolution (New York, Oxford University Press).

Giot, L., Bader, J.S., Brouwer, C., Chaudhuri, A., Kuang, B., Li, Y., Hao, Y.L., Ooi, C.E., Godwin, B., Vitols, E., *et al.* (2003). A protein interaction map of Drosophila melanogaster. *Science* **302**, 1727–1736.

Golding, G.B. and Dean, A.M. (1998). The structural basis of molecular adaptation. *Mol Biol Evol* **15**, 355–369.

Goldman, N. and Yang, Z. (1994). A codon-based model of nucleotide substitution for protein-coding DNA sequences. *Mol Biol Evol* **11**, 725–736.

Good, J.M. and Nachman, M.W. (2005). Rates of protein evolution are positively correlated with developmental timing of expression during mouse spermatogenesis. *Mol Biol Evol* **22**, 1044–1052.

Graur, D. and Li, W.-H. (2000). *Fundamentals of Molecular Evolution*, 2nd edn (Sunderland, Mass., Sinauer Associates).

Gribaldo, S., Casane, D., Lopez, P., and Philippe, H. (2003). Functional divergence prediction from evolutionary analysis: a case study of vertebrate hemoglobin. *Mol Biol Evol* **20**, 1754–1759.

Gu, J., and Gu, X. (2003a). Induced gene expression in human brain after the split from chimpanzee. *Trends Genet* **19**, 63–65.

Gu, J., and Gu, X. (2003b). Natural history and functional divergence of protein tyrosine kinases. *Gene* **317**, 49–57.

Gu, J., and Gu, X. (2004). Further statistical analysis for genome-wide expression evolution in primate brain/liver/fibroblast tissues. *Hum Genomics* **1**, 247–254.

Gu, J., Wang, Y., and Gu, X. (2002a). Evolutionary analysis for functional divergence of Jak protein kinase domains and tissue-specific genes. *J Mol Evol* **54,** 725–733.

Gu, X. (1999). Statistical methods for testing functional divergence after gene duplication. *Molecular Biology and Evolution* **16,** 1664–1674.

Gu, X. (2001a). Mathematical modeling for functional divergence after gene duplication. *J Comput Biol* **8**, 221–234.

Gu, X. (2001b). Maximum-likelihood approach for gene family evolution under functional divergence. *Molecular Biology and Evolution* **18**, 453–464.

Gu, X. (2003). Evolution of duplicate genes versus genetic robustness against null mutations. *Trends Genet* **19**, 354–356.

Gu, X. (2004). Statistical framework for phylogenomic analysis of gene family expression profiles. *Genetics* **167**, 531–542.

Gu, X. (2006). A simple statistical method for estimating Type-II (Cluster-Specific) functional divergence of protein sequences. *Molecular Biology and Evolution* **23**, 1937–1945.

Gu, X. (2007a). Evolutionary framework for protein sequence evolution and gene pleiotropy. *Genetics* **175**, 1813–1822.

Gu, X. (2007b). Stabilizing selection of protein function and distribution of selection coefficient among sites. *Genetica* **130**, 93–97.

Gu, X. (2009). An evolutionary model for the origin of modularity in a complex gene network. *J Exp Zool B Mol Dev Evol* **312**, 75–82.

Gu, X., Fu, Y.X., and Li, W.H. (1995). Maximum likelihood estimation of the heterogeneity of substitution rate among nucleotide sites. *Mol Biol Evol* **12**, 546–557.

Gu, X., Hewett-Emmett, D., and Li, W.H. (1998). Directional mutational pressure affects the amino acid composition and hydrophobicity of proteins in bacteria. *Genetica* **103**, 383–391.

Gu, X., and Huang, W. (2002). Testing the parsimony test of genome duplications: a counterexample. *Genome Res* **12**, 1–2.

Gu, X., Huang, W., Xu, D., and Zhang, H. (2005a). GeneContent: software for whole-genome phylogenetic analysis. *Bioinformatics* **21**, 1713–1714.

Gu, X., and Li, W.H. (1992). Higher rates of amino acid substitution in rodents than in humans. *Mol Phylogenet Evol* **1**, 211–214.

Gu, X., and Li, W.H. (1994). A model for the correlation of mutation rate with GC content and the origin of GC-rich isochores. *J Mol Evol* **38**, 468–475.

Gu, X., and Li, W.H. (1995). The size distribution of insertions and deletions in human and rodent pseudogenes suggests the logarithmic gap penalty for sequence alignment. *J Mol Evol* **40**, 464–473.

Gu, X., and Li, W.H. (1996a). Bias-corrected paralinear and LogDet distances and tests of molecular clocks and phylogenies under nonstationary nucleotide frequencies. *Mol Biol Evol* **13**, 1375–1383.

Gu, X., and Li, W.H. (1996b). A general additive distance with time-reversibility and rate variation among nucleotide sites. *Proc Natl Acad Sci USA* **93**, 4671–4676.

Gu, X., and Li, W.H. (1998). Estimation of evolutionary distances under stationary and nonstationary models of nucleotide substitution. *Proc Natl Acad Sci USA* **95**, 5899–5905.

Gu, X., and Nei, M. (1999). Locus specificity of polymorphic alleles and evolution by a birth-and-death process in mammalian MHC genes. *Mol Biol Evol* **16**, 147–156.

Gu, X., and Su, Z. (2005). Web-based resources for comparative genomics. *Hum Genomics* **2**, 187–190.

Gu, X., and Su, Z. (2007). Tissue-driven hypothesis of genomic evolution and sequence-expression correlations. *Proc Natl Acad Sci USA* **104**, 2779–2784.

Gu, X., Su, Z., and Huang, Y. (2009). Simultaneous expansions of microRNAs and protein-coding genes by gene/genome duplications in early vertebrates. *J Exp Zool B Mol Dev Evol* **312B**, 164–170.

Gu, X., and Vander Velden, K. (2002). DIVERGE: phylogeny-based analysis for functional-structural divergence of a protein family. *Bioinformatics* **18**, 500–501.

Gu, X., Wang, Y., and Gu, J. (2002b). Age distribution of human gene families shows significant roles of both large- and small-scale duplications in vertebrate evolution. *Nat Genet* **31**, 205–209.

Gu, X., and Zhang, H. (2004). Genome phylogenetic analysis based on extended gene contents. *Molecular Biology and Evolution* **21**, 1401–1408.

Gu, X., and Zhang, J.Z. (1997). A simple method for estimating the parameter of substitution rate variation among sites. *Molecular Biology and Evolution* **14**, 1106–1113.

Gu, X., Zhang, Z., and Huang, W. (2005b). Rapid evolution of expression and regulatory divergences after yeast gene duplication. *Proc Natl Acad Sci USA* **102**, 707–712.

Gu, Z., Cavalcanti, A., Chen, F.C., Bouman, P., and Li, W.H. (2002c). Extent of gene duplication in the genomes of Drosophila, nematode, and yeast. *Mol Biol Evol* **19**, 256–262.

Gu, Z., Nicolae, D., Lu, H.H., and Li, W.H. (2002d). Rapid divergence in expression between duplicate genes inferred from microarray data. *Trends Genet* **18**, 609–613.

Gu, Z., Rifkin, S.A., White, K.P., and Li, W.H. (2004). Duplicate genes increase gene expression diversity within and between species. *Nat Genet* **36**, 577–579.

Gu, Z., Steinmetz, L.M., Gu, X., Scharfe, C., Davis, R.W., and Li, W.H. (2003). Role of duplicate genes in genetic robustness against null mutations. *Nature* **421**, 63–66.

Guigo, R. (1998). Assembling genes from predicted exons in linear time with dynamic programming. *J Comput Biol* **5**, 681–702.

Guo, H., Weiss, R.E., Gu, X., and Suchard, M.A. (2007). Time squared: Repeated measures on phylogenies. *Molecular Biology and Evolution* **24**, 352–362.

Guo, H.H., Choe, J., and Loeb, L.A. (2004). Protein tolerance to random amino acid change. *Proc Natl Acad Sci USA* **101**, 9205–9210.

Hahn, M.W., Conant, G.C., and Wagner, A. (2004). Molecular evolution in large genetic networks: does connectivity equal constraint? *J Mol Evol* **58**, 203–211.

Hahn, M.W., De Bie, T., Stajich, J.E., Nguyen, C., and Cristianini, N. (2005). Estimating the tempo and mode of gene family evolution from comparative genomic data. *Genome Res* **15**, 1153–1160.

Hahn, M.W., and Kern, A.D. (2005). Comparative genomics of centrality and essentiality in three eukaryotic protein-interaction networks. *Mol Biol Evol* **22**, 803–806.

Hahn, M.W., Stajich, J.E., and Wray, G.A. (2003). The effects of selection against spurious transcription factor binding sites. *Mol Biol Evol* **20**, 901–906.

Han, J.D., Bertin, N., Hao, T., Goldberg, D.S., Berriz, G.F., Zhang, L.V., Dupuy, D., Walhout, A.J., Cusick, M.E., Roth, F.P., *et al.* (2004). Evidence for dynamically organized modularity in the yeast protein-protein interaction network. *Nature* **430**, 88–93.

Hansen, T., and Martins, E. (1996). Translating Between microevolutionary process and macroevolutionary patterns: the correlation structure of interspecific data. *Evolution* **50**, 1404–1417.

Harrison, R., Papp, B., Pal, C., Oliver, S.G., and Delneri, D. (2007). Plasticity of genetic interactions in metabolic networks of yeast. *P Natl Acad Sci USA* **104**, 2307–2312.

Hartl, D.L., and Taubes, C.H. (1996). Compensatory nearly neutral mutations: selection without adaptation. *J Theor Biol* **182**, 303–309.

Hartl, D.L., and Taubes, C.H. (1998). Towards a theory of evolutionary adaptation. *Genetica* **102–103**, 525–533.

Hartwell, L.H., Hopfield, J.J., Leibler, S., and Murray, A.W. (1999). From molecular to modular cell biology. *Nature* **402**, C47–C52.

Harvey P.H., and Pagel M.D. (1991). The comparative methods in Evolutionary biology. Oxford University Press. Oxford, UK.

Hasegawa, M. and Hashimoto, T. (1993). Ribosomal-Rna trees misleading. *Nature* **361**, 23–23.

Hastings, W.K. (1970). Monte-Carlo sampling methods using Markov chains and their applications. *Biometrika* **57**, 97–109.

Hazkani-Covo, E., Wool, D., and Graur, D. (2005). In search of the vertebrate phylotypic stage: A molecular examination of the developmental hourglass model and von Baer's third law. *J Exp Zool Part B* **304B**, 150–158.

He, S., Gu, X., Mayden, R.L., Chen, W.J., Conway, K.W., and Chen, Y. (2008). Phylogenetic position of the enigmatic genus Psilorhynchus (Ostariophysi: Cypriniformes): evidence from the mitochondrial genome. *Mol Phylogenet Evol* **47**, 419–425.

He, X. and Zhang, J. (2005). Rapid subfunctionalization accompanied by prolonged and substantial neofunctionalization in duplicate gene evolution. *Genetics* **169**, 1157–1164.

He, X. and Zhang, J. (2006). Toward a molecular understanding of pleiotropy. *Genetics* **173**, 1885–1891.

Hedges, S.B., and S., Kumar (Eds.). 2009. The Timetree of Life. Oxford University Press. Oxford, UK.

Hendy, M.D., and Penny, D. (1982). Branch and bound algorithms to determine minimal evolutionary trees. *Math. Biosci.* **59**, 277–290.

Henikoff, S. and Henikoff, J.G. (1992). Amino-Acid Substitution Matrices from Protein Blocks. P *Natl Acad Sci USA* **89**, 10915–10919.

Hennig, W. (1966). *Phylogenetic Systematics* (Urbana, University of Illinois Press).

Hertz, G.Z. and Stormo, G.D. (1999). Identifying DNA and protein patterns with statistically significant alignments of multiple sequences. *Bioinformatics* **15**, 563–577.

Higgins, D.G. and Sharp, P.M. (1988). CLUSTAL: a package for performing multiple sequence alignment on a microcomputer. *Gene* **73**, 237–244.

Higgins, D.G., Thompson, J.D., and Gibson, T.J. (1996). Using CLUSTAL for multiple sequence alignments. *Methods enzymol* **266**, 383–402.

Hirsh, A.E. and Fraser, H.B. (2001). Protein dispensability and rate of evolution. *Nature* **411**, 1046–1049.

Hoekstra, H.E. and Coyne, J.A. (2007). The locus of evolution: evo devo and the genetics of adaptation. *Evolution* **61**, 995–1016.

Holland, P.W., Garcia-Fernandez, J., Williams, N.A., and Sidow, A. (1994). Gene duplications and the origins of vertebrate development. *Development Suppl issue*, 125–133.

Holmquist, R., Goodman, M., Conroy, T., and Czelusniak, J. (1983). The spatial distribution of fixed mutations within genes coding for proteins. *Journal of Molecular Evolution* **19**, 437–448.

Holt, L.J., Tuch, B.B., Villen, J., Johnson, A.D., Gygi, S.P., and Morgan, D.O. (2009). Global analysis of Cdk1 substrate phosphorylation sites provides insights into evolution. *Science* **325**, 1682–1686.

House, C.H. and Fitz-Gibbon, S.T. (2002). Using homolog groups to create a whole-genomic tree of free-living organisms: An update. *Journal of Molecular Evolution* **54**, 539–547.

Huang, W., Fu, Y.X., Chang, B.H., Gu, X., Jorde, L.B., and Li, W.H. (1998). Sequence variation in ZFX introns in human populations. *Mol Biol Evol* **15**, 138–142.

Huang, Y. and Gu, X. (2007). A bootstrap based analysis pipeline for efficient classification of phylogenetically related animal miRNAs. *BMC Genomics* **8**, 66.

Huang, Y., Zheng, Y., Su, Z., and Gu, X. (2009). Differences in duplication age distributions between human GPCRs and their downstream genes from a network prospective. *BMC Genomics* **10 Suppl 1**, S14.

Huelsenbeck, J.P., Ronquist, F., Nielsen, R., and Bollback, J.P. (2001). Evolution - Bayesian inference of phylogeny and its impact on evolutionary biology. *Science* **294**, 2310–2314.

Hughes, A.L. and Nei, M. (1988). Pattern of nucleotide substitution at major histocompatibility complex class I loci reveals overdominant selection. *Nature* **335**, 167–170.

Hughes, A. L. (1994). The Evolution of Functionally Novel Proteins after Gene Duplication. Proceedings of the Royal Society of London Series B-Biological Sciences **256** (1346): 119–124.

Hughes, T., Ekman, D., Ardawatia, H., Elofsson, A., and Liberles, D.A. (2007). Evaluating dosage compensation as a cause of duplicate gene retention in Paramecium tetraurelia. *Genome Biol* **8**, 213.

Huminiecki, L. and Wolfe, K.H. (2004). Divergence of spatial gene expression profiles following species-specific gene duplications in human and mouse. *Genome Res* **14,** 1870–1879.

Huson, D.H., and Steel, M. (2004). Phylogenetic trees based on gene content. *Bioinformatics* **20**, 2044–2049.

Huynen, M.A., and Snel, B. (2000). Gene and context: integrative approaches to genome analysis. *Adv Protein Chem* **54**, 345–379.

Ihmels, J., Collins, S.R., Schuldiner, M., Krogan, N.J., and Weissman, J.S. (2007). Backup without redundancy: genetic interactions reveal the cost of duplicate gene loss. *Mol Syst Biol* **3,** 86.

Imhof, M. and Schlotterer, C. (2001). Fitness effects of advantageous mutations in evolving Escherichia coli populations. *Proc Natl Acad Sci USA* **98**, 1113–1117.

Ispolatov, I., Krapivsky, P.L., and Yuryev, A. (2005). Duplication-divergence model of protein interaction network. *Phys Rev E Stat Nonlin Soft Matter Phys* **71**, 061911.

Jacquier, A. (2009). The complex eukaryotic transcriptome: unexpected pervasive transcription and novel small RNAs. *Nature Reviews Genetics* **10**, 833–844.

Jeong, H., Mason, S.P., Barabasi, A.L., and Oltvai, Z.N. (2001). Lethality and centrality in protein networks. *Nature* **411**, 41–42.

Jeong, H., Neda, Z., and Barabasi, A.L. (2003). Measuring preferential attachment in evolving networks. *Europhys Lett* **61**, 567–572.

Jeong, H., Tombor, B., Albert, R., Oltvai, Z.N., and Barabasi, A.L. (2000). The large-scale organization of metabolic networks. *Nature* **407**, 651–654.

Jiang, C., Gu, J., Chopra, S., Gu, X., and Peterson, T. (2004). Ordered origin of the typical two- and three-repeat Myb genes. *Gene* **326**, 13–22.

Jin, L. and Nei, M. (1990). Limitations of the evolutionary parsimony method of phylogenetic analysis. *Mol Biol Evol* **7**, 82–102.

Johnson, N.L., and Kotz, S. (1969). *Discrete Distributions* (Boston, Houghton Mifflin).

Johnson, N.L. and Kotz, S. (1970). *Continuous Univariate Distributions* (New York, Hougton Mifflin).

Jones, D.T., Taylor, W.R., and Thornton, J.M. (1992). The rapid generation of mutation data matrices from protein sequences. *Comput Appl Biosci* **8**, 275–282.

Jordan, I.K., Bishop, G.R., and Gonzalez, D.S. (2001). Sequence and structural aspects of functional diversification in class I alpha-mannosidase evolution. *Bioinformatics* **17**, 965–976.

Jordan, I.K., Marino-Ramirez, L., and Koonin, E.V. (2005). Evolutionary significance of gene expression divergence. *Gene* **345**, 119–126.

Jordan, I.K., Rogozin, I.B., Wolf, Y.I., and Koonin, E.V. (2002). Essential genes are more evolutionarily conserved than are nonessential genes in bacteria. *Genome Res* **12**, 962–968.

Jordan, I.K., Wolf, Y.I., and Koonin, E.V. (2003). No simple dependence between protein evolution rate and the number of protein-protein interactions: only the most prolific interactors tend to evolve slowly. *Bmc Evol Biol* **3**, 1.

Jordan, I.K., Wolf, Y.I., and Koonin, E.V. (2004). Duplicated genes evolve slower than singletons despite the initial rate increase. *Bmc Evol Biol* **4**, 22.

Jukes, T.H., and Cantor, C.R. (1969). Evolution of protein molecules. In Mammalian protein metabolism (H. N. Munro, ed.), pp.21–132. Academic, New York.

Kacser, H. and Burns, J.A. (1979). MOlecular democracy: who shares the controls? *Biochem Soc Trans* **7**, 1149–1160.

Kafri, R., Bar-Even, A., and Pilpel, Y. (2005). Transcription control reprogramming in genetic backup circuits. *Nature Genetics* **37**, 295–299.

Kafri, R., Levy, M., and Pilpel, Y. (2006). The regulatory utilization of genetic redundancy through responsive backup circuits. *P Natl Acad Sci USA* **103**, 11653–11658.

Kamath, R.S., Fraser, A.G., Dong, Y., Poulin, G., Durbin, R., Gotta, M., Kanapin, A., Le Bot, N., Moreno, S., Sohrmann, M., *et al.* (2003). Systematic functional analysis of the Caenorhabditis elegans genome using RNAi. *Nature* **421**, 231–237.

Kasprzyk, A., Keefe, D., Smedley, D., London, D., Spooner, W., Melsopp, C., Hammond, M., Rocca-Serra, P., Cox, T., and Birney, E. (2004). EnsMart: a generic system for fast and flexible access to biological data. *Genome Res* **14**, 160–169.

Kato, K. (2009). Impact of the next generation DNA sequencers. *Int J Clin Exp Med* **2**, 193–202.

Keightley, P.D. (1994). The distribution of mutation effects on viability in Drosophila melanogaster. *Genetics* **138**, 1315–1322.

Keller, E.F. (2005). Revisiting "scale-free" networks. *Bioessays* **27**, 1060–1068.

Kellis, M., Birren, B.W., and Lander, E.S. (2004). Proof and evolutionary analysis of ancient genome duplication in the yeast Saccharomyces cerevisiae. *Nature* **428**, 617–624.

Kerr, M.K. and Churchill, G.A. (2001). Bootstrapping cluster analysis: Assessing the reliability of conclusions from microarray experiments. *P Natl Acad Sci USA* **98**, 8961–8965.

Khaitovich, P., Enard, W., Lachmann, M., and Paabo, S. (2006a). Evolution of primate gene expression. *Nature Reviews Genetics* **7**, 693–702.

Khaitovich, P., Hellmann, I., Enard, W., Nowick, K., Leinweber, M., Franz, H., Weiss, G., Lachmann, M., and Paabo, S. (2005a). Parallel patterns of evolution in the genomes and transcriptomes of humans and chimpanzees. *Science* **309**, 1850–1854.

Khaitovich, P., Kelso, J., Franz, H., Visagie, J., Giger, T., Joerchel, S., Petzold, E., Green, R.E., Lachmann, M., and Paabo, S. (2006b). Functionality of intergenic transcription: An evolutionary comparison. Plos *Genetics* **2**, 1590–1598.

Khaitovich, P., Muetzel, B., She, X.W., Lachmann, M., Hellmann, I., Dietzsch, J., Steigele, S., Do, H.H., Weiss, G., Enard, W., *et al.* (2004a). Regional patterns of gene expression in human and chimpanzee brains. *Genome Res* **14**, 1462–1473.

Khaitovich, P., Paabo, S., and Weiss, G. (2005b). Toward a neutral evolutionary model of gene expression. *Genetics* **170**, 929–939.

Khaitovich, P., Tang, K., Franz, H., Kelso, J., Hellmann, I., Enard, W., Lachmann, M., and Paabo, S. (2006c). Positive selection on gene expression in the human brain. *Current Biology* **16**, R356–R358.

Khaitovich, P., Weiss, G., Lachmann, M., Hellmann, I., Enard, W., Muetzel, B., Wirkner, U., Ansorge, W., and Paabo, S. (2004b). A neutral model of transcriptome evolution. *PLoS Biol* **2**, E132.

Kimura, M. (1968). Evolutionary rate at the molecular level. *Nature* **217**, 624–626.

Kimura, M. (1979). Model of effectively neutral mutations in which selective constraint is incorporated. *Proc Natl Acad Sci USA* **76**, 3440–3444.

Kimura, M. (1980). A simple method for estimating evolutionary rates of base substitutions through comparative studies of nucleotide sequences. *J Mol Evol* **16**, 111–120.

Kimura, M. (1983). *The Neutral Theory of Molecular Evolution* (Cambridge, Cambridgeshire; New York, Cambridge University Press).

Kimura, M. and Ota, T. (1971). Protein polymorphism as a phase of molecular evolution. *Nature* **229**, 467–469.

King, M.C. and Wilson, A.C. (1975). Evolution at two levels in humans and chimpanzees. *Science* **188**, 107–116.

Kirschner, M. and Gerhart, J. (1998). Evolvability. *P Natl Acad Sci USA* **95**, 8420–8427.

Kishino, H., Miyata, T., and Hasegawa, M. (1990). Maximum likelihood inference of protein phylogeny and the origin of chloroplasts. *Journal of Molecular Evolution* **31**, 151–160.

Knudsen, B. and Miyamoto, M.M. (2001). A likelihood ratio test for evolutionary rate shifts and functional divergence among proteins. *Proc Natl Acad Sci USA* **98**, 14512–14517.

Koehl, P. and Levitt, M. (2002). Protein topology and stability define the space of allowed sequences. *Proc Natl Acad Sci USA* **99**, 1280–1285.

Kondrashov, A.S. (1998). Measuring spontaneous deleterious mutation process. *Genetica* **102–103**, 183–197.

Kondrashov, A.S., Sunyaev, S., and Kondrashov, F.A. (2002a). Dobzhansky-Muller incompatibilities in protein evolution. *Proc Natl Acad Sci USA* **99**, 14878–14883.

Kondrashov, F.A., Rogozin, I.B., Wolf, Y.I., and Koonin, E.V. (2002b). Selection in the evolution of gene duplications. *Genome Biol* **3**, RESEARCH0008.

Koonin, E.V. and Wolf, Y.I. (2006). Evolutionary systems biology: links between gene evolution and function. *Curr Opin Biotechnol* **17**, 481–487.

Korbel, J.O., Snel, B., Huynen, M.A., and Bork, P. (2002). SHOT: a web server for the construction of genome phylogenies. *Trends Genet* **18**, 158–162.

Korf, I., Flicek, P., Duan, D., and Brent, M.R. (2001). Integrating genomic homology into gene structure prediction. *Bioinformatics* **17 Suppl 1**, S140–148.

Koshi, J.M. and Goldstein, R.A. (1996). Probabilistic reconstruction of ancestral protein sequences. *Journal of Molecular Evolution* **42**, 313–320.

Krylov, D.M., Wolf, Y.I., Rogozin, I.B., and Koonin, E.V. (2003). Gene loss, protein sequence divergence, gene dispensability, expression level, and interactivity are correlated in eukaryotic evolution. Genome Res *13*, 2229-2235.

Kumar, S., and Hedges, S.B. (1998). A molecular timescale for vertebrate evolution. *Nature* **392**, 917–920.

Kumar, S., Nei, M., Dudley, J., and Tamura, K. (2008). MEGA: a biologist-centric software for evolutionary analysis of DNA and protein sequences. *Brief Bioinform* **9**, 299–306.

Lake, J.A. (1994). Reconstructing evolutionary trees from DNA and protein sequences: paralinear distances. *Proc Natl Acad Sci USA* **91**, 1455–1459.

Lanave, C., Preparata, G., Saccone, C., and Serio, G. (1984). A new method for calculating evolutionary substitution rates. *Journal of Molecular Evolution* **20**, 86–93.

Lande, R. (1980). The genetic covariance between characters maintained by pleiotropic mutations. *Genetics* **94**, 203–215.

Landgraf, R., Xenarios, I., and Eisenberg, D. (2001). Three-dimensional cluster analysis identifies interfaces and functional residue clusters in proteins. *J Mol Biol* **307**, 1487–1502.

Landry, C.R., Levy, E.D., and Michnick, S.W. (2009). Weak functional constraints on phosphoproteomes. *Trends Genet* **25**, 193–197.

Larget, B., and Simon, D.L. (1999). Markov chain Monte Carlo algorithms for the Bayesian analysis of phylogenetic trees. *Molecular Biology and Evolution* **16**, 750–759.

Lawrence, C.E., Altschul, S.F., Boguski, M.S., Liu, J.S., Neuwald, A.F., and Wootton, J.C. (1993). Detecting subtle sequence signals: a Gibbs sampling strategy for multiple alignment. *Science* **262**, 208–214.

Lawrence, J.G. (1999). Gene transfer, speciation, and the evolution of bacterial genomes. *Curr Opin Microbiol* **2**, 519–523.

Lee, T.I., Rinaldi, N.J., Robert, F., Odom, D.T., Bar-Joseph, Z., Gerber, G.K., Hannett, N.M., Harbison, C.T., Thompson, C.M., Simon, I., *et al.* (2002). Transcriptional regulatory networks in Saccharomyces cerevisiae. *Science* **298**, 799–804.

Lee, Y.H., Ota, T., and Vacquier, V.D. (1995). Positive selection is a general phenomenon in the evolution of abalone sperm lysin. *Molecular Biology and Evolution* **12**, 231–238.

Lewontin, R.C. (1974). *The Genetic Basis of Evolutionary Change* (New York, Columbia University Press).

Li, W.-H. (1997). *Molecular Evolution* (Sunderland, Mass., Sinauer Associates).

Li, W.H. (1993). Unbiased estimation of the rates of synonymous and nonsynonymous substitution. *J Mol Evol* **36,** 96–99.

Li, W.H., and Gu, X. (1996). Estimating evolutionary distances between DNA sequences. *Methods Enzymol* **266**, 449–459.

Li, W.H., Wu, C.I., and Luo, C.C. (1985). A new method for estimating synonymous and nonsynonymous rates of nucleotide substitution considering the relative likelihood of nucleotide and codon changes. *Mol Biol Evol* **2**, 150–174.

Li, W.H., Yang, J., and Gu, X. (2005). Expression divergence between duplicate genes. *Trends Genet* **21**, 602–607.

Liang, H. and Li, W.H. (2007). Gene essentiality, gene duplicability and protein connectivity in human and mouse. *Trends Genet* **23**, 375–378.

Liao, B.Y. and Zhang, J. (2007). Mouse duplicate genes are as essential as singletons. *Trends Genet* **23**, 378-381.

Lichtarge, O., Bourne, H.R., and Cohen, F.E. (1996). An evolutionary trace method defines binding surfaces common to protein families. *J Mol Biol* **257**, 342–358.

Lin, J. and Gerstein, M. (2000). Whole-genome trees based on the occurrence of folds and orthologs: implications for comparing genomes on different levels. *Genome Res* **10**, 808–818.

Liu, C., Bai, B., Skogerbo, G., Cai, L., Deng, W., Zhang, Y., Bu, D., Zhao, Y., and Chen, R. (2005). NONCODE: an integrated knowledge database of non-coding RNAs. *Nucleic Acids Res* **33**, D112–115.

Liu, Q.Q., Yao, Q.H., Wang, H.M., and Gu, M.H. (2004). [Endosperm-specific expression of the ferritin gene in transgenic rice (Oryza sativa L.) results in increased iron content of milling rice]. *Yi Chuan Xue Bao* **31**, 518–524.

Livingstone, C.D. and Barton, G.J. (1996). Identification of functional residues and secondary structure from protein multiple sequence alignment. *Methods Enzymol* **266**, 497–512.

Lockhart, P.J., Steel, M.A., Hendy, M.D., and Penny, D. (1994). Recovering evolutionary trees under a more realistic model of sequence evolution. *Molecular Biology and Evolution* **11**, 605–612.

Loewe, L., Charlesworth, B., Bartolome, C., and Noel, V. (2006). Estimating selection on nonsynonymous mutations. *Genetics* **172**, 1079–1092.

Logsdon, J.M. and Faguy, D.M. (1999). Evolutionary genomics: Thermotoga heats up lateral gene transfer. *Current Biology* **9**, R747–R751.

Lopez, P., Casane, D., and Philippe, H. (2002). Heterotachy, an important process of protein evolution. *Molecular Biology and Evolution* **19**, 1–7.

Lopez, P., Forterre, P., and Philippe, H. (1999). The root of the tree of life in the light of the covarion model. *J Mol Evol* **49**, 496–508.

Lucas, E.S., Finn, S.L., Cox, A., Lock, F.R., and Watkins, A.J. (2009). The impact of maternal high fat nutrition on the next generation: food for thought? *J Physiol* **587**, 3425–3426.

Luo, H., Rose, P., Barber, D., Hanratty, W.P., Lee, S., Roberts, T.M., DAndrea, A.D., and Dearolf, C.R. (1997). Mutation in the Jak kinase JH2 domain hyperactivates Drosophila and mammalian Jak-Stat pathways. *Molecular and Cellular Biology* **17**, 1562–1571.

Lynch, M. (2007a). The evolution of genetic networks by non-adaptive processes. *Nat Rev Genet* **8**, 803–813.

Lynch, M. (2007b). The frailty of adaptive hypotheses for the origins of organismal complexity. *P Natl Acad Sci USA* **104**, 8597–8604.

Lynch, M. (2007c). *The Origins of Genome Architecture* (Sunderland, Mass., Sinauer Associates).

Lynch, M. and Conery, J.S. (2000). The evolutionary fate and consequences of duplicate genes. *Science* **290**, 1151–1155.

Lynch, M. and Hill, W.G. (1986). Phenotypic evolution by neutral mutation. *Evolution* **40**, 915–935.

Lynch, M., O'Hely, M., Walsh, B., and Force, A. (2001). The probability of preservation of a newly arisen gene duplicate. *Genetics* 159, 1789–1804.

MacLean, R.C., Bell, G., and Rainey, P.B. (2004). The evolution of a pleiotropic fitness tradeoff in Pseudomonas fluorescens. *Proc Natl Acad Sci USA* **101**, 8072–8077.

Makova, K.D. and Li, W.H. (2003). Divergence in the spatial pattern of gene expression between human duplicate genes. *Genome Res* **13**, 1638–1645.

Manning, G., Young, S.L., Miller, W.T., and Zhai, Y. (2008). The protist, Monosiga brevicollis, has a tyrosine kinase signaling network more elaborate and diverse than found in any known metazoan. *Proc Natl Acad Sci USA* **105**, 9674–9679.

Mardis, E.R. (2008). The impact of next-generation sequencing technology on genetics. *Trends Genet* **24**, 133–141.

Martin, G. and Lenormand, T. (2006). A general multivariate extension of Fisher's geometrical model and the distribution of mutation fitness effects across species. *Evolution* **60**, 893–907.

Mau, B., Newton, M.A., and Larget, B. (1999). Bayesian phylogenetic inference via Markov chain Monte Carlo methods. *Biometrics* **55**, 1–12.

Mayor, C., Brudno, M., Schwartz, J.R., Poliakov, A., Rubin, E.M., Frazer, K.A., Pachter, L.S., and Dubchak, I. (2000). VISTA : visualizing global DNA sequence alignments of arbitrary length. *Bioinformatics* **16**, 1046–1047.

McDonald, J.H. and Kreitman, M. (1991). Adaptive protein evolution at the Adh locus in Drosophila. *Nature* **351**, 652–654.

McGuigan, K. and Sgro, C.M. (2009). Evolutionary consequences of cryptic genetic variation. *Trends Ecol Evol* **24**, 305–311.

Medina, M. (2005). Genomes, phylogeny, and evolutionary systems biology. *P Natl Acad Sci USA* **102**, 6630–6635.

Mercer, J.F.B., Grimes, A., Ambrosini, L., Lockhart, P., Paynter, J.A., Dierick, H., and Glover, T.W. (1994). Mutations in the Murine Homolog of the Menkes Gene in Dappled and Blotchy Mice. *Nature Genetics* **6**, 374–378.

Messier, W., and Stewart, C.B. (1997). Episodic adaptive evolution of primate lysozymes. *Nature* **385**, 151–154.

Metropolis, N., Rosenbluth, A.W., Rosenbluth, M.N., Teller, A.H., and Teller, E. (1953). Equation of state calculations by fast computing machines. *Journal of Chemical Physics* **21**, 1087–1092.

Meyer, I.M. and Durbin, R. (2002). Comparative ab initio prediction of gene structures using pair HMMs. *Bioinformatics* **18**, 1309–1318.

Milo, R., Shen-Orr, S., Itzkovitz, S., Kashtan, N., Chklovskii, D., and Alon, U. (2002). Network motifs: Simple building blocks of complex networks. *Science* **298**, 824–827.

Mintseris, J. and Weng, Z. (2005). Structure, function, and evolution of transient and obligate protein-protein interactions. *Proc Natl Acad Sci USA* **102**, 10930–10935.

Miyata, T. and Yasunaga, T. (1980). Molecular evolution of mRNA: a method for estimating evolutionary rates of synonymous and amino acid substitutions from homologous nucleotide sequences and its application. *J Mol Evol* **16**, 23–36.

Mooers, A.O. and Schluter, D. (1999). Reconstructing ancestor states with maximum likelihood: Support for one- and two-rate models. *Systematic Biol* **48**, 623–633.

Munro, H.N. and Allison, J.B. (1964). *Mammalian Protein Metabolism* (New York, Academic Press).

Nam, J., Kaufmann, K., Theissen, G., and Nei, M. (2005). A simple method for predicting the functional differentiation of duplicate genes and its application to MIKC-type MADS-box genes. *Nucleic Acids Research* **33**, e12.

Natale, D.A., Shankavaram, U.T., Galperin, M.Y., Wolf, Y.I., Aravind, L., and Koonin, E.V. (2000). Towards understanding the first genome sequence of a crenarchaeon by genome annotation using clusters of orthologous groups of proteins (COGs). *Genome Biol* **1**, RESEARCH0009.

National Biomedical Research Foundation. and Dayhoff, M.O. (1978). Protein segment dictionary 78 : from the *Atlas of Protein Sequence and Structure*, volume 5, and supplements 1, 2, and 3 (Silver Spring, Md.Washington, D.C., National Biomedical Research Foundation ;Georgetown University Medical Center).

Needleman, S.B. and Wunsch, C.D. (1970). A general method applicable to the search for similarities in the amino acid sequence of two proteins. *J Mol Biol* **48**, 443–453.

Nei, M. (1969). Gene duplication and nucleotide substitution in evolution. *Nature* **221**, 40–42.

Nei, M. (1987). *Molecular Evolutionary Genetics* (New York, Columbia University Press).

Nei, M., and Kumar, S. (2000). Molecular evolution and phylogenetics. Oxford University Press. Oxford, UK.

Nei, M. (2005). Selectionism and neutralism in molecular evolution. *Mol Biol Evol* **22**, 2318–2342.

Nei, M. (2007). The new mutation theory of phenotypic evolution. *Proc Natl Acad Sci USA* **104**, 12235–12242.

Nei, M. and Gojobori, T. (1986). Simple methods for estimating the numbers of synonymous and nonsynonymous nucleotide substitutions. *Mol Biol Evol* **3**, 418–426.

Nei, M., Gu, X., and Sitnikova, T. (1997). Evolution by the birth-and-death process in multigene families of the vertebrate immune system. *Proc Natl Acad Sci USA* **94**, 7799–7806.

Nei, M., Niimura, Y., and Nozawa, M. (2008). The evolution of animal chemosensory receptor gene repertoires: roles of chance and necessity. *Nat Rev Genet* **9**, 951–963.

Nielsen, R., Bustamante, C., Clark, A.G., Glanowski, S., Sackton, T.B., Hubisz, M.J., Fledel-Alon, A., Tanenbaum, D.M., Civello, D., White, T.J., *et al.* (2005). A scan for positively selected genes in the genomes of humans and chimpanzees. *PLoS Biol* **3**, e170.

Nielsen, R., Hellmann, I., Hubisz, M., Bustamante, C., and Clark, A.G. (2007). Recent and ongoing selection in the human genome. *Nature Reviews Genetics* **8**, 857–868.

Nielsen, R. and Yang, Z. (2003). Estimating the distribution of selection coefficients from phylogenetic data with applications to mitochondrial and viral DNA. *Mol Biol Evol* **20**, 1231–1239.

Nozawa, M., Kawahara, Y., and Nei, M. (2007). Genomic drift and copy number variation of sensory receptor genes in humans. *Proc Natl Acad Sci USA* **104**, 20421–20426.

Oakley, T.H., Gu, Z., Abouheif, E., Patel, N.H., and Li, W.H. (2005). Comparative methods for the analysis of gene-expression evolution: an example using yeast functional genomic data. *Mol Biol Evol* **22**, 40–50.

Ohno, S. (1970). *Evolution by Gene Duplication* (Berlin, New York, Springer-Verlag).

Ohta, T. (1973). Slightly deleterious mutant substitutions in evolution. *Nature* **246**, 96–98.

Ohta, T. (1993). An examination of the generation-time effect on molecular evolution. *Proc Natl Acad Sci USA* **90**, 10676–10680.

Olsen, G.J., Woese, C.R., and Overbeek, R. (1994). The winds of (evolutionary) change: breathing new life into microbiology. *J Bacteriol* **176**, 1–6.

Orr, H.A. (2005). The genetic theory of adaptation: a brief history. *Nat Rev Genet* **6**, 119–127.

Otto, S.P. (2004). Two steps forward, one step back: the pleiotropic effects of favoured alleles. *Proc Biol Sci* **271**, 705–714.

Ovcharenko, I., Loots, G.G., Hardison, R.C., Miller, W., and Stubbs, L. (2004). zPicture: dynamic alignment and visualization tool for analyzing conservation profiles. *Genome Res* **14**, 472–477.

Pagel, M., Meade, A., and Scott, D. (2007). Assembly rules for protein networks derived from phylogenetic-statistical analysis of whole genomes. *Bmc Evol Biol* **7 Suppl 1**, S16.

Pal, C., Papp, B., and Hurst, L.D. (2001). Highly expressed genes in yeast evolve slowly. *Genetics* **158,** 927–931.

Pal, C., Papp, B., and Hurst, L.D. (2003). Genomic function: Rate of evolution and gene dispensability. *Nature* **421**, 496–497; discussion 497–498.

Pal, C., Papp, B., and Lercher, M.J. (2006a). An integrated view of protein evolution. *Nat Rev Genet* **7**, 337–348.

Pal, C., Papp, B., Lercher, M.J., Csermely, P., Oliver, S.G., and Hurst, L.D. (2006b). Chance and necessity in the evolution of minimal metabolic networks. *Nature* **440**, 667–670.

Pang, K.C., Stephen, S., Engstrom, P.G., Tajul-Arifin, K., Chen, W., Wahlestedt, C., Lenhard, B., Hayashizaki, Y., and Mattick, J.S. (2005). RNAdb–a comprehensive mammalian noncoding RNA database. *Nucleic Acids Res* **33**, D125–130.

Papp, B., Pal, C., and Hurst, L.D. (2003). Dosage sensitivity and the evolution of gene families in yeast. *Nature* **424**, 194–197.

Papp, B., Pal, C., and Hurst, L.D. (2004). Metabolic network analysis of the causes and evolution of enzyme dispensability in yeast. *Nature* **429**, 661–664.

Parisi, G. and Echave, J. (2005). Generality of the structurally constrained protein evolution model: assessment on representatives of the four main fold classes. *Gene* **345**, 45–53.

Parra, G., Agarwal, P., Abril, J.F., Wiehe, T., Fickett, J.W., and Guigo, R. (2003). Comparative gene prediction in human and mouse. *Genome Res* **13**, 108–117.

Pastor-Satorras, R., Smith, E., and Sole, R.V. (2003). Evolving protein interaction networks through gene duplication. *Journal of Theoretical Biology* **222**, 199–210.

Pearson, W.R. (1998). Empirical statistical estimates for sequence similarity searches. *Journal of Molecular Biology* **276**, 71–84.

Pearson, W.R. and Lipman, D.J. (1988). Improved tools for biological sequence comparison. *P Natl Acad Sci USA* **85**, 2444–2448.

Penn, O., Stern, A., Rubinstein, N.D., Dutheil, J., Bacharach, E., Galtier, N., and Pupko, T. (2008). Evolutionary modeling of rate shifts reveals specificity determinants in HIV-1 subtypes. *Plos Comput Biol* **4**, e1000214.

Perler, F., Efstratiadis, A., Lomedico, P., Gilbert, W., Kolodner, R., and Dodgson, J. (1980). The evolution of genes - the chicken preproinsulin gene. *Cell* **20**, 555–566.

Piganeau, G. and Eyre-Walker, A. (2003). Estimating the distribution of fitness effects from DNA sequence data: implications for the molecular clock. *Proc Natl Acad Sci USA* **100**, 10335–10340.

Pollock, D.D., Taylor, W.R., and Goldman, N. (1999). Coevolving protein residues: maximum likelihood identification and relationship to structure. *J Mol Biol* **287**, 187–198.

Poon, A. and Otto, S.P. (2000). Compensating for our load of mutations: freezing the meltdown of small populations. *Evolution* **54**, 1467–1479.

Presser, A., Elowitz, M.B., Kellis, M., and Kishony, R. (2008). The evolutionary dynamics of the Saccharomyces cerevisiae protein interaction network after duplication. *P Natl Acad Sci USA* **105**, 950–954.

Prince, V.E. and Pickett, F.B. (2002). Splitting pairs: the diverging fates of duplicated genes. *Nature Reviews Genetics* **3**, 827–837.

Proulx, S.R. and Phillips, P.C. (2005). The opportunity for canalization and the evolution of genetic networks. *Am Nat* **165**, 147–162.

Pupko, T., Pe'er, I., Hasegawa, M., Graur, D., and Friedman, N. (2002). A branch-and-bound algorithm for the inference of ancestral amino-acid sequences when the replacement rate varies among sites: Application to the evolution of five gene families. *Bioinformatics* **18**, 1116–1123.

Pupko, T., Pe'er, I., Shamir, R., and Graur, D. (2000). A fast algorithm for joint reconstruction of ancestral amino acid sequences. *Molecular Biology and Evolution* **17**, 890–896.

Quackenbush, J. (2001). Computational analysis of microarray data. *Nature Reviews Genetics* **2**, 418–427.

Rain, J.C., Selig, L., De Reuse, H., Battaglia, V., Reverdy, C., Simon, S., Lenzen, G., Petel, F., Wojcik, J., Schachter, V., *et al.* (2001). The protein-protein interaction map of Helicobacter pylori. *Nature* **409**, 211–215.

Rannala, B. and Yang, Z. (1996). Probability distribution of molecular evolutionary trees: a new method of phylogenetic inference. *J Mol Evol* **43**, 304–311.

Rastogi, S. and Liberles, D.A. (2005). Subfunctionalization of duplicated genes as a transition state to neofunctionalization. *Bmc Evol Biol* **5**, 28.

Ravasz, E., Somera, A.L., Mongru, D.A., Oltvai, Z.N., and Barabasi, A.L. (2002). Hierarchical organization of modularity in metabolic networks. *Science* **297**, 1551–1555.

Rifkin, S.A., Houle, D., Kim, J., and White, K.P. (2005). A mutation accumulation assay reveals a broad capacity for rapid evolution of gene expression. *Nature* **438**, 220–223.

Rivas, E. and Eddy, S.R. (2001). Noncoding RNA gene detection using comparative sequence analysis. *BMC Bioinformatics* **2**, 8.

Rocha, E.P. and Danchin, A. (2004). An analysis of determinants of amino acids substitution rates in bacterial proteins. *Mol Biol Evol* **21**, 108–116.

Rodriguez-Caso, C., Medina, M.A., and Sole, R.V. (2005). Topology, tinkering and evolution of the human transcription factor network. *FEBS J* **272**, 6423–6434.

Rodriguez, F., Oliver, J.L., Marin, A., and Medina, J.R. (1990). The general stochastic model of nucleotide substitution. *J Theor Biol* **142**, 485–501.

Roth, C., Rastogi, S., Arvestad, L., Dittmar, K., Light, S., Ekman, D., and Liberles, D.A. (2007). Evolution after gene duplication: Models, mechanisms, sequences, systems, and organisms. *J Exp Zool Part B* **308B**, 58–73.

Roth, F.P., Hughes, J.D., Estep, P.W., and Church, G.M. (1998). Finding DNA regulatory motifs within unaligned noncoding sequences clustered by whole-genome mRNA quantitation. *Nat Biotechnol* **16**, 939–945.

Rotonda, J., Nicholson, D.W., Fazil, K.M., Gallant, M., Gareau, Y., Labelle, M., Peterson, E.P., Rasper, D.M., Ruel, R., Vaillancourt, J.P., *et al.* (1996). The three-dimensional structure of apopain/CPP32, a key mediator of apoptosis. *Nat Struct Biol* **3**, 619–625.

Rzhetsky, A., and Nei, M. (1992). Statistical properties of the ordinary least-squares, generalized least-squares, and minimum-evolution methods of phylogenetic inference. *Journal of Molecular Evolution* **35**, 367–375.

Rzhetsky, A. and Nei, M. (1993). Theoretical foundation of the minimum-evolution method of phylogenetic inference. *Mol Biol Evol* **10**, 1073–1095.

Saitou, N. and Nei, M. (1987). The neighbor-joining method: a new method for reconstructing phylogenetic trees. *Mol Biol Evol* **4**, 406–425.

Salathe, M., Ackermann, M., and Bonhoeffer, S. (2006). The effect of multifunctionality on the rate of evolution in yeast. *Mol Biol Evol* **23**, 721–722.

Sankoff, D. (1975). Minimal mutation trees of sequences. *Siam J Appl Math* **28**, 35–42.

Sankoff, D., Sundaram, G., and Kececioglu, J. (1996). Steiner points in the space of genome rearrangements. *International Journal of the Foundations of Computer Science* **7**, 1–9.

Sattath S., and Tversky A. (1977). Additive similarity trees. Psychometrika **42**, 319–345.

Schluter, D. (1995). Uncertainty in ancient phylogenies. *Nature* **377**, 108–109.

Schluter, D., Price, T., Mooers, A. O., and Ludwig, D. (1997). Likelihood of ancestor states in adaptive radiation. Evolution **51**, 1699–1711

Schug, J. and Overton, G.C. (1997). Modeling transcription factor binding sites with Gibbs Sampling and Minimum Description Length encoding. *Proc Int Conf Intell Syst Mol Biol* **5**, 268–271.

Schwartz, S., Elnitski, L., Li, M., Weirauch, M., Riemer, C., Smit, A., Green, E.D., Hardison, R.C., and Miller, W. (2003). MultiPipMaker and supporting tools: Alignments and analysis of multiple genomic DNA sequences. *Nucleic Acids Res* **31**, 3518–3524.

Schwartz, S., Zhang, Z., Frazer, K.A., Smit, A., Riemer, C., Bouck, J., Gibbs, R., Hardison, R., and Miller, W. (2000). PipMaker–a web server for aligning two genomic DNA sequences. *Genome Res* **10**, 577–586.

Sella, G. and Hirsh, A.E. (2005). The application of statistical physics to evolutionary biology. P*roc Natl Acad Sci USA* **102**, 9541–9546.

Sharp, P.M. and Li, W.H. (1987). The rate of synonymous substitution in enterobacterial genes is inversely related to codon usage bias. *Mol Biol Evol* **4**, 222–230.

Shaw, F.H., Geyer, C.J., and Shaw, R.G. (2002). A comprehensive model of mutations affecting fitness and inferences for Arabidopsis thaliana. *Evolution* **56**, 453–463.

Shen-Orr, S.S., Milo, R., Mangan, S., and Alon, U. (2002). Network motifs in the transcription regulation network of Escherichia coli. *Nature Genetics* **31**, 64–68.

Shi, X.F., Gu, H., Susko, E., and Field, C. (2005). The comparison of the confidence regions in phylogeny. *Molecular Biology and Evolution* **22**, 2285–2296.

Sinha, S., Blanchette, M., and Tompa, M. (2004). PhyME: a probabilistic algorithm for finding motifs in sets of orthologous sequences. *BMC Bioinformatics* **5**, 170.

Skovgaard, M., Kodra, J.T., Gram, D.X., Knudsen, S.M., Madsen, D., and Liberles, D.A. (2006). Using evolutionary information and ancestral sequences to understand the sequence-function relationship in GLP-1 agonists. *Journal of Molecular Biology* **363**, 977–988.

Smith, T.F. and Waterman, M.S. (1981). Identification of Common Molecular Subsequences. *Journal of Molecular Biology* **147**, 195–197.

Snel, B., Bork, P., and Huynen, M.A. (1999). Genome phylogeny based on gene content. *Nature Genetics* **21**, 108–110.

Sogin, M.L., Hinkle, G., and Leipe, D.D. (1993). Universal tree of life. *Nature* **362**, 795–795.

Sole, R.V. and Valverde, S. (2006). Are network motifs the spandrels of cellular complexity? *Trends Ecol Evol* **21**, 419–422.

Soyer, O.S. and Bonhoeffer, S. (2006). Evolution of complexity in signaling pathways. *Proc Natl Acad Sci USA* **103**, 16337–16342.

Spencer M., Susko E., Roger A.J. Modelling prokaryote gene content. Evol Bioinform Online (2006) **2**, 165–186.

Steel, M. (1994). Recovering a tree from the leaf colourations it generates under a Markov model. *Appl Math Lett* **7**, 19–23.

Stekel, D.J., Git, Y., and Falciani, F. (2000). The comparison of gene expression from multiple cDNA libraries. *Genome Res* **10**, 2055–2061.

Stone, J.R. and Wray, G.A. (2001). Rapid evolution of cis-regulatory sequences via local point mutations. *Mol Biol Evol* **18**, 1764–1770.

Storey, J.D. (2002). A direct approach to false discovery rates. *J Roy Stat Soc B* **64,** 479–498.

Storey, J.D. and Tibshirani, R. (2003). Statistical significance for genomewide studies. *P Natl Acad Sci USA* **100**, 9440–9445.

Studier, J.A., and Keppler, K.J. (1988). A note on the neighbor-joining algorithm of Saitou and Nei. *Mol Biol Evol* **5**, 729–731.

Su, A.I., Cooke, M.P., Ching, K.A., Hakak, Y., Walker, J.R., Wiltshire, T., Orth, A.P., Vega, R.G., Sapinoso, L.M., Moqrich, A., *et al.* (2002). Large-scale analysis of the human and mouse transcriptomes. *Proc Natl Acad Sci USA* **99**, 4465–4470.

Su, Z. and Gu, X. (2008). Predicting the proportion of essential genes in mouse duplicates based on biased mouse knockout genes. *J Mol Evol* **67**, 705–709.

Su, Z., Huang, Y., and Gu, X. (2007). Tissue-driven hypothesis with Gene Ontology (GO) analysis. *Ann Biomed Eng* **35**, 1088–1094.

Su, Z., Wang, J., Yu, J., Huang, X., and Gu, X. (2006). Evolution of alternative splicing after gene duplication. *Genome Res* **16**, 182–189.

Su, Z., Zeng, Y., and Gu, X. (2009). A preliminary analysis of gene pleiotropy estimated from protein sequences. *J Exp Zoolog B Mol Dev Evol* **314B**,115–122.

Sueoka, N. (1988). Directional mutation pressure and neutral molecular evolution. *Proc Natl Acad Sci USA* **85**, 2653–2657.

Sullivan, J., Holsinger, K.E., and Simon, C. (1995). Among-site rate variation and phylogenetic analysis of 12S rRNA in sigmodontine rodents. *Mol Biol Evol* **12**, 988–1001.

Suzuki, Y. and Gojobori, T. (1999). A method for detecting positive selection at single amino acid sites. *Mol Biol Evol* **16**, 1315–1328.

Tagle, D.A., Koop, B.F., Goodman, M., Slightom, J.L., Hess, D.L., and Jones, R.T. (1988). Embryonic epsilon and gamma globin genes of a prosimian primate (Galago crassicaudatus). Nucleotide and amino acid sequences, developmental regulation and phylogenetic footprints. *J Mol Biol* **203**, 439–455.

Tajima, F. and Nei, M. (1982). Biases of the estimates of DNA divergence obtained by the restriction enzyme technique. *J Mol Evol* **18**, 115–120.

Tajima, F. and Nei, M. (1984). Estimation of evolutionary distance between nucleotide sequences. *Mol Biol Evol* **1**, 269–285.

Tamura, K. and Nei, M. (1993). Estimation of the number of nucleotide substitutions in the control region of mitochondrial DNA in humans and chimpanzees. *Mol Biol Evol* 10, 512–526.

CSH Tan, A., Pasculescu, W.A., Lim, T., Pawson, G.D., Bader, R. Linding (2009). Positive Selection of Tyrosine Loss in Metazoan Evolution. *Science* **325**, 1686–1688.

Tanay, A., Regev, A., and Shamir, R. (2005). Conservation and evolvability in regulatory networks: the evolution of ribosomal regulation in yeast. *Proc Natl Acad Sci USA 1*02, 7203–7208.

Thompson, J.D., Higgins, D.G., and Gibson, T.J. (1994). Clustal-W - improving the sensitivity of progressive multiple sequence alignment through sequence weighting, position-specific gap penalties and weight matrix choice. *Nucleic Acids Research* 22, 4673–4680.

Thompson, W., Rouchka, E.C., and Lawrence, C.E. (2003). Gibbs Recursive Sampler: finding transcription factor binding sites. *Nucleic Acids Res* **31**, 3580–3585.

Torgerson, D.G., Whitty, B.R., and Singh, R.S. (2005). Sex-specific functional specialization and the evolutionary rates of essential fertility genes. *J Mol Evol* **61**, 650–658.

Tourasse, N.J. and Gouy, M. (1997). Evolutionary distances between nucleotide sequences based on the distribution of substitution rates among sites as estimated by parsimony. *Molecular Biology and Evolution* **14**, 287–298.

True, J.R. and Haag, E.S. (2001). Developmental system drift and flexibility in evolutionary trajectories. *Evol Dev* **3**, 109–119.

Tsong, A.E., Tuch, B.B., Li, H., and Johnson, A.D. (2006). Evolution of alternative transcriptional circuits with identical logic. *Nature* **443**, 415–420.

Turelli, M. (1985). Effects of pleiotropy on predictions concerning mutation-selection balance for polygenic traits. *Genetics* **111**, 165–195.

Tusher, V.G., Tibshirani, R., and Chu, G. (2001). Significance analysis of microarrays applied to the ionizing radiation response. *P Natl Acad Sci USA* **98**, 5116–5121.

Uzzell, T. and Corbin, K.W. (1971). Fitting discrete probability distributions to evolutionary events. *Science* **172**, 1089–1096.

Vazquez, A., Dobrin, R., Sergi, D., Eckmann, J.P., Oltvai, Z.N., and Barabasi, A.L. (2004). The topological relationship between the large-scale attributes and local interaction patterns of complex networks. *P Natl Acad Sci USA* **101**, 17940–17945.

von Mering, C., Krause, R., Snel, B., Cornell, M., Oliver, S.G., Fields, S., and Bork, P. (2002). Comparative assessment of large-scale data sets of protein-protein interactions. *Nature* **417**, 399–403.

Wagner, A. (1999). Redundant gene functions and natural selection. *J Evolution Biol* **12**, 1–16.

Wagner, A. (2000a). Decoupled evolution of coding region and mRNA expression patterns after gene duplication: Implications for the neutralist-selectionist debate. *P Natl Acad Sci USA* **97**, 6579–6584.

Wagner, A. (2000b). Robustness against mutations in genetic networks of yeast. *Nature Genetics* **24**, 355–361.

Wagner, A. (2000c). The role of population size, pleiotropy and fitness effects of mutations in the evolution of overlapping gene functions. *Genetics* **154**, 1389–1401.

Wagner, A. (2001). The yeast protein interaction network evolves rapidly and contains few redundant duplicate genes. *Molecular Biology and Evolution* **18**, 1283–1292.

Wagner, G. P., and Mezey J. (2004). The role of genetic architecture constrains in the origin of variational modularity. In: Modularity in development and evolution (Schlosser G, Wagner GP, editors), p. 338–358. Chicago: University of Chicago Press.

Wagner, A. (2005a). *Robustness and Evolvability in Living Systems* (Princeton, N.J., Princeton University Press).

Wagner, A. (2005b). Robustness, evolvability, and neutrality. *FEBS Lett* **579**, 1772–1778.

Wagner, A. (2008). Gene duplications, robustness and evolutionary innovations. *Bioessays* **30**, 367–373.

Wagner, G.P. (1989). Multivariate mutation-selection balance with constrained pleiotropic effects. *Genetics* **122**, 223–234.

Wagner, G.P., Pavlicev, M., and Cheverud, J.M. (2007). The road to modularity. *Nat Rev Genet* **8**, 921–931.

Wakeley, J. (1993). Substitution rate variation among sites in hypervariable region 1 of human mitochondrial DNA. *J Mol Evol* **37**, 613–623.

Wall, D.P., Hirsh, A.E., Fraser, H.B., Kumm, J., Giaever, G., Eisen, M.B., and Feldman, M.W. (2005). Functional genomic analysis of the rates of protein evolution. *Proc Natl Acad Sci USA* **102**, 5483–5488.

Wallace, J.L. (1999). Selective COX-2 inhibitors: is the water becoming muddy? *Trends Pharmacol Sci* **20**, 4–6.

Wang, Y. and Gu, X. (2000). Evolutionary patterns of gene families generated in the early stage of vertebrates. *J Mol Evol* **51**, 88–96.

Wang, Y. and Gu, X. (2001). Functional divergence in the caspase gene family and altered functional constraints: statistical analysis and prediction. *Genetics* **158**, 1311–1320.

Washietl, S., Hofacker, I.L., and Stadler, P.F. (2005). Fast and reliable prediction of noncoding RNAs. *Proc Natl Acad Sci USA* **102**, 2454–2459.

Waterman, M.S. and Vingron, M. (1994). Rapid and accurate estimates of statistical significance for sequence data-base searches. *P Natl Acad Sci USA* **91**, 4625–4628.

Waxman, D. and Peck, J.R. (1998). Pleiotropy and the preservation of perfection. *Science* **279**, 1210–1213.

Welch, J.J. and Waxman, D. (2003). Modularity and the cost of complexity. *Evolution* **57**, 1723–1734.

West-Eberhard, M.J. (2005a). Developmental plasticity and the origin of species differences. *Proc Natl Acad Sci USA* **102 Suppl 1**, 6543–6549.

West-Eberhard, M.J. (2005b). Phenotypic accommodation: adaptive innovation due to developmental plasticity. *J Exp Zool B Mol Dev Evol* **304**, 610–618.

Wheeler, W.C., De Laet, J., Gladstein, D.S. (2002). POY: The Optimization of Alignment Characters. Version 3.0.4. Program and Documentation. New York, NY. Available at ftp.amnh.org/pub/molecular. Documentation by D. Janies and W.C. Wheeler.

Williams, C.S., Mann, M., and DuBois, R.N. (1999). The role of cyclooxygenases in inflammation, cancer, and development. *Oncogene* **18**, 7908–7916.

Williams, E.J. and Hurst, L.D. (2000). The proteins of linked genes evolve at similar rates. *Nature* **407**, 900–903.

Wilson, A.C., Carlson, S.S., and White, T.J. (1977). Biochemical evolution. *Annu Rev Biochem* **46**, 573–639.

Wilson, K.P., Black, J.A., Thomson, J.A., Kim, E.E., Griffith, J.P., Navia, M.A., Murcko, M.A., Chambers, S.P., Aldape, R.A., Raybuck, S.A., *et al.* (1994). Structure and mechanism of interleukin-1 beta converting enzyme. *Nature* **370**, 270–275.

Wingender, E., Chen, X., Fricke, E., Geffers, R., Hehl, R., Liebich, I., Krull, M., Matys, V., Michael, H., Ohnhauser, R., *et al.* (2001). The TRANSFAC system on gene expression regulation. *Nucleic Acids Res* **29**, 281–283.

Winzeler, E.A., Shoemaker, D.D., Astromoff, A., Liang, H., Anderson, K., Andre, B., Bangham, R., Benito, R., Boeke, J.D., Bussey, H., *et al.* (1999). Functional characterization of the S-cerevisiae genome by gene deletion and parallel analysis. *Science* **285**, 901–906.

Wolf, Y.I. (2006). Coping with the quantitative genomics 'elephant': the correlation between the gene dispensability and evolution rate. *Trends Genet* **22**, 354–357.

Wolf, Y.I., Carmel, L., and Koonin, E.V. (2006). Unifying measures of gene function and evolution. *Proc Biol Sci* **273**, 1507–1515.

Wolf, Y.I., Rogozin, I.B., Grishin, N.V., and Koonin, E.V. (2002). Genome trees and the tree of life. *Trends Genet* **18**, 472–479.

Wolfe, K.H. and Shields, D.C. (1997). Molecular evidence for an ancient duplication of the entire yeast genome. *Nature* **387**, 708–713.

Workman, C.T. and Stormo, G. (2000). ANN-Spec: a method for discovering transcription factor binding sites with improved specificity. Paper presented at: Pacific Symposium on Biocomputing.

Wray, G.A., Hahn, M.W., Abouheif, E., Balhoff, J.P., Pizer, M., Rockman, M.V., and Romano, L.A. (2003). The evolution of transcriptional regulation in eukaryotes. *Molecular Biology and Evolution* **20**, 1377–1419.

Wright, S. (1968). Evolution and the Genetics of Populations. Vol.1. University of Chicago Press, Chicago USA.

Wu, C.I., and Li, W.H. (1985). Evidence for higher rates of nucleotide substitution in rodents than in man. *Proc Natl Acad Sci USA* **82**, 1741–1745.

Wu, S. and Gu, X. (2002). Multiple genome rearrangement by reversals. *Pac Symp Biocomput*, 259–270.

Wu, S. and Gu, X. (2003). Algorithms for multiple genome rearrangement by signed reversals. *Pac Symp Biocomput*, 363–374.

Wu, S., Gu, X. (2005). Gene Network: Model, dynamics and simulation. Lecture Notes on Computer Science: **3595**, pp. 12–21.

Wuchty, S., Barabasi, A.L., and Ferdig, M.T. (2006). Stable evolutionary signal in a Yeast protein interaction network. *Bmc Evol Biol* **6**, 8.

Wuchty, S., Oltvai, Z.N., and Barabasi, A.L. (2003). Evolutionary conservation of motif constituents in the yeast protein interaction network. *Nature Genetics* **35**, 176–179.

Wyckoff, G.J., Malcom, C.M., Vallender, E.J., and Lahn, B.T. (2005). A highly unexpected strong correlation between fixation probability of nonsynonymous mutations and mutation rate. *Trends Genet* **21**, 381–385.

Xia, X. and Xie, Z. (2001). DAMBE: software package for data analysis in molecular biology and evolution. *J Hered* **92**, 371–373.

Xu, G., Ma, H., Nei, M., and Kong, H. (2009). Evolution of F-box genes in plants: different modes of sequence divergence and their relationships with functional diversification. *Proc Natl Acad Sci USA* **106**, 835–840.

Yanai, I., Graur, D., and Ophir, R. (2004). Incongruent expression profiles between human and mouse orthologous genes suggest widespread neutral evolution of transcription control. *Omics* **8**, 15–24.

Yanai, I., Korbel, J.O., Boue, S., McWeeney, S.K., Bork, P., and Lercher, M.J. (2006). Similar gene expression profiles do not imply similar tissue functions. *Trends Genet* **22**, 132–138.

Yang, J., Gu, Z., and Li, W.H. (2003). Rate of protein evolution versus fitness effect of gene deletion. *Mol Biol Evol* **20**, 772–774.

Yang, J., Su, A.I., and Li, W.H. (2005). Gene expression evolves faster in narrowly than in broadly expressed mammalian genes. *Mol Biol Evol* **22**, 2113–2118.

Yang, Z. (1993). Maximum-likelihood estimation of phylogeny from DNA sequences when substitution rates differ over sites. *Mol Biol Evol* **10**, 1396–1401.

Yang, Z. (1994a). Estimating the pattern of nucleotide substitution. *J Mol Evol* **39**, 105–111.

Yang, Z. (1994b). Maximum likelihood phylogenetic estimation from DNA sequences with variable rates over sites: approximate methods. *J Mol Evol* **39**, 306–314.

Yang, Z. (1997). PAML: a program package for phylogenetic analysis by maximum likelihood. *Comput Appl Biosci* **13**, 555–556.

Yang, Z. 2006. *Computational Molecular Evolution.* Oxford University Press, Oxford, England.

Yang, Z. and Kumar, S. (1996). Approximate methods for estimating the pattern of nucleotide substitution and the variation of substitution rates among sites. *Molecular Biology and Evolution* **13,** 650–659.

Yang, Z., Kumar, S., and Nei, M. (1995). A new method of inference of ancestral nucleotide and amino acid sequences. *Genetics* **141**, 1641–1650.

Yang, Z. and Rannala, B. (1997). Bayesian phylogenetic inference using DNA sequences: a Markov Chain Monte Carlo Method. *Mol Biol Evol* **14**, 717–724.

Yeger-Lotem, E., Sattath, S., Kashtan, N., Itzkovitz, S., Milo, R., Pinter, R.Y., Alon, U., and Margalit, H. (2004). Network motifs in integrated cellular networks of transcription-regulation and protein-protein interaction. *Proc Natl Acad Sci USA* 101, 5934–5939.

Yu, H., Kim, P.M., Sprecher, E., Trifonov, V., and Gerstein, M. (2007). The importance of bottlenecks in protein networks: correlation with gene essentiality and expression dynamics. *Plos Comput Biol* **3**, e59.

Zhang, H. and Gu, X. (2004). Maximum likelihood for genome phylogeny on gene content. *Stat Appl Genet Mol Biol* **3**, Article31.

Zhang, H., Zhong, Y., Hao, B., and Gu, X. (2009). A simple method for phylogenomic inference using the information of gene content of genomes. *Gene* **441**, 163–168.

Zhang, J., and Gu, X. (1998). Correlation between the substitution rate and rate variation among sites in protein evolution. Genetics *149*, 1615–1625.

Zhang, J., Kumar, S., and Nei, M. (1997). Small-sample tests of episodic adaptive evolution: a case study of primate lysozymes. *Mol Biol Evol* **14**, 1335–1338.

Zhang, J., Rosenberg, H.F., and Nei, M. (1998). Positive Darwinian selection after gene duplication in primate ribonuclease genes. *Proc Natl Acad Sci USA* **95**, 3708–3713.

Zhang, L. and Li, W.H. (2004). Mammalian housekeeping genes evolve more slowly than tissue-specific genes. *Mol Biol Evol* **21**, 236–239.

Zhang, P., Gu, Z., and Li, W.H. (2003). Different evolutionary patterns between young duplicate genes in the human genome. *Genome Biol* **4**, R56.

Zhang, X.S. and Hill, W.G. (2003). Multivariate stabilizing selection and pleiotropy in the maintenance of quantitative genetic variation. *Evolution* **57**, 1761–1775.

Zhang, Z., Gu, J., and Gu, X. (2004). How much expression divergence after yeast gene duplication could be explained by regulatory motif evolution? *Trends Genet* **20**, 403–407.

Zhao, F., Xuan, Z., Liu, L., and Zhang, M.Q. (2005). TRED: a Transcriptional Regulatory Element Database and a platform for in silico gene regulation studies. *Nucleic Acids Res* **33**, D103–107.

Zharkikh, A. (1994). Estimation of evolutionary distances between nucleotide sequences. *J Mol Evol* **39**, 315–329.

Zheng, Y., Xu, D.P., and Gu, X. (2007). Functional divergence after gene duplication and sequence-structure relationship: A case study of G-protein alpha subunits. *J Exp Zool Part B* **308B**, 85–96.

Zhou, H., Gu, J., Lamont, S.J., and Gu, X. (2007). Evolutionary analysis for functional divergence of the toll-like receptor gene family and altered functional constraints. *J Mol Evol* **65**, 119–123.

Zou, Y., Su, Z., Yang, J., Zeng, Y., and Gu, X. (2009). Uncovering genetic regulatory network divergence between duplicate genes using yeast eQTL landscape. *J Exp Zool B Mol Dev Evol* **312**, 722–733.

Zuckerkandl, E. (1976). Evolutionary processes and evolutionary noise at the molecular level. II. A selectionist model for random fixations in proteins. *J Mol Evol* **7**, 269–311.

Index

Note: page numbers in italics refer to Figures and Tables